Singularities
—Kagoshima 2017

Singularities
—Kagoshima 2017

Proceedings of the 5th Franco-Japanese-Vietnamese Symposium on Singularities
Kagoshima, Japan, 27 October – 3 November 2017

Editors-in-Chief

Masaharu Ishikawa
Keio University, Japan

Shoji Yokura
Kagoshima University, Japan

World Scientific

NEW JERSEY • LONDON • SINGAPORE • BEIJING • SHANGHAI • HONG KONG • TAIPEI • CHENNAI • TOKYO

Published by

World Scientific Publishing Co. Pte. Ltd.
5 Toh Tuck Link, Singapore 596224
USA office: 27 Warren Street, Suite 401-402, Hackensack, NJ 07601
UK office: 57 Shelton Street, Covent Garden, London WC2H 9HE

Library of Congress Control Number: 2020938567

British Library Cataloguing-in-Publication Data
A catalogue record for this book is available from the British Library.

SINGULARITIES — KAGOSHIMA 2017
Proceedings of the 5th Franco-Japanese-Vietnamese Symposium on Singularities
The 5th Franco-Japanese-Vietnamese Symposium on Singularities

ISBN 978-981-120-602-3 (hardcover)
ISBN 978-981-120-603-0 (ebook for institutions)
ISBN 978-981-120-604-7 (ebook for individuals)

For any available supplementary material, please visit
https://www.worldscientific.com/worldscibooks/10.1142/11437#t=suppl

Cover Photo

The photo on the cover shows Sakurajima, one of the world's most active volcanoes and the symbol of Kagoshima. It was taken by Nobuo Ichikawa (provided by Sakurajima Volcano Research Center, Disaster Prevention Research Institute, Kyoto University).

Preface

The Fifth Franco-Japanese-Vietnamese Symposium on Singularities (The 5th FJV 2017) was held at Kagoshima University, in Kagoshima, Japan during the period of October 27–November 3, 2017. It was the 5th symposium aiming at research communication between French, Japanese and Vietnamese mathematicians working in Singularity Theory. The symposium presented recent advances and perspectives in Geometry and Topology of Singularities, Algebraic Geometry of Singular Varieties and Applications of Singularity Theory. The homepage* of the symposium is

http://www.math.tohoku.ac.jp/~ishikawa/FJV2017/

The first FJV symposium was held at Nice, France in 2013, followed by Sapporo, Japan in 2014, Hanoi, Vietnam in 2015, Chambery, France in 2016, Kagoshima, Japan in 2017 and Nha Trang, Vietnam, 2018. These symposiums were main activities of the GDRI = Groupe de Recherche International sur les Singularités/International Research Group on Singularities. The homepage of the GDRI is

https://www.i2m.univ-amu.fr/GDRI-Singularities/

For the first time in the series of Franco-Japanese-Vietnamese, Franco-Japanese, Franco-Vietnamese or Japanese-Vietnamese conferences, a mini-school, which was mainly for young mathematicians, was organized during the period of October 27–28 before the main conference (October 29–November 3). The lecturers of the mini-school were Ha Huy Vui (Hanoi, Vietnam), Ilia Itenberg (Paris, France) and Wojciech Kucharz (Krakow, Poland), who each gave three one-hour lectures:

(1) Ha Huy Vui: *Łojasiewicz inequality — Smooth and global versions*
(2) Ilia Itenberg: *Tropical homology and topology of real algebraic varieties*
(3) Wojciech Kucharz: *On continuous rational functions*

*It can also be found at http://user.keio.ac.jp/~ishikawa/FJV2017/

During the main conference sixteen 50-minute talks and eight 30-minute talks were presented and a poster session was also organized and ten posters were presented.

The present volume contains thirteen articles, which are extended versions of the above three lectures of the mini-school and recent works and surveys by some speakers and participants of the FJV 2017 Kagoshima. We note that all the papers in this volume have been carefully refereed.

The 5th FJV 2017 was financially supported by le Centre National de la Recherche Scientifique (CNRS), GDRI, the Vietnam Academy of Science and Technology (VAST) and the Japan Society for the Promotion of Science (JSPS), especially by KAKENHI Grant Numbers JP17H06128 (Osamu Saeki), JP25247004 (Kyoji Saito), JP17H02838 (Toru Ohmoto), JP16K05140 (Masaharu Ishikawa), JP16H03936 (Shoji Yokura). Without these financial supports the symposium would not have been in the position to be successfully organized. On behalf of all the people involved in this symposium we would like to express our deep appreciation to these organizations.

The Scientific and Organizing Committees deserve our great gratitude for creating a well-run, productive and fruitful symposium and the Editorial Board for producing this nice volume of the proceedings of the symposium. Some students of Kagoshima University also helped with the organization of the symposium. It is a great pleasure to thank all of them for their excellent work. Our thanks also go to the editorial staff of World Scientific Publishers for their patience and efficient help with the publication process of the volume.

Finally, on behalf of the Scientific and Organizing Committees, we would like to thank all the participants for their attendance, which made both the mini-school and the conference very fruitful and successful!

Editors-in-Chief
Masaharu Ishikawa Yokohama, Japan
Shoji Yokura Kagoshima, Japan
January, 2020

Editorial board

of the proceedings: Singularities — Kagoshima 2017

Organization

of the 5th Franco-Japanese-Vietnamese Symposium on Singularities (FJV2017 Kagoshima)

Scientific Committee

Nguyen Viet Dung (Chairman)	– VAST, Vietnam
Alexandru Dimca	– Nice University, France
Goulwen Fichou	– Rennes University, France
Toshitake Kohno	– The University of Tokyo, Japan
Krzysztof Kurdyka	– Savoie University, France
Toru Ohmoto	– Hokkaido University, Japan
Patrick Popescu-Pampu	– Lille University, France
Tiến-Sơn Phạm	– Dalat University, Vietnam
Osamu Saeki	– Kyushu University, Japan
Kiyoshi Takeuchi	– University of Tsukuba, Japan

Organizing Committee

Vincent Blanlœil	– Strasbourg University, France
Hiroaki Ishida	– Kagoshima University, Japan
Masaharu Ishikawa	– Tohoku University, Japan
Hiroyuki Nakaoka	– Kagoshima University, Japan
Vu The Khoi	– VAST, Vietnam
Shoji Yokura	– Kagoshima University, Japan

Contents

PART A

Articles from Mini-school

Smooth and global versions of the Łojasiewicz inequality

Ha Huy Vui

Thang Long Institute of Mathematics and Applied Sciences,
Hanoi, Vietnam
E-mail: hhvui@math.ac.vn

The classical Łojasiewicz inequality says that if f is a germ of an analytic function at $0 \in \mathbb{R}^n$, then there exist constants $c, \alpha > 0$ and a neighborhood U of 0 such that, for all $x \in U$, we have

$$|f(x)| \geq c\,\mathrm{dist}(x, f^{-1}(0))^{\alpha}.$$

In this paper we report on some recent results of [DHT], [H], [HNP] giving some analogs of this inequality for smooth functions, both of local and global cases.

Keywords: Lojasiewicz inequality, atypical value

2010 MSC: Primary 32S20; Secondary 32S30, 14P25, 14H50

Let $\mathrm{dist}(x, X)$ denote the distance from a point x to a subset X of $\mathbb{R}^n$. The classical Łojasiewicz inequality says that if f is a germ of an analytic function at $0 \in \mathbb{R}^n$, then there exist $c, \alpha > 0$ and a neighborhood U of 0 such that, for all $x \in U$, we have

$$|f(x)| \geq c\,\mathrm{dist}(x, f^{-1}(0))^{\alpha}. \tag{1}$$

Put

$$\mathcal{L}_0(f) := \inf\{\alpha > 0 : \ \exists c, \alpha, U \ \text{ s.t. } \ (1) \text{ holds}\}.$$

We will call $\mathcal{L}_0(f)$ the *Łojasiewicz exponent* of f at the origin.

Inequality (1) was born to solve a big problem of analysis ([L, Ho]). Later it becomes a very useful tool for studying many questions in analysis, geometry and algebra (See [T] and references therein).

The author is partially supported by Vietnam National Foundation for Science and Technology Development (NAFOSTED) under grant number 101.04-2019.316.

In general the Łojasiewicz inequality does not hold for germs of smooth functions, since smooth functions may have zeros of infinite order. Also, (1) cannot be true globally. In fact, if instead of a bounded neighborhood U, we take $U = \mathbb{R}^n$, then there exists a polynomial $f\colon \mathbb{R}^n \to \mathbb{R}$, having a sequence $\{x^k\} \subset \mathbb{R}^n, \|x^k\| \to \infty$, such that $|f(x^k)| \le M < +\infty$ and $\mathrm{dist}(x^k, f^{-1}(0)) \to \infty$, hence, f does not satisfy (1) for all $x \in \mathbb{R}^n$.

In these notes we report on some recent results of [DHT], [H], [HNP] giving some analogs of the classical Łojasiewicz inequality, both for local and global cases.

1. The case of complex variables

It is useful to start with the case of complex variables.

Theorem 1.1 ([RT]). *Let $f\colon (\mathbb{C}^n, 0) \to (\mathbb{C}, 0)$ be a germ of a holomorphic function and m the multiplicity of f at the origin. Then there exist $c > 0$ and a neighborhood U of $0 \in \mathbb{C}^n$ such that*

$$|f(x)| \ge c\,\mathrm{dist}(x, f^{-1}(0))^m$$

for all $x \in U$. Moreover, the Łojasiewicz exponent of f at 0 is equal to m.

Theorem 1.2. *Let $f\colon \mathbb{C}^n \to \mathbb{C}$ be a polynomial of degree d. Then there exists $c > 0$ such that for all $x \in \mathbb{C}^n$, we have*

$$|f(x)| \ge c\,\mathrm{dist}(x, f^{-1}(0))^d.$$

Remark 1.3. One can show that for every polynomial f of degree d, there exist a sequence $\{x^k\} \subset \mathbb{C}^n$, $\|x^k\| \to \infty$ and a constant $M > 0$ such that

$$\frac{1}{M}\mathrm{dist}(x^k, f^{-1}(0))^d \le |f(x^k)| \le M\mathrm{dist}(x^k, f^{-1}(0))^d.$$

Hence, by Theorem 1.2, the degree can be considered as the Łojasiewicz exponent of the global Łojasiewicz inequality for a complex polynomial.

Proof of Theorem 1.1. By the Weierstrass preparation theorem we may assume that in some neighborhood U of the origin, f is of the form

$$f(x) = u(x)(x_n^m + a_1(x')x_n^{m-1} + \cdots + a_m(x')),$$

where $x' = (x_1, \ldots, x_{n-1})$; $a_1(x'), \ldots, a_m(x')$ are germs of holomorphic functions at the origin of $\mathbb{C}^{n-1}$, $a_1(0) = \cdots = a_m(0) = 0$; $u(x)$ is a germ of a holomorphic function at the origin of $\mathbb{C}^n$, $u(0) \ne 0$.

Choosing U of the form

$$U = \{(x', x_n) \in \mathbb{C}^{n-1} \times \mathbb{C} : \|x'\| < \varepsilon_1, |x_n| < \varepsilon_2\}$$

where ε_1 is sufficiently small in comparing with ε_2, we have

- If $\|x'\| < \varepsilon_1$ and $\tau_1(x'), \ldots, \tau_m(x')$ denote the roots of $f(x', x_n)$ considered as a polynomial in x_n, then $|\tau_i(x')| < \varepsilon_2$, $i = 1, \ldots, m$;
- There exists $\delta > 0$ such that if $(x', \tau(x')) \in U$ and $f(x', \tau(x')) = 0$, then $|\tau(x')| \le \delta\|x'\|$.

Then, for $x \in U$, we have

$$\begin{aligned} |f(x)| &= u(x)\prod_{i=1}^{m}(x_n - \tau_i(x')) \\ &\ge \frac{|u(0)|}{2}(\min_{i=1,\ldots,m}|x_n - \tau_i(x')|^m) \\ &\ge \frac{|u(0)|}{2}\mathrm{dist}(x, f^{-1}(0))^m. \end{aligned}$$

Since, for some $M > 0$

$$\begin{aligned} \frac{1}{M}\mathrm{dist}((0, x_n), f^{-1}(0))^m &\le |f(0, x_n)| = u(0)|x_n|^m \\ &\le M\mathrm{dist}((0, x_n), f^{-1}(0))^m \end{aligned}$$

for all $x \in U$, we conclude that $\mathcal{L}_0(f) = m$. □

The proof of Theorem 1.2 is similar to that of Theorem 1.1.

2. The case of smooth functions

We begin with the case when $f\colon (\mathbb{R}^n, 0) \to (\mathbb{R}, 0)$ is a germ of a smooth function at the origin.

Let m be the multiplicity of f at 0. Then f can be written in the form

$$f(x_1, \ldots, x_{n-1}, x_n) = f(x', x_n) = u(x)[x_n^m + a_1(x')x_n^{m-1} + \cdots + a_m(x')]$$

in some neighborhood U of the origin of $\mathbb{R}^n$. Here $u(x), a_i(x')$ are germs of smooth functions (at $0 \in \mathbb{R}^n$ and $0' \in \mathbb{R}^{n-1}$, respectively).

Put

$$V := \{x \in U : \ f(x) = 0\}$$

$$V_1 := \{x \in U : \ \frac{\partial f}{\partial x_n}(x) = 0\}.$$

Theorem 2.1 ([H]). *(Analog of Theorem 1.1 for smooth case) There exist a neighborhood Ω of the origin of $\mathbb{R}^n$ and a constant $c > 0$ such that*

$$|f(x)| \ge c\,\mathrm{dist}(x, V \cup V_1)^m$$

for all $x \in \Omega$.

Now, let $f\colon \mathbb{R}^n \to \mathbb{R}$ be a smooth function defined on $\mathbb{R}^n$.

Theorem 2.2 ([HNP]). *(Analog of Theorem 1.2 for smooth case) Assume that there exist* $d \in \mathbb{N}$ *and* $M > 0$ *such that*

$$\frac{\partial^d f}{\partial x_n^d}(x) \geq M$$

for all $x \in \mathbb{R}^n$. *Then, there exists* $c > 0$ *such that for all* $x \in \mathbb{R}^n$, *we have*

$$|f(x)| \geq c \operatorname{dist}(x, V \cup V_1)^d,$$

where

$$V := \{x \in \mathbb{R}^n : \ f(x) = 0\} \quad \text{and} \quad V_1 \colon = \{x \in \mathbb{R}^n : \ \frac{\partial f}{\partial x_n}(x) = 0\}.$$

The proofs of these results are based on the following

Lemma 2.3 (Van der Corput lemma [G]). *Let* $\mathcal{U}(t)$ *be a real-valued* C^k*-function,* $k \in \mathbb{N}$, *that satisfies the* k^{th} *derivative* $\mathcal{U}^{(k)}(t) \geq 1$ *for all* $t \in I$, *where* $I = [a, b]$ *or* $I = \mathbb{R}$. *Then there exists a finite constant* $c(k)$, *depending only on* k *such that the following estimate is valid for all* $\alpha > 0$:

$$\operatorname{mes}\{t \in I : \ |\mathcal{U}(t)| \leq \alpha\} \leq c(k)\alpha^{\frac{1}{k}},$$

where $\operatorname{mes}\{t \in I : \ |\mathcal{U}(t)| \leq \alpha\}$ *denotes the Lebesgue measure of the set* $\{t \in I : \ |\mathcal{U}(t)| \leq \alpha\}$.

Proof of Theorem 2.1. Without loss of generality, we may assume that

$$f(x', x_n) = x_n^m + a_1(x')x_n^{m-1} + \cdots + a_m(x'),$$

where $a_i(0') = 0, i = 1, \ldots, m$.

Fix $(0', a) \in U$. We have

$$|f(0', a)| = |f(0', -a)| = |a|^m.$$

Hence, there exists $b > 0$ such that if $\|x'\| \leq b$ then

$$\min\{|f(x', a)|, |f(x', -a)|\} \geq \frac{|a|^m}{2}.$$

Put

$$\Omega = \{(x', x_n) : \ \|x'\| \leq b, |x_n| \leq |a|\}.$$

Clearly $\Omega \subset U$ if a and b are sufficiently small.

We will show that there exists $c > 0$ such that for all $x \in \Omega$, we have

$$|f(x)| \geq c\,\mathrm{dist}(x, V \cup V_1)^m. \tag{2}$$

Let $x_0 = (x'_0, x^0_n) \in \Omega$.

If $|f(x_0)| \geq \dfrac{|a|^m}{2}$ then

$$|f(x_0)| \geq \frac{1}{2}\mathrm{dist}(x_0, f^{-1}(0))^m$$

and (2) holds. Assume that $|f(x_0)| < \dfrac{|a|^m}{2}$.

Set

$$\sum(x'_0) := \{x_n \in [-a, a] : \ |f(x'_0, x_n)| \leq |f(x_0)|\}.$$

Since $\dfrac{\partial^m f}{\partial x_n^m}(x'_0, x_n) = m!$, by Lemma 2.3, we have

$$\mathrm{mes}\sum(x'_0) \leq \frac{c(m)}{(m!)^{1/m}}|f(x_0)|^{\frac{1}{m}} \tag{3}$$

where $c(m)$ is a positive constant, which does not depend on x_0.

$\sum(x'_0)$ is a semi-algebraic closed subset of $\mathbb{R}$, hence $\sum(x'_0)$ is a finite union of some closed intervals $[a_i, b_i], i = 1, \ldots, s$ and some isolated points $c_1, \ldots, c_k$.

Since $|f(x_0)| < \dfrac{|a|^m}{2}$ and $|x'_0| \leq b$, the points a and $-a$ do not belong to $\sum(x'_0)$. Hence, we have

$$|f(x'_0, a_i)| = |f(x'_0, b_i)| = |f(x'_0, c_j)| = |f(x_0)|, \ i = 1, \ldots, s; \ j = 1, \ldots, k.$$

Clearly, $x^0_n \in \{a_i, b_i, c_j, \ i = 1, \ldots, s; \ j = 1, \ldots, k\}$. If $x^0_n = c_j$ then $\dfrac{\partial f}{\partial x_n}(x'_0, c_j) = 0$, i.e. $(x_0, c_j) \in V_1$ and (2) holds automatically. Assume that $x^0_n \in \{a_i, b_i, i = 1, \ldots, s\}$, says, $x^0_n = a_1$. Since $|f(x'_0, a_1)| = |f(x'_0, b_1)|$, it is easy to see that there exists either $t_1 \in [a_1, b_1]$ such that $f(x'_0, t_1) = 0$, i.e. $(x'_0, t_1) \in V$, or $t_0 \in [a_1, b_1]$ such that $\dfrac{\partial f}{\partial x_n}(x'_0, t_0) = 0$, i.e. $(x'_0, t_0) \in V_1$. Using (3), in the first case we have

$$\frac{c(m)}{(m!)^{1/m}}|f(x_0)|^{\frac{1}{m}} \geq |a_1 - t_1| \geq \mathrm{dist}(x_0, V),$$

and in the second case we have

$$\frac{c(m)}{(m!)^{1/m}}|f(x_0)|^{\frac{1}{m}} \geq \mathrm{mes}\sum(x'_0) \geq |a_1 - t_0| \geq \mathrm{dist}(x_0, V_1).$$

Hence (2) holds. Theorem 2.1 is proved. □

The proof of Theorem 2.2 is similar to that of Theorem 2.1.

Using Theorems 2.1 and 2.2, we can prove that the set V_1 can be considered as the testing set for the existence of the Łojasiewicz inequality for smooth functions.

Theorem 2.4 ([H]). *Let $f\colon (\mathbb{R}^n, 0) \to (\mathbb{R}, 0)$ be a germ of a smooth function at the origin, written in the form*

$$f(x_1, \ldots, x_{n-1}, x_n) = f(x', x_n) = u(x)[x_n^m + a_1(x')x_n^{m-1} + \cdots + a_m(x')],$$

where m is the multiplicity of f at the origin. Assume that there exist $c', \alpha > 0$ and a neighborhood Ω of $0 \in \mathbb{R}^n$ such that

$$|f(x)| \geq c' \operatorname{dist}(x, V)^\alpha$$

for all $x \in V_1 \cap \Omega$. Then there exists $c > 0$ such that

$$|f(x)| \geq c \operatorname{dist}(x, V)^{\max\{m,\alpha\}}$$

for all $x \in \Omega$.

Theorem 2.5. *Let $f\colon \mathbb{R}^n \to \mathbb{R}$ be a smooth function. Assume that*

i) *There are $d \in \mathbb{N}$ and $M > 0$ such that $\dfrac{\partial^d f}{\partial x_n^d}(x) \geq M$ for all $x \in \mathbb{R}^n$;*

ii) *There exist $c, \alpha > 0$ such that*

$$|f(x)| \geq c \operatorname{dist}(x, V)^\alpha$$

for all $x \in V_1$.

Then there exists $c > 0$ such that

$$|f(x)| \geq c \min\{\operatorname{dist}(x, V)^d, \operatorname{dist}(x, V)^\alpha\}$$

for all $x \in \mathbb{R}^n$.

For the proof of Theorem 2.4, we refer to ([H], lemma 2.1). The proof of Theorem 2.5 is similar to that of Theorem 2.4.

If there exist $c', \alpha > 0$ and a neighborhood Ω such that

$$|f(x)| \geq c' \operatorname{dist}(x, V)^\alpha$$

for all $x \in V_1 \cap \Omega$, we put

$$\mathcal{L}_0(f; V_1) = \inf\{\alpha > 0 : \exists c', \alpha > 0 \text{ and } \Omega \text{ s.t. } \forall x \in \Omega \cap V_1,\ |f(x)| \geq c' \operatorname{dist}(x, V)^\alpha\}.$$

We call $\mathcal{L}_0(f; V_1)$ the *Łojasiewicz exponent at* 0 *of f, w.r.t. V_1.*

Corollary 2.6. *Let $f\colon (\mathbb{R}^n, 0) \to (\mathbb{R}, 0)$ be a germ of a smooth function, for which the Łojasiewicz inequality holds. Then we have*

$$\mathcal{L}_0(f) = \max\{m, \mathcal{L}_0(f, V_1)\},$$

where m is the multiplicity of f at the origin.

Remark 2.7. If $f\colon (\mathbb{R}^n, 0) \to (\mathbb{R}, 0)$ is a germ of an analytic function of multiplicity m at $0 \in \mathbb{R}$, then

$$\mathcal{L}_0(f) = \max\{m, \mathcal{L}_0(f, V_1)\}.$$

3. Case $n = 2$

Let $f\colon (\mathbb{R}^2, 0) \to (\mathbb{R}, 0)$ be a germ of a smooth function. In this section, based on Theorem 2.5 and the smooth Puiseux theorem of V. Rychkov [R], we give a concrete sufficient condition for the Łojasiewicz inequality to hold for f. Then, under this condition, we compute the Łojasiewicz exponent of f.

Let $\mathbb{C}[[x]]$ be the ring of formal series with coefficients in $\mathbb{C}$. We use the following rings of germs of $\mathbb{C}$-valued functions:

- $C^\infty((x))$ and $C^\infty((x, y))$ are the rings of germs of C^∞ functions at the origin of $\mathbb{R}$ and $\mathbb{R}^2$, respectively.
- $C_+((x))$ and $C_+^\infty((x))$ are the rings of one-sided germs, consisting of functions $g(x)$ defined in a right neighborhood of zero of the form $[0, \epsilon)$, where $\epsilon > 0$ can depend on $g(x)$, which are continuous, respectively C^∞, up to zero.
- $A_+((x^\gamma)), \gamma \geq 0$, is the subring of $C_+((x))$ consisting of germs $g(x)$ for which there exists a series $\widetilde{g}(x) \in \mathbb{C}[[x^\gamma]]$, $\widetilde{g}(x) = \sum_{n=0}^\infty c_n x^{n\gamma}$, such that $g(x) \sim \widetilde{g}(x)$ in the sense that for any N

$$g(x) - \sum_{n=0}^{N} c_n x^{n\gamma} = O(x^{(N+1)\gamma}), \quad x \longrightarrow 0;$$

  such a $\widetilde{g}(x)$ is uniquely determined and is called the *asymptotic expansion* of $g(x)$.

The ring of germs of $\mathbb{R}$-valued functions will be denoted by adding an $\mathbb{R}$ to the above notation, e.g. $\mathbb{R}A_+((x))$, $\mathbb{R}C_+((x))$.

Let $f(x, y) \in \mathbb{R}C^\infty((x, y))$ and $\overline{f}(x) = \sum c_{ij} x^i y^j$ be the formal Taylor series of f at the origin.

Set

$$\text{supp}(f) = \{(i,j) \in \mathbb{R}^2 : \ c_{ij} \neq 0\}.$$

Let $\Gamma(f)$ be the convex hull of the set

$$\{(i,j) + \mathbb{R}_+^2 : \ (i,j) \in \text{supp}(f)\}.$$

Then $\Gamma(f)$ is the *Newton polygon* of f. Let E be a compact edge of the boundary of $\Gamma(f)$ joining points (A_E, B_E) and (A'_E, B'_E), where $B'_E > B_E$.

We set

$$n_E = B'_E - B_E, \quad \gamma_E = \frac{A_E - A'_E}{B'_E - B_E}.$$

If m is the multiplicity of f at the origin, for studying the Łojasiewicz inequality of f, we may assume that f is of the form

$$f(x,y) = y^m + a_{m-1}(x)y^{m-1} + \cdots + a_0(x),$$

where $a_i(x) \in C^\infty((x)), a_i(0) = 0, i = 0, \ldots, m-1$.

It follows from the smooth Puiseux theorem ([R]) that:

i) In the region $x > 0, f(x,y)$ admits a factorization

$$f(x,y) = \prod_{i=1}^{B}(y - Y_i(x)) \prod_E \prod_{i=1}^{n_E}(y - Y_{E_i}(x)) \tag{4}$$

where B is the y-coordinate of the infinite horizontal edge of $\Gamma(f)$, $Y_i(x) \in C_+((x))$ and $Y_i(x) = O(x^N)$ as $x \to 0$ for any $N > 0$, $Y_{E_i} \in A_+((x^{\frac{1}{m!}}))$ with asymptotic expansions of the form $Y_{E_i}(x) = c_{E_i}x^{\gamma_E} + \cdots$ as $x \to 0$, where $c_{E_i} \neq 0$ and E runs through all compact edges of $\Gamma(f)$.

ii) In the region $x > 0$, $\dfrac{\partial f}{\partial y}(x,y)$ admits a factorization

$$\frac{\partial f}{\partial y}(x,y) = \prod_{i=1}^{B'}(y - Y'_i(x)) \prod_{E'} \prod_{i=1}^{n_{E'}}(y - Y_{E'_i}(x)) \tag{5}$$

where B' is the y-coordinate of the infinite horizontal edge of $\Gamma(\dfrac{\partial f}{\partial y})$, $Y'_i(x) \in C_+((x))$ and $Y'_i(x) = O(x^N)$, as $x \to 0$, for any $N > 0$, $Y_{E'_i}(x) \in A_+((x^{\frac{1}{(m-1)!}}))$ with asymptotic expansion $Y_{E'_i}(x) = c_{E'_i}x^{\gamma_{E'}} + \cdots$ as $x \to 0$, where $c_{E'_i} \neq 0$ and E' runs through all compact edges of $\Gamma(\dfrac{\partial f}{\partial y})$.

Notation 3.1. Let $\mathcal{P}_{\mathbb{C}}(f)$ denote the sets of all functions $Y_i(x)$ and $Y_{E_i}(x)$ in (4) and $\mathcal{P}_{\mathbb{C}}(\frac{\partial f}{\partial y})$ denote the set of all functions $Y_i'(x), Y'_{E_i'}(x)$ in (5).

Set

$$\mathcal{P}_{\mathbb{R}}(f) := \mathcal{P}_{\mathbb{C}}(f) \cap (\mathbb{R}C_+((x)) \cup \mathbb{R}A_+((x^{\frac{1}{m!}})))$$

and

$$\mathcal{P}_{\mathbb{R}}(\frac{\partial f}{\partial y}) := \mathcal{P}_{\mathbb{C}}(\frac{\partial f}{\partial y}) \cap (\mathbb{R}C_+((x)) \cup \mathbb{R}A_+((x^{\frac{1}{(m-1)!}}))).$$

Then, as germs of sets at the origin, we have

$$\begin{aligned} V^+ &:= V \cap \{(x,y) \in \mathbb{R}^2 : \ x \geq 0\} \\ &= \bigcup_{Y \in \mathcal{P}_{\mathbb{R}}(f)} \{(x,y) \in \mathbb{R}^2 : \ x \geq 0, y = Y(x)\} \end{aligned}$$

and

$$\begin{aligned} V_1^+ &:= V_1 \cap \{(x,y) \in \mathbb{R}^2 : \ x \geq 0\} \\ &= \bigcup_{Y' \in \mathcal{P}_{\mathbb{R}}(\frac{\partial f}{\partial y})} \{(x,y) \in \mathbb{R}^2 : \ x \geq 0, y = Y'(x)\}. \end{aligned}$$

For two real-valued functions $\varphi > 0$ and $\psi > 0$, we write $\varphi \asymp \psi$ if $\frac{\varphi(x)}{\psi(x)}$ lies between two positive constants for x near $0 \in \mathbb{R}$. If $\varphi(x) \asymp x^s$ for some s, we set $v(\varphi) := s$. If $\varphi(x) = O(x^N)$ as $x \to 0$ for any $N > 0$, we set $v(\varphi) := \infty$. If φ and ψ are two functions, we set $\theta(\varphi, \psi) := v(|\varphi - \psi|)$ and call it the *contact order* of φ and ψ.

For each $Y' \in \mathcal{P}_{\mathbb{R}}(\frac{\partial f}{\partial y})$ we denote

$$\theta_{Y'} = \sum_{Y \in \mathcal{P}_{\mathbb{C}}(f)} \theta(Y', Y).$$

Clearly $\theta_{Y'} \in \mathbb{Q} \cup \{\infty\}$.

We set

$$\delta_{Y'} := \begin{cases} 1 & \text{if } \ \mathcal{P}_{\mathbb{R}}(f) = \emptyset, \\ \max\{\theta(Y, Y') : Y \in \mathcal{P}_{\mathbb{R}}(f)\} & \text{if } \ \mathcal{P}_{\mathbb{R}}(f) \neq \emptyset. \end{cases}$$

Definition 3.2. We say that f satisfies the *finite contact condition* if for every $Y' \in \mathcal{P}_{\mathbb{R}}(\frac{\partial f}{\partial y}) \setminus \mathcal{P}_{\mathbb{R}}(f)$, we have $\theta_{Y'} < \infty$.

Theorem 3.3 ([H]). *Let* $f(x,y) = y^m + a_{m-1}(x)y^{m-1} + \cdots + a_0(x) \in \mathbb{R}C^\infty((x,y))$*, where* m *is the multiplicity of* f *at* $0 \in \mathbb{R}^2$*. Assume that* f *satisfies the finite contact condition. Then the number*

$$L_+(f;V_1) = \max\{\frac{\theta_{Y'}}{\delta_{Y'}}\}$$

where Y' *runs through all elements of* $\mathcal{P}_\mathbb{R}(\frac{\partial f}{\partial y}) \setminus \mathcal{P}_\mathbb{R}(f)$*, has the property that there exist* $c > 0$ *and* $\epsilon > 0$ *such that*

$$|f(x,y)| \geq c\,\mathrm{dist}((x,y),V)^{L_+(f;V_1)}$$

for all $(x,y) \in V_1 \cap \{(x,y) \in \mathbb{R}^2 :\ x \geq 0\} \cap B_\epsilon$*, where* $B_\epsilon = \{(x,y) \in \mathbb{R}^2 : \|(x,y)\| < \epsilon\}$ *and* ϵ *sufficiently small. Moreover,* $L_+(f;V_1)$ *is the smallest number with this property.*

Corollary 3.4. *Assume that both* $f(x,y)$ *and* $\widetilde{f}(x,y) = f(-x,y)$ *satisfy the finite contact condition. Then*

$$\mathcal{L}_0(f;V_1) = \max\{L_+(f;V_1), L_+(\widetilde{f};V_1)\}.$$

Theorem 3.5 ([H]). *Let* $f(x,y) = y^m + a_{m-1}(x)y^{m-1} + \cdots + a_0(x) \in \mathbb{R}C^\infty((x,y))$*, where* m *is the multiplicity of* f *at* $0 \in \mathbb{R}^2$*. Assume that both* $f(x,y)$ *and* $f(-x,y)$ *satisfy the finite contact condition. Then*

$$\mathcal{L}_0(f) = \max\{\mathcal{L}_0(f;V_1), m\}.$$

Proof. Theorem 3.5 follows from Theorem 2.5 and Corollary 3.4. □

The proof of Theorem 3.3 follows from the definitions of $\theta_{Y'}$ and $\delta_{Y'}$ and the following.

Lemma 3.6 ([H]). *Let* $Y(x) \in \mathcal{P}_\mathbb{R}(f)$ *and* X_Y *be the germ at the origin of the set* $\{(x,y) \in \mathbb{R}^2 :\ x \geq 0, y = Y(x)\}$*. Let* $Y'(x) \in \mathcal{P}_\mathbb{R}(\frac{\partial f}{\partial y})$*. Then*

$$\mathrm{dist}((x,Y'(x)),X_Y) \asymp |Y'(x) - Y(x)|.$$

Remark 3.7. Let $f(x,y)$ be a germ of an analytic function at $0 \in \mathbb{R}^2$, then $\mathcal{P}_\mathbb{C}(f), \mathcal{P}_\mathbb{R}(f)$ (resp. $\mathcal{P}_\mathbb{C}(\frac{\partial f}{\partial y}), \mathcal{P}_\mathbb{R}(\frac{\partial f}{\partial y})$) are the set of complex and real Puiseux series of $f = 0$ (resp. of $\frac{\partial f}{\partial y} = 0$). The finite contact condition holds automatically.

Theorem 3.5 gives a method of computation of the Łojasiewicz exponent of an analytic germ. This method of computation of $\mathcal{L}_0(f)$ for analytic germs is different from that of [K] in the following aspect: instead of using the so-called *real approximations* of Puiseux series of $f = 0$ to express $\mathcal{L}_0(f)$ as in [K], we use real Puiseux series of $\dfrac{\partial f}{\partial y}$ and $f = 0$.

4. Global Łojasiewicz inequality for real polynomials

It is not difficult to find the criterion of the existence of the global Łojasiewicz inequality for real polynomials.

Definition 4.1 ([DHT]). Let $f : \mathbb{R}^n \to \mathbb{R}$ be a polynomial.

i) A sequence $\{x^k\} \subset \mathbb{R}^n$ is said to be *of the first type* of f, if

$$\|x^k\| \to \infty,$$

$$|f(x^k)| \to 0,$$

$$\operatorname{dist}(x^k, f^{-1}(0)) \geq \delta > 0, \text{ for some } \delta > 0.$$

ii) A sequence $\{x^k\} \subset \mathbb{R}^n$ is said to be *of the second type* of f, if

$$\|x^k\| \to \infty,$$

$$\operatorname{dist}(x^k, f^{-1}(0)) \to +\infty,$$

$$|f(x^k)| \leq M < +\infty,$$

where M is a positive constant.

Let f be of the form

$$f(x_1, \dots, x_n) = a_0 x_n^d + a_1(x')x_n^{d-1} + \dots + a_d(x'),$$

where d is the degree of f and $a_0, a_1(x'), \dots, a_d(x')$ are polynomials in $x' = (x_1, \dots, x_{n-1})$, $\deg a_i(x') \leq i, i = 0, \dots, d$.

Set

$$V_1 : = \{x \in \mathbb{R}^n : \frac{\partial f}{\partial x_n}(x) = 0\}.$$

Definition 4.2.

i) A sequence $\{x^k\}$ is *of the first type* of f w.r.t. V_1, if

$$\{x^k\} \subset V_1,$$

$$\|x^k\| \to \infty,$$

$$|f(x^k)| \to 0$$

and

$$\mathrm{dist}(x^k, f^{-1}(0)) \geq \delta > 0,$$

where δ is a positive constant.

ii) A sequence $\{x^k\}$ is *of the second type* of f w.r.t. V_1, if

$$\{x^k\} \subset V_1,$$

$$\|x^k\| \to \infty,$$

$$\mathrm{dist}(x^k, f^{-1}(0)) \to +\infty,$$

$$|f(x^k)| \leq M < +\infty,$$

where M is a positive constant.

Theorem 4.3 ([DHT], see also [HD]). *Let*

$$f(x_1, ..., x_n) = a_0 x_n^d + a_1(x')x_n^{d-1} + \cdots + a_d(x')$$

be a polynomial of degree d and

$$V_1 := \{x \in \mathbb{R}^n : \frac{\partial f}{\partial x_n}(x) = 0\}.$$

Then, the following statements are equivalent

(i) *There are no sequences of the first or the second types of f;*

(ii) *There exist $c > 0, \alpha > 0$ and $\beta > 0$ such that*

$$|f(x)|^\alpha + |f(x)|^\beta \geq c\,\mathrm{dist}(x, f^{-1}(0)),$$

for all $x \in \mathbb{R}^n$;

(iii) *There are no sequences of the first or the second types of f w.r.t. V_1.*

Proof. (i)$\Rightarrow$ (ii). For each $t \in (0, \infty)$, set

$$\varphi(t) = \sup_{x \in |f|^{-1}(t)} \mathrm{dist}(x, f^{-1}(0)).$$

It follows from (i) that $\varphi(t)$ is well-defined and $\varphi(t)$ is a semialgebraic function. Moreover, we can show that

$$\varphi(t) \to 0, \quad \text{as} \quad t \to 0;$$

$$\varphi(t) \to \infty, \quad \text{as} \quad t \to \infty.$$

For any $0 < \delta < M < +\infty$, $\varphi(t)$ is bounded on $[\delta, M]$. Then, for δ sufficiently small and M sufficiently large, we have

$$\varphi(t) = c_1 t^\alpha + o(t^\alpha),\ t \in [0, \delta],$$

$$\varphi(t) = c_2 t^\beta + o(t^\beta),\ t \in [M, \infty].$$

Hence, there exists δ sufficiently small such that

$$|f(x)|^\alpha \geq \frac{c_1}{2}\operatorname{dist}(x, f^{-1}(0)) \tag{6}$$

for all $x \in \mathbb{R}^n$, $|f(x)| \leq \delta$ and there is $M \gg 1$ such that

$$|f(x)|^\beta \geq \frac{c_2}{2}\operatorname{dist}(x, f^{-1}(0)) \tag{7}$$

for all $x \in \mathbb{R}^n$, $|f(x)| \geq M$.

Clearly, $\operatorname{dist}(x, f^{-1}(0))$ is bounded on the set $\{x \in \mathbb{R}^n : \delta \leq |f(x)| \leq M\}$. This, together with Inequalities (6) and (7) implies (ii).

(ii)$\Rightarrow$ (iii): is straightforward.

(iii)$\Rightarrow$ (i): The proof of this uses Lemma 2.3 (see [HD] for the details). □

We call (6) and (7) the *Łojasiewicz inequality near* $f^{-1}(0)$ *and far* $f^{-1}(0)$, respectively.

Set

$$\alpha(f) := \sup\{\alpha > 0 : \exists \delta \in (0,1), c_1 > 0 \text{ s.t. (6) holds}\}$$

and

$$\beta(f) := \inf\{\beta > 0 : \exists M \in (1, +\infty), c_2 > 0 \text{ s.t. (7) holds}\}.$$

Then, under condition (i) (or (iii)) of Theorem 4.3, we have

$$|f(x)|^{\alpha(f)} + |f(x)|^{\beta(f)} \geq c \operatorname{dist}(x, f^{-1}(0)) \tag{8}$$

for all $x \in \mathbb{R}^n$.

In this inequality, the exponents $\alpha(f)$ and $\beta(f)$ are best possible.

Remark 4.4. If $\alpha(f) \geq \beta(f)$ then Inequality (8) can be replaced by one of the following inequalities

$$|f(x)|^{\alpha(f)} \geq c \operatorname{dist}(x, f^{-1}(0))$$

for all $x \in \mathbb{R}^n$; or

$$|f(x)|^{\beta(f)} \geq c \operatorname{dist}(x, f^{-1}(0))$$

for all $x \in \mathbb{R}^n$.

In these inequalities the exponents $\alpha(f)$ and $\beta(f)$ are best possible.

Note that the condition $\alpha(f) \geq \beta(f)$ holds for polynomials of complex variables, by Theorems 1.1 and 1.2, and also for real polynomials of one variables.

Nevertheless, in general one may have $\alpha(f) < \beta(f)$, as in the following example

$$f(x,y) = y^{2n} + (y - x^n)^2$$

where we have $f^{-1}(0) = \{(0,0)\}$, $\operatorname{dist}((x,y), f^{-1}(0)) = \|(x,y)\|$.

By ([H], Example 2.13 and [K], Example 3.8) we have $\mathcal{L}_0(f) = 2n^2$.

In this example, we have $\alpha(f) = \dfrac{1}{\mathcal{L}_0(f)} = \dfrac{1}{2n^2}$.

Clearly, $f(x,y) \to +\infty$ as $\|(x,y)\| \to +\infty$. Hence if $|f(x,y)| \gg 1$, we have $\|(x,y)\| > 1$.

It follows then from

$$|f(x,y)|^{\beta(f)} \geq c \operatorname{dist}((x,y), f^{-1}(0)) = c\|(x,y)\|$$

that

$$(\|(x,y)\|^{2n})^{\beta(f)} \geq c_1 |f(x,y)|^{\beta(f)} \geq c_1 c \|(x,y)\|$$

for some $c_1 > 0$.

Hence, since $\|(x,y)\| > 1$, we set

$$2n\beta(f) > 1$$

or

$$\beta(f) > \frac{1}{2n}.$$

Thus $\beta(f) > \dfrac{1}{2n} > \dfrac{1}{2n^2} = \alpha(f)$.

Thus, in order to describe the global Łojasiewicz inequality, in general we have to use both exponents $\alpha(f)$ and $\beta(f)$.

5. Global Łojasiewicz inequality and topology at infinity of real polynomial

Set

$$F^1(f) := \{t \in \mathbb{R} : \exists \{x^k\} \subset \mathbb{R}^n, \|x^k\| \to \infty, f(x^k) \to t,$$
$$\operatorname{dist}(x^k, f^{-1}(t)) \geq \delta > 0\}.$$

Let

$$F(f) := \{t \in \mathbb{R} : \exists \{x^k\} \subset \mathbb{R}^n, \|x^k\| \to \infty, \|\nabla f(x^k)\| \to 0, f(x^k) \to t\}$$

be the Fedoryuk set of f.

Theorem 5.1. *We have*

$$F^1(f) \subset F(f).$$

For the proof, we use

Theorem 5.2 ([E], Theorem 1.1). *Let X be a complete metric space, and $F\colon X \to \mathbb{R} \cup \{+\infty\}$ a l.s.c. function, not $\equiv +\infty$, bounded from below. For every point $u \in X$ satisfying*

$$\inf F \leq F(u) \leq \inf F + \epsilon, (\epsilon > 0)$$

and every $\lambda > 0$, there exists some point $v \in X$ such that

$$F(v) \leq F(u)$$

$$\operatorname{dist}(u, v) \leq \lambda$$

$$\forall w \neq v, F(w) > F(v) - \frac{\epsilon}{\lambda} \operatorname{dist}(v, w).$$

Proof of Theorem 5.1. Let $t_0 \in F^1(f)$ and $\{x^k\}$ be a sequence such that

$$\|x^k\| \to \infty$$

$$f(x^k) - t_0 > 0, f(x^k) \to t_0,$$

$$\operatorname{dist}(x^k, f^{-1}(t_0)) \geq \delta.$$

Put $X = \{x \in \mathbb{R}^n : f(x) \geq t_0\}$.

Then $f\colon X \to \mathbb{R}$ is a continuous function bounded from below on a complete metric space.

Put

$$\epsilon_k = f(x^k) - t_0,$$

$$\lambda_k = \sqrt{\epsilon_k}.$$

It follows from Theorem 5.2, there exists $y^k \in X$ such that

(i) $f(y^k) \leq f(x^k)$
(ii) $\operatorname{dist}(x^k, y^k) \leq \sqrt{\epsilon_k}$

(iii) $f(x) \geq f(y^k) - \sqrt{\epsilon_k}\text{dist}(x, y^k)$ for all $x \in \mathbb{R}^n$.

Since $\text{dist}(x^k, y^k) \to 0$ and $\text{dist}(x^k, f^{-1}(0)) > \delta$, we may assume that

$$B(y^k, \frac{\delta}{2}) := \{x \in \mathbb{R}^n : \|x - y^k\| \leq \frac{\delta}{2}\} \subset X \setminus f^{-1}(0).$$

Now, it follows from (iii) that

$$f(x) \geq f(y^k) - \sqrt{\epsilon_k}\text{dist}(x, y^k) \text{ for all } x \in B(y^k, \frac{\delta}{2})$$

which implies that

$$\|\nabla f(y^k)\| \leq \sqrt{\epsilon_k}.$$

We have $\|y^k\| \to \infty$, $\|\nabla f(y^k)\| \to 0$ and $f(y^k) \to t_0$. Therefore $t_0 \in F(f)$. □

We recall

Definition 5.3. Let $f\colon \mathbb{R}^n \to \mathbb{R}$ be a polynomial function. A value $t_0 \in \mathbb{R}$ will be called a *typical value of f at infinity*, if there are a compact set K and an open interval $(-\epsilon + t_0, t_0 + \epsilon)$ such that the restriction $f\colon f^{-1}((-\epsilon + t_0, t_0 + \epsilon)) \setminus K \to (-\epsilon + t_0, t_0 + \epsilon)$ is a trivial fibration of class C^∞.

Otherwise, t_0 will be called an *atypical value of f at infinity.*

Problem. Give a criterion of the atypical values of f.

This is an open problem in general. It is proved in [TZ] that a regular value of a polynomial of two variables f is atypical if and only if either the characteristic of the fiber $f^{-1}(t)$ is not constant in a neighborhood of t_0 or there is a connected component of $f^{-1}(t)$ which vanishes at infinity as t tends to t_0.

Definition 5.4. Let $f\colon \mathbb{R}^n \to \mathbb{R}$ be a polynomial. We say that f *has a vanishing component at infinity as t tends to t_0* if for every $\delta > 0$ and every $R > 0$ there exists $t \in (t_0 - \delta, t_0 + \delta)$, such that the fiber $f^{-1}(t)$ admits a connected component W_t with $W_t \cap B_R = \emptyset$.

Theorem 5.5 ([DHT]). *Let $f\colon \mathbb{R}^2 \to \mathbb{R}$ be a polynomial of two variable, $t_0 \in F^1(f)$. Then f has a vanishing component at infinity as t tends to t_0.*

Proof. We use the argument of [DKL]. We may assume that f is of the form

$$f(x, y) = a_0 y^d + a_1(x) y^{d-1} + \cdots + a_d(x),$$

where d is the degree of f.

Let $V = f^{-1}(0)$ and $V_1 = (\frac{\partial f}{\partial y})^{-1}(0)$.

Let $t_0 \in F^1(f)$, i.e. there exists a sequence of the first type

$$\{x^k\} \subset \mathbb{R}^2, \|x^k\| \to \infty,$$

$$f(x^k) \to t_0, \operatorname{dist}(x^k, f^{-1}(0)) \geq \delta > 0.$$

It follows from Theorem 4.3 that we can assume that $\{x^k\} \subset V_1$, and moreover $\{x^k\}$ is contained in some branch Γ at infinity of V_1.

Let $(x, y) \in \Gamma$ if and only if $y = \gamma(x)$.

We have then

$$f(\gamma(x)) \to t_0, \quad \text{as} \quad x \to \infty,$$

$$\operatorname{dist}((x, \gamma(x)), V) \geq \delta > 0.$$

For $\epsilon > 0$, let

$$T_\epsilon(V \cup V_1) = \{(x, y) \in \mathbb{R}^2 : \operatorname{dist}((x, y), V \cup V_1) \leq \epsilon\}.$$

Clearly, there exist R sufficiently large and ϵ sufficiently small such that

$$(T_\epsilon(\Gamma) \setminus B_R) \cap T_\epsilon(V) = \emptyset.$$

Since, by Theorem 2.2,

$$|f(x, y)| \geq c \operatorname{dist}((x, y), V \cup V_1)^d$$

for all $(x, y) \in \mathbb{R}^2$, with some constant $c > 0$, there exists ρ such that

$$f^{-1}((t_0 - \rho, t_0 + \rho)) \subset T_\epsilon(V \cup V_1).$$

Let U_ρ be a connected component of $f^{-1}((t_0 - \rho, t_0 + \rho)) \setminus B_R$ containing the branch Γ.

One can show that $U_\rho \subset T_\epsilon(\Gamma \setminus B_R)$ and $\overline{U_\rho} \cap T_\epsilon(V) = \emptyset$.

Now, let W_x be a connected component of $f^{-1}(f(x, \gamma(x)))$ containing the point $(x, \gamma(x))$. It is not difficult to prove that for any $r > 0$, $W_x \cap B_r = \emptyset$ if $|x|$ is sufficiently large. □

The following example of [DKL] shows that the conclusion of Theorem 5.5 is no longer true for $n > 2$.

Let

$$f(x, y, z) = (y^2 + (xy - 1)^2)z.$$

We have $V = \{f = 0\} = \{z = 0\}$

$$f(k, \frac{1}{k}, 1) \to 0$$

$$\mathrm{dist}((k, \frac{1}{k}, 1), V) = 1.$$

So $s^k = (k, \frac{1}{k}, 1)$ is a sequence of the first type and $0 \in F^1(f)$. On the other hand, let $\epsilon \neq 0$, then $f^{-1}(\epsilon)$ has only one connected component which is the graph $z = \dfrac{\epsilon}{y^2 + (xy-1)^2}$. Hence, f is a trivial C^∞ fibration. Consequently, f does not have a vanishing component as $t \to 0$.

Acknowledgments

The author is thankful to the referee for reading the manuscript carefully and giving his valuable suggestions to improve the presentation of the paper.

References

[DHT] S. T. Dinh, H. V. Hà, N. T. Thao, *Łojasiewicz inequality for polynomial function on non-compact domains,* Internat. J. Math. **23** (2012), no. 4 [28 pages]

[DKL] S. T. Dinh, K. Kurdyka, O. Le Gal, *Łojasiewicz inequality on non-compact domains and singularities at infinity,* Internat. J. Math. **24** (2013), no. 10 [8 pages].

[E] I. Ekeland, *On the variational principle,* J. Math. Anal. Appl. **47** (1974), 324–353.

[G] L. Grafakos, *Classical Fourier Analysis,* Springer, 2008.

[H] H. V. Ha, *Computation of the Łojasiewicz exponent for a germ of a smooth function in two variables,* Studia Math. **240** (2018), 161–176.

[HD] H. V. Ha, D. V. Doat, *On the global Łojasiewicz inequality for polynomial functions,* Preprint, Institute of Mathematics, VAST, 2017.

[HNP] H. V. Ha, H. V. Ngai, T. S. Pham, *A global smooth version of the classical Łojasiewicz inequality,* J. Math. Anal. Appl. **421** (2015), 1559–1572.

[Ho] L. Hörmander, *On the division of distributions by polynomials,* Ark. Mat. **3** (1958), 555–568.

[K] C. T. Kuo, *Computation of Łojasiewicz exponent of $f(x, y)$,* Comment. Math. Helv. **49** (1974), 201–213.

[L] S. Łojasiewicz, *Sur le problème de la division,* Studia Math. **18** (1959) 87–136.

[RT] J. J. Risler, D. Trotman, *Bilipschitz invariance of the multiplicity,* Bulletin of the London Math. Soc. **29** (1997), no. 2, 200–204.

[R] V. S. Rychkov, *Sharp L^2 bounds for oscillatory integral operators with C^∞ phase,* Math. Z. **236** (2001), 461–489.

[T] B. Teissier, *Variétés polaires 1*, Inventiones Math. **40** (1977), 267–292.
[TZ] M. Tibar, A. Zaharia, *Asymptotic behaviour of families of real curves,* manuscripta mathematica **99** (1999), 383–393.

Topology of real algebraic varieties and tropical homology

Ilia Itenberg

Institut de Mathématiques de Jussieu - Paris Rive Gauche,
Sorbonne Université, 4 place Jussieu, 75252 Paris Cedex 5, France
E-mail: ilia.itenberg@imj-prg.fr

This text is based on the notes of the lectures given by the author at the 5th Franco-Japanese-Vietnamese Symposium on Singularities in Kagoshima.

2010 MSC: 14P25, 14N99

1. Topology of algebraic hypersurfaces in $\mathbb{R}P^n$

1.1. *Hilbert 16-th problem*

The questions concerning the topology of real algebraic hypersurfaces in $\mathbb{R}P^n$ were included by D. Hilbert in the 16-th problem of his famous list [11]. By an *algebraic hypersuface* in $\mathbb{R}P^n$ we mean a real homogeneous polynomial in $n+1$ variables which is considered up to multiplication by a non-zero real constant. Such a polynomial has a zero locus $\mathbb{R}X \subset \mathbb{R}P^n$ and a zero locus $\mathbb{C}X \subset \mathbb{C}P^n$. The former zero locus is called the *real part* of the hypersurface. The *degree* of an algebraic hypersurface in $\mathbb{R}P^n$ is the degree of a defining polynomial. All hypersurfaces considered in these notes are *non-singular*, that is, their defining polynomials do not have critical points in $\mathbb{C}^{n+1} \setminus \{0\}$.

Fix positive integers n and d; what kind of topology can have the real part $\mathbb{R}X$ of a non-singular hypersurface X of degree d in $\mathbb{R}P^n$? In the case of curves ($n = 2$), this question was solved by A. Harnack [10]. Each connected component of a non-singular curve of degree d in $\mathbb{R}P^2$ is a circle smoothly embedded in $\mathbb{R}P^2$. According to the Harnack theorem, the number of these connected components can take any integer value between 0 (respectively, 1) and $(d-1)(d-2)/2+1$ if d is even (respectively, odd). The upper bound

$$b_0(\mathbb{R}X) \leq \frac{(d-1)(d-2)}{2} + 1,$$

where $b_0(\mathbb{R}X)$ is the number of connected components of the real part $\mathbb{R}X$ a non-singular curve X of degree d in $\mathbb{R}P^2$ is called the *Harnack inequality.*

The question about possible topology of the pair $(\mathbb{R}P^2, \mathbb{R}X)$, where X is a non-singular curve of degree d in $\mathbb{R}P^2$, is much more difficult; the answer is known only for $d \leq 7$.

In the case $n = 3$, the answer to the question on possible topology of the real part of a non-singular surface of degree d is known only for $d \leq 4$ (in fact, for non-singular surfaces of degree $d \leq 4$, several finer classifications are available). For non-singular surfaces of degree 5 in $\mathbb{R}P^3$, even the maximal possible number of connected components of the real part is not known (this maximal number is at least 23 and at most 25; see [15], [18]).

1.2. *Betti numbers*

A natural question, *a priori* simpler than the one formulated above, concerns the possible values of the Betti numbers of the real part of a non-singular hypersurface of degree d in $\mathbb{R}P^n$. In these notes, we consider Betti numbers with $\mathbb{Z}/2\mathbb{Z}$-coefficients: by the p-th Betti number of a topological space Y we mean $b_p(Y) = \dim_{\mathbb{Z}/2\mathbb{Z}} H_p(Y; \mathbb{Z}/2\mathbb{Z})$.

A generalization of the Harnack inequality is the *Smith-Thom inequality* for total Betti numbers (see [9] and [23]): for any algebraic hypersurface X in $\mathbb{R}P^n$, one has

$$\sum_{p=0}^{n-1} b_p(\mathbb{R}X) \leq \sum_{p=0}^{2n-2} b_p(\mathbb{C}X).$$

In fact, the inequality is valid for any (quasi-projective) real algebraic variety of dimension $n-1$ (and even in a more general situation); see, for example, [6] and [28]. A real algebraic variety realizing equality in the Smith-Thom inequality is said to be *maximal*.

Sharp upper bounds for individual Betti numbers of the real part of a non-singular hypersurface of degree d in $\mathbb{R}P^n$ are, in general, difficult to obtain. O. Viro [25] proposed the following conjecture (related to the so-called Ragsdale conjecture for real algebraic curves; see [20]): for any smooth projective real algebraic surface X with simply connected complex point set $\mathbb{C}X$, one has

$$b_1(\mathbb{R}X) \leq h^{1,1}(\mathbb{C}X),$$

where, as before, $\mathbb{R}X$ denotes the real part of X. Counter-examples to the Ragsdale conjecture [12] give rise to counter-examples to the Viro conjecture, in particular, among algebraic surfaces in $\mathbb{R}P^3$ (see [13], [4], [5]). Nevertheless, the Viro conjecture is true for the *primitive T-surfaces* (see [13]), *i.e.*, for algebraic surfaces in $\mathbb{R}P^3$ that are close to non-singular tropical limit.

In these notes, we consider the question on possible values of Betti numbers of real parts of primitive T-hypersurfaces of degree d in $\mathbb{R}P^n$ (see Section 2.1 for the definitions). The following upper bound (adapting the Viro conjecture to the case of primitive T-hypersurfaces in $\mathbb{R}P^n$) was proved very recently by A. Renaudineau and K. Shaw [21] (see Section 4.2 for details): *for any integer* $0 \leq p \leq n-1$ *and any primitive* T*-hypersurface* X *of degree* d *in* $\mathbb{R}P^n$, *we have*

$$b_p(\mathbb{R}X) \leq \begin{cases} h^{p,n-1-p}(\mathbb{C}X), & \text{if } p = (n-1)/2, \\ h^{p,n-1-p}(\mathbb{C}X) + 1, & \text{otherwise.} \end{cases}$$

The case $n = 3$ is treated in [13]. A proof of an asymptotic version of the above inequalities for primitive T-hypersurfaces can be found in [16].

2. Combinatorial patchworking

2.1. *Construction*

The special class of real algebraic hypersurfaces we are interested in is described by the *combinatorial patchworking.* This construction is a particular case of so-called *Viro method.* Let n and d be positive integers, and let $T^n(d)$ be the simplex in $\mathbb{R}^n$ with vertices $(0, 0, \dots, 0)$, $(0, 0, \dots, 0, d)$, $(0, \dots, 0, d, 0)$, $\dots, (d, 0, \dots, 0)$. We shorten the notation of $T^n(d)$ to T, when n and d are unambiguous and call $T^n(d)$ the *standard* n*-simplex of size* d. Take a triangulation τ of T with straight edges and integer vertices (*i.e.*, vertices having integer coordinates). Suppose that a distribution of signs at the vertices of τ is given. The sign (plus or minus) at the vertex with coordinates $(i_1, \dots, i_n)$ is denoted by $\sigma_{i_1,\dots,i_n}$.

Denote by T_* the union of all symmetric copies of T under reflections and compositions of reflections with respect to coordinate hyperplanes. Extend the triangulation τ to a symmetric triangulation τ_* of T_*, and the distribution of signs $\sigma_{i_1,\dots,i_n}$ to a distribution at the vertices of the extended triangulation by the following rule: passing from a vertex to its mirror image with respect to a coordinate hyperplane we preserve the sign if the distance from the vertex to the plane is even, and change the sign if the distance is odd.

If an n-simplex of the triangulation of T_* has vertices of different signs, select a piece of hyperplane being the convex hull of the middle points of the edges having endpoints of opposite signs. Denote by Γ the union of the selected pieces. It is a piecewise-linear hypersurface contained in T_*. It is not a simplicial subcomplex of T_*, but can be deformed by an isotopy preserving τ_* to a subcomplex Υ of the first barycentric subdivision τ_*' of τ_*. Each n-simplex of τ_*' has a unique vertex belonging to τ_*. Denote by τ_*^+ the union of all the n-simplices of τ_*' containing positive vertices of τ_* and by τ_*^- the union of all the remaining n-simplices. The subcomplex Υ is the intersection of τ_*^+ and τ_*^-.

Identify by the symmetry with respect to the origin the faces of T_*. The quotient space $\widetilde{T}$ is homeomorphic to the real projective space $\mathbb{R}P^n$. Denote by $\widetilde{\Gamma}$ the image of Γ in $\widetilde{T}$. (An example for $n = 2$ and $d = 10$ is shown in Figure 1.)

A triangulation τ of T is said to be *convex* if there exists a convex piecewise-linear function $\nu : T \longrightarrow \mathbb{R}$ whose domains of linearity coincide with the n-simplices of τ.

Theorem 1 (O. Viro). (see [24], [26]) *If τ is convex, there exists a non-singular hypersurface X of degree d in $\mathbb{R}P^n$ and a homeomorphism $\mathbb{R}P^n \to \widetilde{T}$ mapping the real part $\mathbb{R}X$ of X onto $\widetilde{\Gamma}$.*

Fig. 1. Construction of a counter-example to the Ragsdale conjecture

Consider a family (depending on positive real parameter t) of polynomials

$$P_t(x_0, \ldots, x_n) = \sum_{(i_1,\ldots,i_n)\in V} \sigma_{i_1,\ldots i_n} t^{-\nu(i_1,\ldots i_n)} x_0^{d-i_1-\ldots-i_n} x_1^{i_1} \ldots x_n^{i_n},$$

where V is the set of vertices of τ and ν is a convex piecewise-linear function certifying that the triangulation τ is convex. If t is big enough, the polynomial P_t is called *Viro polynomial* and defines a hypersurface that satisfies the properties described in Theorem 1. The hypersurface X defined by a Viro polynomial is called a *T-hypersurface.* If the triangulation τ is *primitive* (that is, each n-simplex of τ is of volume $\frac{1}{n!}$), then X is called a *primitive T-hypersurface.*

2.2. *Euler characteristic of primitive T-hypersurfaces*

A smooth real algebraic variety X is said to be *distinguished* if the Euler characteristic of the real part $\mathbb{R}X$ of X is equal to the signature of the complex point set $\mathbb{C}X$ of X. If X is of even (complex) dimension, then by the signature of $\mathbb{C}X$ we mean the signature of the intersection form in the middle homology; if the dimension of X is odd, we say that the signature of $\mathbb{C}X$ is 0. An example of distinguished real algebraic varieties is provided by primitive T-hypersurfaces in $\mathbb{R}P^n$ (of course, this statement is immediate if n is even).

Theorem 2 (B. Bertrand). *Let n and d be positive integers. Then, any primitive T-hypersurface of degree d in $\mathbb{R}P^n$ is a distinguished real algebraic variety.*

In the case $n = 3$ this statement was proved in [13]. The general statement was proved by B. Bertrand [3]. A simpler proof can be found in [1].

3. Tropical hypersurfaces in $\mathbb{R}^n$

The construction of combinatorial patchworking is directly related to tropical geometry (for tropical reformulations of the combinatorial patchworking, see [7] and [21]). We recall here several basic tropical notions.

3.1. *Tropical operations*

Tropical geometry can be seen as an algebraic geometry based on so-called *tropical numbers.* The set $\mathbb{T}$ of tropical numbers is the set $\mathbb{R}$ of real numbers completed with negative infinity: $\mathbb{T} = \mathbb{R} \cup \{-\infty\}$. Thus, $\mathbb{T}$ has a natural topology and is homeomorphic to a (closed) half-line. The tropical numbers are equipped with two arithmetic operations: *tropical addition* and *tropical multiplication*. The tropical addition is the operation of taking maximum, and the tropical multiplication is the ordinary addition:

$$\text{“}a + b\text{”} = \max\{a, b\}, \quad \text{“}a \cdot b\text{”} = a + b.$$

(We use the quotation marks when operations are tropical.) The set $\mathbb{T}$ equipped with these two operations is a *semi-field*, which means that

- $(\mathbb{T}, \text{“}+\text{”})$ is a commutative monoid (with the neutral element $-\infty$),
- $(\mathbb{T}^\times, \text{“}\cdot\text{”})$, where $\mathbb{T}^\times = \mathbb{T} \setminus \{-\infty\}$, is an abelian group (with the neutral element 0),
- and the tropical multiplication is distributive with respect to the tropical addition:

  $$\text{“}a \cdot (b + c)\text{”} = \text{“}(a \cdot b) + (a \cdot c)\text{”}$$

  for any $a, b, c \in \mathbb{T}$.

This semi-field is called *tropical*.

3.2. *Maslov's dequantization*

The tropical arithmetic operations can be seen as results of deformation of standard addition and multiplication. For any real number $t > 1$, consider the base t logarithmic map

$$\log_t : \mathbb{R}_{\geq 0} \to \mathbb{T}.$$

This map establishes a bijection between the set $\mathbb{R}_{\geq 0}$ of non-negative real numbers and the set $\mathbb{T}$. Thus, $\log_t$ allows one to introduce new arithmetic operations on $\mathbb{T}$:

$$a \oplus_t b = \log_t(t^a + t^b), \quad a \odot_t b = a + b.$$

These operations degenerate to the tropical operations when t tends to $+\infty$.

3.3. *Tropical hypersurfaces in* $\mathbb{R}^n$

Let n be a positive integer. In the tropical framework, the space $\mathbb{R}^n = (\mathbb{T}^\times)^n$ can be seen as tropical analog of the complex torus $(\mathbb{C}^\times)^n$, where $\mathbb{C}^\times = \mathbb{C} \setminus \{0\}$.

Let $V \subset (\mathbb{Z}_{\geq 0})^n$ be a finite set. Consider a *tropical polynomial*

$$p(u_1, \ldots, u_n) = \text{“} \sum_{(i_1,\ldots,i_n)\in V} a_{i_1,\ldots,i_n} u_1^{i_1} \ldots u_n^{i_n} \text{”},$$

where all coefficients $a_{i_1,\ldots,i_n}$ are tropical numbers (assume that all these coefficients are different from $-\infty$). Such a tropical polynomial gives rise to a *tropical polynomial function*

$$f_p : (u_1, \ldots, u_n) \mapsto \max_{(i_1,\ldots,i_n)\in V} \{a_{i_1,\ldots,i_n} + i_1 u_1 + \ldots + i_n u_n\}.$$

This is a convex piecewise-linear function. It is defined on $\mathbb{R}^n = (\mathbb{T}^\times)^n$ and it takes values in $\mathbb{R} = \mathbb{T}^\times$.

Denote by $\mathcal{X}_p$ the *corner locus* in $\mathbb{R}^n$ of the function f_p, that is, the subset of $\mathbb{R}^n$ formed by the points where the function f_p is not locally affine-linear. Figure 2 shows an example of such a corner locus appearing in the case of a tropical polynomial of degree 1 (with three monomials).

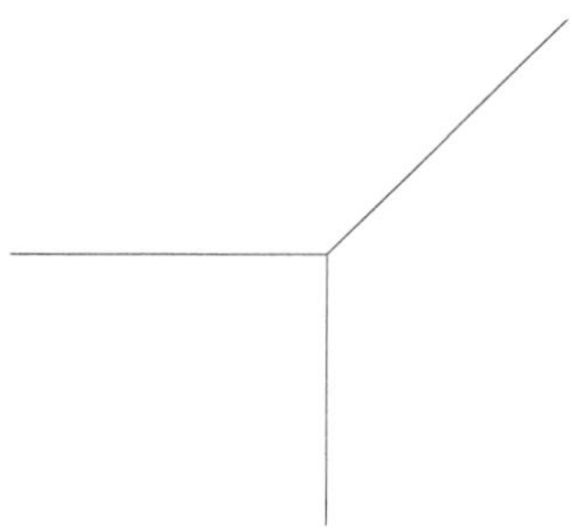

Fig. 2. Tropical line

The graph of f_p has a natural structure of polyhedral complex, so the corner locus $\mathcal{X}_p$ is also a polyhedral complex. It defines a subdivision Φ_p of $\mathbb{R}^n$. This subdivision is dual to a certain subdivision Ψ_p of the convex hull Δ_p of $V \subset \mathbb{R}^n$; the convex hull Δ_p is called the *Newton polytope* of p; this definition makes sense for any field or semi-field of coefficients. The latter subdivision can be described as follows. For any strata δ of Φ_p, take a point $(u_1^0, \ldots, u_n^0)$ in the relative interior of δ and evaluate at $(u_1^0, \ldots, u_n^0)$ the affine-linear functions $a_{i_1,\ldots,i_n} + i_1 u_1 + \ldots + i_n u_n$, $(i_1, \ldots, i_n) \in V$. The convex hull of the points of V that correspond to the affine-linear functions $a_{i_1,\ldots,i_n} + i_1 u_1 + \ldots + i_n u_n$ realizing the maximum is the polytope dual to δ. If δ is of dimension ℓ, then the dual polytope is of dimension $n - \ell$.

The dual polytopes form a subdivision Ψ_p of Δ_p (see, for example, [7] for details).

The *tropical hypersurface* defined by p in $\mathbb{R}^n$ is the polyhedral complex $\mathcal{X}_p$ whose $(n-1)$-dimensional facets are equipped with positive integer weights equal to integer lengths of the dual segments (the *integer length* of a segment with integer endpoints is the number of its integer points diminished by 1). We use the same notation $\mathcal{X}_p$ for this tropical hypersurface. The *degree* and *Newton polytope* of $\mathcal{X}_p$ are, respectively, the degree and Newton polytope of the defining tropical polynomial p. Figure 3 contains examples of tropical curves of small degrees in $\mathbb{R}^2$ (only the weights different from 1 are indicated); Figure 4 presents examples of dual subdivisions.

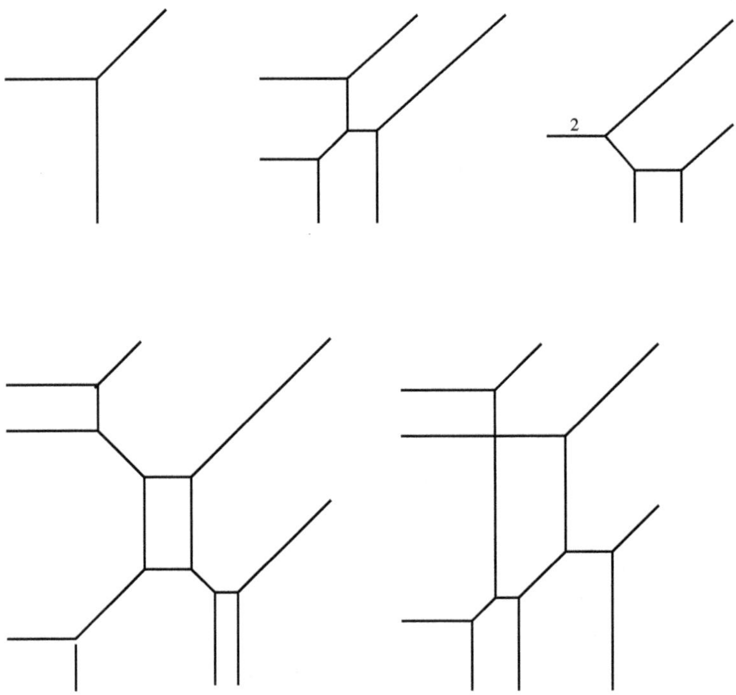

Fig. 3. Examples of tropical curves of degrees 1, 2 and 3 in $\mathbb{R}^2$

The subdivision Ψ_p can be described in a different way. Consider the convex hull in $\mathbb{R}^{n+1}$ of the graph of the function $(i_1,\dots,i_n) \mapsto -a_{i_1,\dots,i_n}$ defined on V. The projection to Δ_p of the lower part of the boundary of this convex hull gives the subdivision Ψ_p. In particular, Ψ_p is convex in the sense of Section 2.1: there exists a convex piecewise-linear function $\nu : \Delta_p \longrightarrow \mathbb{R}$ whose domains of linearity coincide with the n-dimensional polytopes of Ψ_p. The tropical polynomial function

$$f_p : (u_1, \ldots, u_n) \mapsto \max_{(i_1,\ldots,i_n)\in V} \{a_{i_1,\ldots,i_n} + i_1 u_1 + \ldots + i_n u_n\}$$

is the *Legendre transform* of the function $(i_1, \ldots, i_n) \mapsto -a_{i_1,\ldots,i_n}$ defined on V.

The tropical hypersurface $\mathcal{X}_p$ is said to be *non-singular*, if the dual subdivision Ψ_p is a primitive triangulation. In this case, the weights of all $(n-1)$-dimensional facets of $\mathcal{X}_p$ are equal to 1, but the latter property does not guarantee that the tropical hypersurface is non-singular.

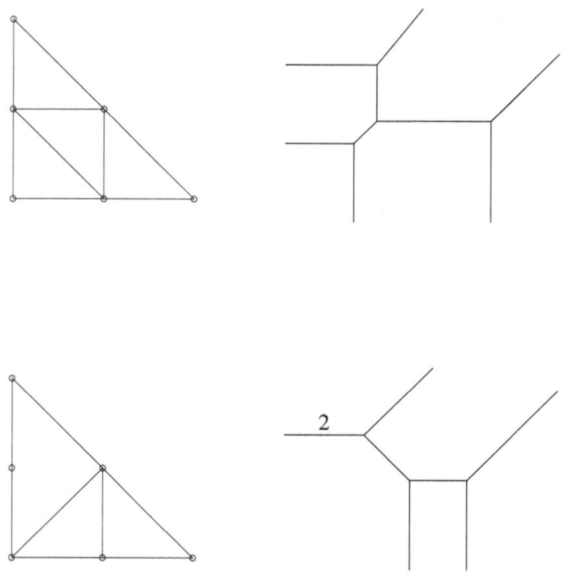

Fig. 4. Examples of tropical conics and their dual subdivisions

3.4. *Geometric interpretation*

Tropical hypersurfaces in $\mathbb{R}^n$ are $(n-1)$-dimensional weighted polyhedral complexes. Any face of such a tropical hypersurface $\mathcal{X}_p$ has rational direction (that is, direction defined over $\mathbb{Q}$). With any face we associate the intersection of the direction of the face with the integral lattice; the intersection is a lattice whose rank coincides with the dimension of the face. Furthermore, at any face δ of $\mathcal{X}_p$ of dimension $n-2$, the following *balancing condition* is satisfied: the vector

$$\sum_{\kappa} w_\kappa v_\kappa$$

belongs to the direction of δ, where the sum is taken over all $(n-1)$-dimensional faces κ of $\mathcal{X}_p$ that are adjacent to δ, the positive integer w_κ is the weight of κ, and v_κ is a vector belonging to the direction of κ such that

- v_κ together with a basis of the lattice associated with δ form a basis of the lattice associated with κ,
- v_κ points towards the $(n-1)$-dimensional half-space determined by κ.

This statement can be deduced, for example, from the duality formulated in Section 3.3. The other way around, it is easy to see that any $(n-1)$-dimensional finite weighted polyhedral complex in $\mathbb{R}^n$ (the weights are positive integers associated with the faces of dimension $n-1$) satisfying the two conditions above (all faces have rational directions and the balancing condition holds at each face of dimension $n-2$) is a tropical hypersurface, that is, it can be defined by a tropical polynomial.

3.5. *Limits of amoebas*

Another description of tropical hypersurfaces in $\mathbb{R}^n$ is *via* limits of amoebas. For any real number $t > 1$, consider the base t logarithmic map $\mathrm{Log}_t : (\mathbb{C}^\times)^n \to \mathbb{R}^n$ defined by

$$(z_1, \ldots, z_n) \mapsto (\log_t |z_1|, \ldots, \log_t |z_n|).$$

If Z is an algebraic variety in $(\mathbb{C}^\times)^n$, the *amoeba* (more precisely, *t-amoeba*) of Z is $\mathrm{Log}_t(\mathbb{C}Z) \subset \mathbb{R}^n$, where $\mathbb{C}Z$ is the complex point set of Z. Any tropical hypersurface $\mathcal{X}$ in $\mathbb{R}^n$ can be approximated by a family $\mathrm{Log}_t(\mathbb{C}Z_t)$, $t \to +\infty$, of amoebas of algebraic hypersurfaces having the same Newton polytope as $\mathcal{X}$. Indeed, if

$$p(u_1, \ldots, u_n) = \text{``} \sum_{(i_1,\ldots,i_n)\in V} a_{i_1,\ldots,i_n} u_1^{i_1} \ldots u_n^{i_n} \text{''}$$

is a tropical polynomial defining the tropical hypersurface $\mathcal{X}$, then an approximating family Z_t is given by the Viro polynomials

$$P_t(z_1, \ldots, z_n) = \sum_{(i_1,\ldots,i_n)\in V} t^{a_{i_1,\ldots,i_n}} z_1^{i_1} \ldots z_n^{i_n}.$$

If $t \to +\infty$, the amoebas $\mathrm{Log}_t(\mathbb{C}Z_t)$ converge to $\mathcal{X}$ on any compact subset of $\mathbb{R}^n$ (with respect to the Hausdorff metric on closed subsets, $\mathbb{R}^n$ being equipped with the Euclidean metric). This statement can be viewed as a version of the Mikhalkin-Rullgård theorem (see, [17, 22]). The coefficients of monomials of P_t can be multiplied by arbitrary non-zero complex numbers $\alpha_{i_1,\ldots,i_n}$. This does not change the limit of amoebas of the family: only the powers of t are important for a description of the limit. In Section 3.7, we will deal with real coefficients, and since we will be interested in the real part of the limiting object, we will have to take into account the signs of the coefficients of monomials.

3.6. *Hypersurfaces in tropical projective spaces*

Let n be a positive integer. The *tropical projective space* $\mathbb{T}P^n$ is defined as the quotient of $\mathbb{T}^{n+1} \setminus \{(-\infty, \ldots, -\infty)\}$ by the equivalence relation

$$(u_0, \ldots, u_n) \sim (u_0 + \lambda, \ldots, u_n + \lambda)$$

for any $(u_0, \ldots, u_n) \in \mathbb{T}^{n+1} \setminus \{(-\infty, \ldots, -\infty)\}$ and any $\lambda \in \mathbb{R} = \mathbb{T}^\times$. The space $\mathbb{T}P^n$ is homeomorphic to an n-dimensional simplex. Each point of $\mathbb{T}P^n$ has a *sedentarity*: this is the collection of indices of the coordinates that take value $-\infty$ at this point. All points of an open face of $\mathbb{T}P^n$ have the same sedentarity; the latter is called the sedentarity of the face. Each closed face of $\mathbb{T}P^n$ is naturally identified with the tropical projective space of the corresponding dimension. For any pair (A, B) of disjoint closed faces of $\mathbb{T}P^n$ such that the sum of dimensions of A and B is equal to $n-1$ (in such a situation, we write $A = B^{\mathrm{op}}$), there is a natural projection

$$\mathrm{pr}_{A,B} : \mathbb{T}P^n \setminus A \to B.$$

One can, of course, describe tropical projective spaces by gluing affine charts: each of the $n+1$ affine charts of $\mathbb{T}P^n$ is defined by the condition $u_j \neq -\infty$, where j is one of the integers $0, \ldots, n$. For example, the tropical projective line $\mathbb{T}P^1$ can be obtained in the following way: take two copies of $\mathbb{T}$ with coordinates $u_1^{(1)}$ and $u_1^{(2)}$, and glue these copies along $\mathbb{T}^\times$ using the identification $u_1^{(2)} = -u_1^{(1)}$. Similarly, the tropical projective space $\mathbb{T}P^n$ can be constructed by gluing $n+1$ copies of $\mathbb{T}^n$. The intersection of all these $n+1$ charts (that is, the interior of the simplex $\mathbb{T}P^n$) can be identified with $\mathbb{R}^n$ (once one of the charts is chosen; we always assume that the chosen chart is the one defined by $u_0 \neq -\infty$).

If $x \in \mathbb{T}P^n$ is a point, denote by $\mathcal{D}(x) \subset \mathbb{T}P^n$ the open strata containing x. The tangent space $\mathcal{T}(x)$ of $\mathcal{D}(x)$ at x is identified with $\mathbb{R}^k$, where k is the dimension of $\mathcal{D}(x)$.

Let d be a positive integer, and let $\mathcal{X} \subset \mathbb{R}^n$ be a (non-singular) tropical hypersurface of degree d having the standard n-simplex of size d as Newton polytope. The closure $\overline{\mathcal{X}} \subset \mathbb{T}P^n$ of $\mathcal{X}$ is called a *regular (non-singular) tropical hypersurface of degree d in $\mathbb{T}P^n$*. Naturally extending the map $\mathrm{Log}_t : (\mathbb{C}^\times)^n \to \mathbb{R}^n$ to the logarithmic map from $\mathbb{C}P^n$ to $\mathbb{T}P^n$, we can represent $\overline{\mathcal{X}}$ as limit of amoebas of hypersurfaces of degree d in $\mathbb{C}P^n$. These considerations can be generalized to other convex lattice polytopes and the corresponding toric varieties (see, for example, [7]).

3.7. *Combinatorial patchworking revisited*

A proof of the combinatorial patchworking theorem (Theorem 1) can be obtained along the following lines (see [27] for details). Consider a Viro polynomial

$$P_t(x_0, \ldots, x_n) = \sum_{(i_1,\ldots,i_n)\in V} \sigma_{i_1,\ldots i_n} t^{-\nu(i_1,\ldots i_n)} x_0^{d-i_1-\ldots-i_n} x_1^{i_1} \ldots x_n^{i_n},$$

where t is sufficiently big (we use the notation of Section 2.1). The statement is proved orthant by orthant, so let us study the zero set of P_t in the open positive orthant of the affine chart $x_0 \neq 0$ with the affine coordinates

$$y_1 = x_1/x_0, \quad \ldots, \quad y_n = x_n/x_0.$$

(The other orthants can be treated using the obvious coordinate changes.) So, from now on, we assume that $y_1 > 0, \ldots, y_n > 0$. The equality $P_t(y_1, \ldots, y_n) = 0$ can be rewritten as

$$P_t^+(y_1, \ldots, y_n) = P_t^-(y_1, \ldots, y_n),$$

where P_t^+ (respectively, P_t^-) is formed by the monomials of P_t that have positive (respectively, negative) coefficients.

When t tends to $+\infty$, the graphs of the polynomials P_t^+ and P_t^- in the logarithmic coordinates

$$u_1 = \log_t y_1, \quad \ldots, \quad u_n = \log_t y_n, \quad u_{n+1} = \log_t y_{n+1}$$

tend, respectively, to the graphs

$$u_{n+1} = p^+(u_1, \ldots, u_n) \quad \text{and} \quad u_{n+1} = p^-(u_1, \ldots, u_n)$$

of the tropical polynomials

$$p^{\pm}(u_1, \ldots, u_n) = \text{“} \sum_{(i_1,\ldots,i_n)\in V^{\pm}} -\nu(i_1, \ldots i_n)u_1^{i_1} \ldots u_n^{i_n} \text{”},$$

where V^+ (respectively, V^-) is the set of vertices of τ that have positive (respectively, negative) sign.

The projection that forgets the coordinate u_{n+1} sends the intersection of the graphs of p^+ and p^- to a part $\mathcal{X}_{\text{positive}}$ of the tropical hypersurface $\mathcal{X}$ defined in $\mathbb{R}^n$ (with the coordinates u_1, …, u_n) by the tropical polynomial

$$p(u_1, \ldots, u_n) = \text{“} \sum_{(i_1,\ldots,i_n)\in V} -\nu(i_1, \ldots i_n)u_1^{i_1} \ldots u_n^{i_n} \text{”}.$$

To describe $\mathcal{X}_{\text{positive}}$, recall that each connected component of the complement of $\mathcal{X} \subset \mathbb{R}^n$ corresponds to a vertex of τ. Since the vertices of τ have signs, the connected components of the complement of $\mathcal{X} \subset \mathbb{R}^n$ are divided into two classes: positive and negative. The part $\mathcal{X}_{\text{positive}}$ is the common boundary of the union of positive components and the union of negative components.

Since P_t^+ and P_t^- tend, respectively, to the graphs of the tropical polynomials p^+ and p^-, it can be shown that the projection (that forgets the coordinate u_{n+1}) of the intersection of the graphs of P_t^+ and P_t^- is isotopic (in the stratified sense) to $\mathcal{X}_{\text{positive}}$. Furthermore, under a natural identification of $\mathbb{T}P^n$ and the simplex T, the closure of $\mathcal{X}_{\text{positive}}$ in $\mathbb{T}P^n$ is isotopic (again in the stratified sense) to the intersection of the piecewise-linear hypersurface Γ with the simplex T (see Section 2.1 for notation).

4. Tropical homology

4.1. *Definition and properties*

This section contains a short introduction to tropical homology. We restrict our attention to the case of regular non-singular tropical hypersurfaces in $\mathbb{T}P^n$. For a general and more detailed presentation of tropical homology and cohomology, the reader is referred to [14] and [7].

Let k be a non-negative integer, and let $\Sigma = \bigcup \varsigma \subset \mathbb{R}^k$ be a rational polyhedral fan (that is, each cone of this fan is generated by a finite collection of integer vectors). For each cone $\varsigma \subset \Sigma$, we denote by $< \varsigma >_{\mathbb{Z}}$ the integral lattice in the vector subspace linearly spanned by ς.

For any non-negative integer p, the group ${}^{\mathbb{Z}}\mathcal{F}_p(\Sigma)$ is the subgroup of $\wedge^p\mathbb{Z}^k$ generated by the elements $v_1 \wedge \cdots \wedge v_p$, where $v_1, \ldots, v_p \in < \varsigma >_{\mathbb{Z}}$ for some cone $\varsigma \in \Sigma$. It is important that all p vectors $v_1, \ldots, v_p$ come from the *same* cone.

Let $\mathcal{X}$ be a regular non-singular tropical hypersurface in $\mathbb{T}P^n$. The polyhedral decomposition of $\mathcal{X}$ into faces gives it a natural cell structure. Let $x \in \mathcal{X}$ be a point; denote by I its sedentarity, and denote by $|I|$ the number of elements in I. We define $\Sigma(x)$, the *tangent cone to* $\mathcal{X}$ *at* x, to be the cone in $\mathcal{T}(x) \cong \mathbb{R}^{n-|I|}$ consisting of vectors $r \in \mathcal{T}(x)$ such that $x + \varepsilon r \in \mathcal{X} \cap \mathcal{D}(x)$ for a sufficiently small $\varepsilon > 0$ (depending on r).

For any abelian group G and any non-negative integer p, we define the coefficient group ${}^G\mathcal{F}_p(x)$ to be ${}^{\mathbb{Z}}\mathcal{F}_p(\Sigma(x)) \otimes G$. We will be particularly interested in the cases $G = \mathbb{Q}$ and $G = \mathbb{Z}/2\mathbb{Z}$. For any point x' belonging to the same open strata θ of $\mathcal{X}$ as x, the groups ${}^G\mathcal{F}_p(x')$ and ${}^G\mathcal{F}_p(x)$ are naturally identified; so we use the notation ${}^G\mathcal{F}_p(\theta)$ for these groups. The geometric meaning of the groups ${}^G\mathcal{F}_p(x)$ is as follows. The cone $\Sigma(x)$ can be seen as limit of t-amoebas of a constant family $\mathcal{L}_t = \mathcal{L}$, where $\mathcal{L}$ is a hypersurface defined in $(\mathbb{C}^\times)^{n-|I|}$ by a polynomial whose Newton polytope is dual (in the sense of Section 3.3) to the open face containing x. The monomials of such a defining polynomial are in a natural bijection with the regions of $\mathbb{T}P^n \setminus \mathcal{X}$ that are adjacent to x. Up to monomial coordinate change, the hypersurface $\mathcal{L}$ is a hyperplane in $(\mathbb{C}^\times)^{n-|I|}$ (that is, the complement of a hyperplane arrangement in $\mathbb{C}P^{n-|I|-1}$). The group ${}^G\mathcal{F}_p(x)$ is isomorphic to $H_p(\mathbb{C}\mathcal{L}; G)$.

Let θ_1 and θ_2 be adjacent open faces of $\mathcal{X}$ such that $\dim\theta_2 = \dim\theta_1 - 1$. There are two options: either θ_1 and θ_2 have the same sedentarity, or the sedentarity of θ_2 is equal to the sedentarity of θ_1 increased by 1. In the first case, denote by $\partial_{\theta_1,\theta_2} : {}^G\mathcal{F}_p(\theta_1) \to {}^G\mathcal{F}_p(\theta_2)$ the inclusion map. In the second case, denote by $\partial_{\theta_1,\theta_2} : {}^G\mathcal{F}_p(\theta_1) \to {}^G\mathcal{F}_p(\theta_2)$ the morphism induced by the restriction to θ_1 of the projection $\mathrm{pr}_{\theta_2^{\mathrm{op}},\theta_2}$.

The groups ${}^G\mathcal{F}_p$ can play the role of coefficient system. Considering the singular simplices *compatible with the stratification of* $\mathcal{X}$ (that is, the singular simplices such that the image of each open face of the standard simplex is contained in a face of $\mathcal{X}$) and using the coefficient maps $\partial_{\theta_1,\theta_2} : {}^G\mathcal{F}_p(\theta_1) \to {}^G\mathcal{F}_p(\theta_2)$, we obtain a simplical chain complex $\mathcal{C}(\mathcal{X};\, {}^G\mathcal{F}_p)$ and *tropical homology groups* $H_q(\mathcal{X};\, {}^G\mathcal{F}_p)$, $q = 0, 1, \ldots$.

Consider a family of Viro polynomials

$$P_t(x_0, \ldots, x_n) = \sum_{(i_1,\ldots,i_n) \in V} \alpha_{i_1,\ldots i_n} t^{-\nu(i_1,\ldots i_n)} x_0^{d-i_1-\ldots-i_n} x_1^{i_1} \ldots x_n^{i_n},$$

such that the amoebas of the corresponding hypersurfaces $Z_t \subset \mathbb{C}P^n$ approximate $\mathcal{X}$. A corollary of a more general statement proved in [14] is the following statement: *for any non-negative integers p and q, the dimension (over $\mathbb{Q}$) of the group $H_q(\mathcal{X}; {}^{\mathbb{Q}}\mathcal{F}_p)$ is equal to the Hodge number $h^{p,q}(Z_t)$ of Z_t for sufficiently big t.*

4.2. *Inequalities for Betti numbers*

As already mentioned in Section 1.2, in the case of primitive T-hypersurface X in $\mathbb{R}P^n$, there are the following upper bounds for individual Betti numbers $b_p(\mathbb{R}X)$.

Theorem 3 (A. Renaudineau and K. Shaw). *Let n and d be positive integers. Then, for any integer $0 \le p \le n-1$ and any primitive T-hypersurface X of degree d in $\mathbb{R}P^n$, we have*

$$b_p(\mathbb{R}X) \le \begin{cases} h^{p,n-1-p}(\mathbb{C}X), & \text{if } p = (n-1)/2, \\ h^{p,n-1-p}(\mathbb{C}X) + 1, & \text{otherwise.} \end{cases}$$

Remark 4.1. The inequalities of Theorem 3 are false in general (for non-singular real algebraic hypersurfaces in $\mathbb{R}P^n$). For example,

- the real part of a non-singular cubic surface in $\mathbb{R}P^3$ can have two connected components, while the geometric genus $h^{2,0}(Y_3)$ of such a surface Y_3 is zero;
- the real part of a non-singular quartic surface in $\mathbb{R}P^3$ can have 10 connected components, while the geometric genus $h^{2,0}(Y_4)$ of such a surface Y_4 is equal to 1.

Furthermore, the inequalities of Theorem 3 are false even for T-hypersurfaces in $\mathbb{R}P^n$ (without assumption of primitivity). Examples (for the first Betti number in the case of T-surfaces in $\mathbb{R}P^3$) can be found in [13].

The idea of the proof of Theorem 3 is as follows (for details, the reader is referred to [21]). The hypersurface X is produced by the combinatorial patchworking; let

$$P_t(x_0, \dots, x_n) = \sum_{(i_1,\dots,i_n)\in V} \sigma_{i_1,\dots i_n} t^{-\nu(i_1,\dots i_n)} x_0^{d-i_1-\dots-i_n} x_1^{i_1} \dots x_n^{i_n},$$

be the corresponding family of Viro polynomials, and let $\mathcal{X}$ be the limiting tropical hypersurface of degree d in $\mathbb{T}P^n$. On $\mathcal{X}$, one can introduce a real version $\mathcal{S}_p$ of the coefficient system ${}^{\mathbb{Z}/2\mathbb{Z}}\mathcal{F}_p$, $p = 0, 1, \dots$. For each point $x \in \mathcal{X}$, we can choose a hypersurface $\mathcal{L}$ in $(\mathbb{C}^\times)^{n-|I|}$, where I is the sedentarity of x (see Section 4.1), in such a way that a defining polynomial of $\mathcal{L}$ is real and the signs of its coefficients coincide with the signs of the corresponding regions of $\mathbb{T}P^n \setminus \mathcal{X}$ (and, thus, with the signs of the corresponding monomials of P_t). For any non-negative integer p, we put $\mathcal{S}_p(x) = H_p(\mathbb{R}\mathcal{L}; \mathbb{Z}/2\mathbb{Z})$, where $\mathbb{R}\mathcal{L}$ is the real part of $\mathcal{L}$. Thus, $\mathcal{S}_p(x) = 0$ if $p > 0$, and the dimension (over $\mathbb{Z}/2\mathbb{Z}$) of $\mathcal{S}_0(x)$ is equal to the number of connected components of $\mathbb{R}\mathcal{L}$. The definition of the morhisms $\partial_{\theta_1,\theta_2}$ given in Section 4.1 can be easily adapted to the groups $\mathcal{S}_0(x)$, and we obtain a simplicial chain complex $\mathcal{C}(\mathcal{X}, \mathcal{S}_0)$ and homology groups $H_q(\mathcal{X}, \mathcal{S}_0)$ for any non-negative integer q. One has $H_q(\mathbb{R}X; \mathbb{Z}/2\mathbb{Z}) \simeq H_q(\mathcal{X}; \mathcal{S}_0)$.

For $\mathcal{L}$, or for the corresponding complement of hyperplane arrangement, one can consider the so-called *Viro homomorphisms* bv_p, $p = 0, 1, \ldots$ (see [8] for the definition of these homomorphisms). The complement of any hyperplane arrangement is a maximal variety in the sense of the Smith-Thom inequality (see [19]); thus, the target space of bv_p is $H_p(\mathbb{C}\mathcal{L}; \mathbb{Z}/2\mathbb{Z}) = {}^{\mathbb{Z}/2\mathbb{Z}}\mathcal{F}_p(x)$, see [8] for explanations. The homomorphism bv_0 is defined on $\mathcal{S}_0(x) = H_0(\mathbb{R}\mathcal{L}; \mathbb{Z}/2\mathbb{Z})$, and for each $p \geq 1$, the homomorphism bv_p is defined on the kernel K_p of bv_{p-1}. We obtain a filtration

$$\mathcal{S}_0(x) = K_0 \supset K_1 \supset \ldots \supset K_{n-|I|-1} \supset K_{n-|I|} = 0.$$

The maximality of $\mathcal{L}$ implies that all the homomorphisms bv_p are isomorphisms: for any integer $0 \leq p \leq n - |I| - 1$, one has $K_p/K_{p+1} \simeq {}^{\mathbb{Z}/2\mathbb{Z}}\mathcal{F}_p(x)$.

The above filtration gives rise to a filtration of the simplicial chain complex $\mathcal{C}(\mathcal{X}; \mathcal{S}_0)$ and produce a spectral sequence that calculates the homology groups $H_q(\mathbb{R}X; \mathbb{Z}/2\mathbb{Z}) \simeq H_q(\mathcal{X}; \mathcal{S}_0)$. The first page of the spectral sequence is formed by the groups $H_q(\mathcal{X}, {}^{\mathbb{Z}/2\mathbb{Z}}\mathcal{F}_p)$. Thus, we obtain the inequalities

$$b_p(\mathbb{R}X) \leq \sum_{q=0}^{n-1} \dim_{\mathbb{Z}/2\mathbb{Z}} H_q(\mathcal{X}, {}^{\mathbb{Z}/2\mathbb{Z}}\mathcal{F}_p).$$

As it was shown by Ch. Arnal, A. Renaudineau and K. Shaw [2], for any non-negative integers p and q, one has

$$\dim_{\mathbb{Z}/2\mathbb{Z}} H_q(\mathcal{X}, {}^{\mathbb{Z}/2\mathbb{Z}}\mathcal{F}_p) = \dim_{\mathbb{Q}} H_q(\mathcal{X}, {}^{\mathbb{Q}}\mathcal{F}_p).$$

Thus, one can use the interpretation given in [14] for tropical homology groups (see Section 4.1) in order to obtain the inequality

$$b_p(\mathbb{R}X) \leq \sum_{q=0}^{n-1} h^{p,q}(\mathbb{C}X).$$

It remains to notice that $h^{p,q}(\mathbb{C}X) = 0$ if $p \neq q$ and $p + q \neq n - 1$; in addition $h^{p,p}(\mathbb{C}X) = 1$ if $p \neq (n-1)/2$.

References

[1] Ch. Arnal, *Patchwork combinatoire et topologie d'hypersurfaces algébriques réelles*, Master thesis, Université Pierre et Marie Curie, Paris, 2017.

[2] Ch. Arnal, A. Renaudineau, and K. Shaw, *Lefschetz section theorems for tropical hypersurfaces*. Preprint arXiv:1907.06420, 2019.

[3] B. Bertrand, *Euler characteristic of primitive T-hypersurfaces and maximal surfaces*, J. Inst. Math. de Jussieu **9** (2010), no. 1, 1–27.

[4] F. Bihan, *Une sextique de l'espace projectif réel avec un grand nombre d'anses*, Revista Matematica Complutense **14** (2001), no. 2, 439–461.

[5] F. Bihan, *Asymptotic behaviour of Betti numbers of real algebraic surfaces*, Comm. Math. Helv. **78** (2003), 227–244.

[6] G. Bredon, *Introduction to Compact Transformation Groups*, Academic Press, 1972.

[7] E. Brugallé, I. Itenberg, G. Mikhalkin, and K. Shaw, *Brief introduction to tropical geometry*, Proceedings of the Gökova Geometry-Topology Conference 2014 (GGT), 2015, 1–75.

[8] A. Degtyarev and V. Kharlamov, *Topological properties of real algebraic varieties: Rokhlin's way*, Rus. Math. Surv. **55** (2000), no. 4, 735–814.

[9] E. E. Floyd, *On periodic maps and the Euler characteristics of associated spaces*, Trans. Amer. Math. Soc. **72** (1952), 138–147.

[10] A. Harnack, *Über Vieltheiligkeit der ebenen algebraischen Curven*, Math. Ann. **10** (1876), 189–199.

[11] D. Hilbert, *Mathematische probleme*, Nachrichten von der Gesellschaft der Wissenschaften zu Göttingen, 1900.

[12] I. Itenberg, *Contre-exemples à la conjecture de Ragsdale*, C. R. Acad. Sci. Paris, Sér. I, Math. **317** (1993), no. 3, 277–282.

[13] I. Itenberg, *Topology of real algebraic T-surfaces*, Rev. Mat. Univ. Complut. Madrid **10**, Real algebraic and analytic geometry (Segovia, 1995), 1997, pp. 131–152.

[14] I. Itenberg, L. Katzarkov, G. Mikhalkin, and I. Zharkov, *Tropical homology*, Math. Ann., 2018. DOI: 10.1007/s00208-018-1685-9.

[15] I. Itenberg and V. Kharlamov, *Towards the maximal number of components of a nonsingular surface of degree* 5 *in* $\mathbb{R}P^3$, Topology of real algebraic varieties and related topics, volume 173 of Amer. Math. Soc. Transl. Ser. 2, Amer. Math. Soc., Providence, RI, 1996, pp. 111–118.

[16] I. Itenberg and O. Viro, *Asymptotically maximal real algebraic hypersurfaces of projective space*, Proceedings of Gökova Geometry-Topology Conference 2006 (GGT), 2007, 91–105.

[17] G. Mikhalkin, *Decomposition into pairs-of-pants for complex algebraic hypersurfaces*, Topology **43** (2004), no. 5, 1035–1065.

[18] S. Orevkov, *Real quintic surface with* 23 *components*, C. R. Acad. Sci. Paris, Sér. I, Math. **333** (2001), no. 2, 115–118.

[19] P. Orlik and H. Terao, *Arrangements of hyperplanes*, Grundlehren der Mathematischen Wissenschaften **300**, Springer-Verlag, Berlin, 1992.

[20] V. Ragsdale, *On the arrangement of the real branches of plane algebraic curves*, Amer. J. Math. **28** (1906), 377–404.

[21] A. Renaudineau and K. Shaw, *Bounding the Betti numbers of real hypersurfaces near the tropical limit.* Preprint arXiv:1805.02030, 2018.

[22] H. Rullgård, *Stratification des espaces de polynômes de Laurent et la structure de leurs amibes*, C. R. Acad. Sci. Paris, Sér I, Math. **331** (2000), 355–358.

[23] R. Thom, *Sur l'homologie des variétés algébriques réelles*, Diff. and Comb. Top., Symp. in honor of M. Morse, Princeton Univ. Press, 1965, pp. 255–265.

[24] O. Viro, *Gluing of algebraic hypersurfaces, smoothing of singularities and construction of curves*, Proc. Leningrad Int. Topological Conf. (Leningrad, Aug. 1983), Nauka, Leningrad, 1983, pp. 149–197 (in Russian).

[25] O. Viro, *Progress in the topology of real algebraic varieties over the last six years*, Rus. Math. Surv. **41** (1986), no. 3, 55–82.

[26] O. Viro, *Patchworking real algebraic varieties.* Preprint, Uppsala University, 1994.

[27] O. Viro, *Dequantization of real algebraic geometry on logarithmic paper*, European Congress of Mathematics, Vol. I (Barcelona, 2000), volume 201 of Progr. Math., Birkhäuser, Basel, 2001, pp. 135–146.

[28] G. Wilson, *Hilbert's sixteenth problem*, Topology **17** (1978), no. 1, 53–73.

On continuous rational functions

Wojciech Kucharz

Institute of Mathematics
Faculty of Mathematics and Computer Science,
Jagiellonian University,
Łojasiewicza 6
30-348 Kraków, Poland
E-mail: Wojciech.Kucharz@im.uj.edu.pl

The aim of this article is to present some recent developments in real algebraic geometry. We focus on continuous functions that admit a rational representation. Such functions have some remarkable algebraic and geometric properties.

Keywords: Real algebraic variety, regular function, rational function, regulous function, regulous variety

2010 MSC: 14P05, 14P99, 26C15, 57R22, 58A07

0. Introduction

The purpose of the present article is to provide an introduction to *regulous geometry*, which is a new branch of real algebraic geometry developed in [4, 9–16, 21–38, 43, 46, 47] and focusing on functions that admit a rational representation. The name *regulous* is derived from *regular* and *continuous*. We do not attempt to write a comprehensive survey, but instead concentrate on a few ideas that are in some form present in all the papers cited above.

Throughout this paper, by a function on a set Ω we always mean a real-valued function. Given a collection F of functions on Ω, we set

$$Z(F) = \{x \in \Omega \colon f(x) = 0 \text{ for all } f \in F\}$$

and write $Z(f_1, \dots, f_r)$ for $Z(F)$ if $F = \{f_1, \dots, f_r\}$. For foundations of real algebraic geometry we refer to [6].

A locally ringed space $(X, \mathcal{R}_X)$, where $\mathcal{R}_X$ is a sheaf of functions on X, is called an *affine real algebraic variety* if, for some positive integer n, it is isomorphic to an algebraic subset of $\mathbb{R}^n$ endowed with the Zariski topology and the sheaf of regular functions. The class of affine real algebraic varieties is identical with the class of quasiprojective real algebraic varieties [6,

Proposition 3.2.10, Theorem 3.4.4]. Consequently, many useful constructions can be performed within the class of affine real algebraic varieties, blowing-up being an important example. Algebraic subvarieties of X are always assumed to be Zariski closed subsets. The set of smooth points

$$X^{\mathrm{sm}} := \{x \in X\colon \text{ the local ring } \mathcal{R}_{X,x} \text{ is regular}\}$$

is Zariski open dense in X, hence the singular locus

$$\mathrm{Sing}(X) := X \setminus X^{\mathrm{sm}}$$

is Zariski closed nowhere dense in X.

It will be convenient to consider regular functions in a more general setting than usual. We say that a function $f\colon W \to \mathbb{R}$, defined on some subset W of X, is *regular at a point* $x \in W$ if there exist a Zariski open neighborhood $X_x \subseteq X$ of x and a regular function $\varphi_x\colon X_x \to \mathbb{R}$ such that $f|_{W\cap X_x} = \varphi_x|_{W\cap X_x}$. Thus, assuming that X is an algebraic subset of $\mathbb{R}^n$, the function f is regular at x if and only if there exist two polynomial functions p, q on $\mathbb{R}^n$ with $q(x) \neq 0$ and $f = p/q$ on $W \setminus Z(q)$. We say that f is a *regular function* if it is regular at every point in W.

Any rational function R on X determines a regular function

$$R\colon X \setminus \mathrm{Pole}(R) \to \mathbb{R},$$

where $\mathrm{Pole}(R)$ stands for the polar set of R.

Convention. Any real algebraic variety is endowed with the Zariski topology (Z-topology) and carries also the Euclidean topology (E-topology) which is induced by the standard metric on $\mathbb{R}$. Unless explicitly stated otherwise, we will always work with the Euclidean topology.

1. Regulous functions on $\mathbb{R}^n$

In this section k stands for a fixed nonnegative integer.

Definition 1.1. A function f, defined on an open subset $U \subseteq \mathbb{R}^n$, is said to be *k-regulous* if it is of class $\mathcal{C}^k$ and there exist two polynomial functions p, q on $\mathbb{R}^n$ such that $Z(q) \neq \mathbb{R}^n$ and $f = p/q$ on $U \setminus Z(q)$; a 0-regulous function is called *regulous*.

Example 1.2. The function f_k on $\mathbb{R}^2$ defined by

$$f_k(x,y) = \begin{cases} \frac{x^{3+k}}{x^2+y^2} & \text{for } (x,y) \neq (0,0) \\ 0 & \text{for } (x,y) = (0,0) \end{cases}$$

is k-regulous but it is not $(k+1)$-regulous.

For any open set $U \subseteq \mathbb{R}^n$, the collection $\mathcal{R}^k(U)$ of all k-regulous functions on U forms a ring (commutative with 1) under pointwise addition and multiplication.

Proposition 1.3. *The ring $\mathcal{R}^k(\mathbb{R}^n)$ is not Noetherian if $n \geq 2$.*

Proof. For any nonnegative integer m, the function $g_m \colon \mathbb{R}^n \to \mathbb{R}$ defined by

$$g_m(x_1, x_2, \ldots, x_n) = \begin{cases} \frac{x_1^{3+k}}{x_1^2+(x_2-m)^2} & \text{if } (x_1, x_2) \neq (0, m) \\ 0 & \text{otherwise} \end{cases}$$

is k-regulous. Let $I_m \subseteq \mathcal{R}^k(\mathbb{R}^n)$ be the ideal generated by $g_0, \ldots, g_m$. It suffices to prove that $I_m \neq I_{m+1}$ for all m. Suppose to the contrary that $I_m = I_{m+1}$ for some m, hence

$$g_{m+1} = \sum_{i=0}^{m} h_i g_i,$$

where $h_i \in \mathcal{R}^k(\mathbb{R}^n)$ for $i = 0, \ldots, m$. Evaluating at $(x_1, m+1, 0, \ldots, 0)$, we get

$$x_1^{1+k} = \sum_{i=0}^{m} h_i(x_1, m+1, 0, \ldots, 0) \frac{x_1^{3+k}}{x_1^2 + (m+1-i)^2}.$$

Consequently,

$$x_1^{1+k} \left(1 - \sum_{i=0}^{m} h_i(x_1, m+1, 0, \ldots, 0) \frac{x_1^2}{x_1^2 + (m+1-i)^2} \right) = 0.$$

Clearly, the last equality cannot hold for all x_1 close to $0 \in \mathbb{R}$, a contradiction. □

Proposition 1.3 comes from [12, Proposition 4.16]; its proof is an adaptation of the original argument given by Kurdyka [39, p. 462] in the context of arc-analytic semialgebraic functions. Actually, according to [9], no maximal ideal of the ring $\mathcal{R}^k(\mathbb{R}^n)$ is finitely generated if $n \geq 2$.

Although Proposition 1.3 is a negative statement, the ring $\mathcal{R}^k(\mathbb{R}^n)$ does have some remarkable properties. Fichou, Huisman, Mangolte, and Monnier [12, Théorème 5.24] obtained the following.

Theorem 1.4. *Let I be an ideal of the ring $\mathcal{R}^k(\mathbb{R}^n)$. If a function f in $\mathcal{R}^k(\mathbb{R}^n)$ vanishes on $Z(I)$, then f^N belongs to I for some positive integer N.*

Theorem 1.4 has exactly the same formulation as the classical Hilbert's Nullstellensatz for the ring of polynomials in n variables over an algebraically closed field. Let us recall that a variant of the Nullstellensatz is available for the ring of polynomials over $\mathbb{R}$, but it has a completely different formulation and plays an entirely different role [6, Corollary 4.1.8].

After some preparation, we will give in Section 3 a somewhat simplified version of the original proof of Theorem 1.4.

Remark 1.5. A function on $\mathbb{R}^n$ that is regulous and of class $\mathcal{C}^\infty$ is actually regular [24, Proposition 2.1]. Therefore one gains no new insight by considering such functions.

2. Regulous functions on varieties

Let X be an affine real algebraic variety and let $f\colon W \to \mathbb{R}$ be a function defined on some subset W of X.

Definition 2.1. Let $Y \subseteq X$ be the Zariski closure of W. A rational function R on Y is said to be a *rational representation* of f if there exists a Zariski open dense subset $Y^0 \subseteq Y$ such that

$$Y^0 \subseteq Y \setminus \mathrm{Pole}(R) \text{ and } f|_{W\cap Y^0} = R|_{W\cap Y^0}.$$

By a *stratification* of X, we mean a finite collection of pairwise disjoint Zariski locally closed subsets of X whose union is X.

Proposition 2.2. *For $f\colon W \to \mathbb{R}$, the following conditions are equivalent:*

(a) *For every algebraic subvariety $Z \subseteq X$, the restriction $f|_{W\cap Z}$ admits a rational representation.*

(a′) *For every irreducible algebraic subvariety $Z \subseteq X$, the restriction $f|_{W\cap Z}$ admits a rational representation.*

(b) *There exists a sequence of algebraic subvarieties*

$$X = X_0 \supseteq X_1 \supseteq \ldots \supseteq X_{m+1} = \emptyset$$

such that the restriction of f to $W \cap (X_i \setminus X_{i+1})$ is a regular function for $i = 0, \ldots, m$.

(c) *There exists a stratification $\mathcal{S}$ of X such that for every stratum $S \in \mathcal{S}$ the restriction of f to $W \cap S$ is a regular function.*

Proof. The argument is straightforward. □

Recall that $f\colon W \to \mathbb{R}$ is a *semialgebraic function* if its graph is a semialgebraic subset of $X \times \mathbb{R}$; in that case, $W \subseteq X$ is a semialgebraic subset by the Tarski–Seidenberg theorem [6, Theorem 2.2.1].

A subset $A \subseteq X$ is said to be *constructible* if it belongs to the Boolean algebra generated by the Zariski closed subsets of X; or equivalently if A is the finite union of Zariski locally closed subsets of X.

Corollary 2.3. *Assume that for* $f\colon W \to \mathbb{R}$ *the equivalent conditions* (a), (a′), (b), (c) *are satisfied. Then we have:*

(i) *If* $W \subseteq X$ *is a semialgebraic set, then* f *is a semialgebraic function.*
(ii) *If* $W \subseteq X$ *is a constructible set, then so is the zero set* $Z(f) \subseteq X$.

Proof. Obvious. □

Definition 2.4. We say that $f\colon W \to \mathbb{R}$ is a *regulous function* if it is continuous and the equivalent conditions (a), (a′), (b), (c) are satisfied.

In some papers, regulous functions are called *hereditarily rational* [21, 22] or *stratified-regular* [28, 34, 35, 46]. A continuous function that admits a rational representation is often called simply a *continuous rational function* [24–27, 29, 33].

Clearly, any regulous function is continuous and admits a rational representation. The converse holds in an important special case.

Theorem 2.5. *Let* X *be an affine real algebraic variety,* $W \subseteq X^{sm}$ *an open subset, and* $f\colon W \to \mathbb{R}$ *a continuous function that admits a rational representation. Then* f *is a regulous function.*

Proof. Let $Z \subseteq X$ be an irreducible algebraic subvariety. The issue is to prove that the restriction $f|_{W\cap Z}$ admits a rational representation. It suffices to consider X irreducible. We use induction on $n := \dim X$, starting with $n = 0$ when there is nothing to prove. If $n \geq 1$ and $\dim Z < n$, we can choose a codimension 1 irreducible algebraic subvariety $Z_1 \subseteq X$ with $Z \subseteq Z_1$. Hence, by the induction hypothesis, the proof is reduced to the case $\operatorname{codim}_X Z = 1$. By replacing X with a Zariski open subset, we may assume that both X and Z are smooth. The nontrivial case is $W \cap Z \neq \emptyset$. Since f admits a rational representation, there exist a rational function R on X and a Zariski open dense subset $X^0 \subseteq X$ such that

$$X^0 \subseteq X \setminus \operatorname{Pole}(R) \text{ and } f|_{W\cap X^0} = R|_{W\cap X^0}.$$

If Z is not contained in $\mathrm{Pole}(R)$, then the restriction R_Z of R is a well-defined rational function on Z. Clearly, R_Z is a rational representation of $f|_{W\cap Z}$ and the proof is complete.

Suppose then that $Z \subseteq \mathrm{Pole}(R)$. This leads to a contradiction as we will now show. Regarding $\mathbb{R}$ as a subset of the real projective line $\mathbb{P}^1(\mathbb{R})$, we obtain a rational map $\tilde{R}\colon X \to \mathbb{P}^1(\mathbb{R})$ determined by R. The polar set $\mathrm{Pole}(\tilde{R}) \subseteq X$ of $\tilde{R}$ has codimension at least 2; see [20, p. 129, Theorem 2.17]. Since $f\colon W \to \mathbb{R}$ is continuous and $W \cap X^0$ is dense in W, we get

$$f(x) = \tilde{R}(x) \text{ for all } x \in W \cap (X \setminus \mathrm{Pole}(\tilde{R})).$$

It follows that

$$W \cap (X \setminus \mathrm{Pole}(R)) = W \cap (X \setminus \mathrm{Pole}(\tilde{R})),$$

hence $W \cap Z \subseteq \mathrm{Pole}(\tilde{R})$. Since $W \cap Z$ is Zariski dense in Z, we get $Z \subseteq \mathrm{Pole}(\tilde{R})$, a contradiction. □

Remark 2.6. It follows from Theorem 2.5 that Definitions 1.1 and 2.4 are compatible.

Theorem 2.5 is a variant of [22, Proposition 8]; the assumption $W \subseteq X^{\mathrm{sm}}$ cannot be omitted as shown in the following example.

Example 2.7. Consider the algebraic surface

$$S := \{(x, y, z) \in \mathbb{R}^3 \colon x^3 - (1 + z^2)y^3 = 0\}$$

and the function $f\colon S \to \mathbb{R}$ defined by

$$f(x, y, z) = (1 + z^2)^{\frac{1}{3}}.$$

Note that S is irreducible, $\mathrm{Sing}(S) = z$-axis and $f(x, y, z) = x/y$ on $S \setminus (z\text{-axis})$. In particular, f is continuous and has a rational representation. However, f is not regulous since $f|_{z\text{-axis}}$ does not admit a rational representation.

Example 2.7 comes from the paper of Kollár and Nowak [22]. As it is amply demonstrated in [22], continuous functions that admit a rational representation behave in a rather unusual way and have many undesirable properties. By introducing regulous functions, we obtain a more useful class of functions.

Theorem 2.8. *Let X be an affine real algebraic variety and let $f\colon W \to \mathbb{R}$ be a regulous function defined on some subset W of X. Let $\gamma\colon (-1, 1) \to X$ be an analytic arc with $\gamma((-1, 1)) \subseteq W$. Then the composite function $f \circ \gamma\colon (-1, 1) \to \mathbb{R}$ is analytic.*

Proof. We may assume that X is an algebraic subset of $\mathbb{R}^n$. The Zariski closure $Z \subseteq X$ of $\gamma((-1,1))$ is an irreducible algebraic set (possibly with $\dim Z > 1$). By assumption, the restriction $f|_{W \cap Z}$ admits a rational representation. Note that the Zariski closure of $W \cap Z$ in X coincides with Z. Hence there exist an algebraic subset Z_1 of Z and two polynomial functions p, q on $\mathbb{R}^n$ with the following properties: $\dim Z_1 < \dim Z$, $Z \cap Z(q) \subseteq Z_1$, and

$$f = p/q \quad \text{on} \quad W \cap (Z \setminus Z_1).$$

Clearly, $I_1 := \gamma^{-1}(Z_1)$ is a discrete subset of $I = (-1,1)$, and

$$f(\gamma(t)) = \frac{p(\gamma(t))}{q(\gamma(t))} \quad \text{for} \quad t \in I \setminus I_1.$$

It follows that $f \circ \gamma$ is a meromorphic function on $(-1,1)$. Thus $f \circ \gamma$ is analytic, being meromorphic and continuous. $\square$

Theorem 2.8 means that regulous functions are *arc-analytic*. The latter class of functions, introduced by Kurdyka [39], has numerous applications in algebraic/analytic geometry and singularity theory; see [1–3, 40, 42, 44] and the references therein.

3. The constructible topology

To begin with we establish basic properties of constructible sets.

Lemma 3.1. *Let X be an affine real algebraic variety, $A \subseteq X$ a constructible subset, and $Y \subseteq X$ the Zariski closure of A. Then $Y^{sm} \subseteq \overline{A}$, where $\overline{A}$ denotes the closure of A in X (in the E-topology).*

Proof. The set A can be written as a finite union

$$A = \bigcup_{i \in I} (Z_i \setminus Z_i'),$$

where $Z_i' \subseteq Z_i$ are Zariski closed subsets of X with $Z_i' \neq Z_i$ and Z_i irreducible for all $i \in I$. Note that $Z_i \subseteq Y$ because $Z_i \setminus Z_i' \subseteq Y$ and the Zariski closure of $Z_i \setminus Z_i'$ is equal to Z_i. Let $Y_1, \ldots, Y_r$ be the irreducible components of Y. Then $Z_i \subseteq Y_j$ for some j. Setting $I_j := \{i \in I \colon Z_i \subseteq Y_j\}$, we get

$$Y_j = \bigcup_{i \in I_j} Z_i$$

for all $j = 1, \ldots, r$. Set

$$Y_j^{'} := \bigcup_{i \in I_j} Z_i^{'},$$

then $\dim Y_j^{'} < \dim Y_j$ and $Y_j \setminus Y_j^{'} \subseteq A$. Since $Y_j^{\mathrm{sm}} \setminus Y_j^{'}$ is E-dense in Y_j^{sm}, we get $Y_j^{\mathrm{sm}} \subseteq \overline{A}$. Hence

$$Y^{\mathrm{sm}} \subseteq Y_1^{\mathrm{sm}} \cup \ldots \cup Y_r^{\mathrm{sm}} \subseteq \overline{A},$$

which completes the proof. □

Recall that a topological space X is called *Noetherian* if any descending chain of its closed subsets

$$A_1 \supseteq A_2 \supseteq A_3 \supseteq \ldots$$

is stationary, that is, there exists a positive integer m such that $A_i = A_m$ for all $i \geq m$.

Proposition 3.2. *Let X be an affine real algebraic variety. Then the collection of all constructible E-closed subsets of X forms the family of closed sets for a Noetherian topology on X.*

Proof. It suffices to prove that any chain

$$A_1 \supseteq A_2 \supseteq A_3 \supseteq \ldots$$

of constructible E-closed subsets of X is stationary. To this end, let $Y_i \subseteq X$ be the Zariski closure of A_i. We use induction on $\dim Y_1$. The case $\dim Y_1 = 0$ is obvious since then Y_1 is a finite set.

Suppose that $\dim Y_1 \geq 1$. The chain

$$Y_1 \supseteq Y_2 \supseteq Y_3 \supseteq \ldots$$

of Zariski closed subsets is stationary, hence there exists a positive integer l such that $Y_i = Y_l$ for all $i \geq l$. Setting $Y := Y_l$, we get in view of Lemma 3.1 a sequence

$$A_l \setminus Y^{\mathrm{sm}} \supseteq A_{l+1} \setminus Y^{\mathrm{sm}} \supseteq A_{l+2} \setminus Y^{\mathrm{sm}} \supseteq \ldots$$

of constructible E-closed subsets of X. This sequence is stationary by the induction hypothesis since

$$A_l \setminus Y^{\mathrm{sm}} \subseteq \mathrm{Sing}(Y) \text{ and } \dim \mathrm{Sing}(Y) < \dim Y \leq \dim Y_1.$$

It follows that the chain

$$A_1 \supseteq A_2 \supseteq A_3 \supseteq \ldots$$

is stationary. □

The topology on X described in Proposition 3.2 is called the *constructible topology* (*C-topology*).

On X we have

$$\text{Z-topology} \subseteq \text{C-topology} \subseteq \text{E-topology},$$

where the inclusions are proper except some trivial cases.

The following example is instructive.

Example 3.3. The Cartan umbrella

$$S := \{(x, y, z) \in \mathbb{R}^3 \colon x^3 - z(x^2 + y^2) = 0\}$$

is an irreducible algebraic surface with $\operatorname{Sing}(S) = z$-axis. The surface S is connected and $S \setminus (z\text{-axis})$ is not dense in S. The set

$$A := (S \setminus (z\text{-axis})) \cup \{(0, 0, 0)\}$$

is C-closed but not Z-closed in $\mathbb{R}^3$.

The following classical result on semialgebraic functions will be useful; see [6, Proposition 2.6.4] for the proof.

Proposition 3.4. *Let $f \colon A \to \mathbb{R}$ be a continuous semialgebraic function defined on a locally closed semialgebraic subset $A \subseteq \mathbb{R}^n$. Let $g \colon A \setminus Z(f) \to \mathbb{R}$ be a continuous semialgebraic function. Then for any integer N large enough, the function $h \colon A \to \mathbb{R}$ defined by*

$$h = f^N g \text{ on } A \setminus Z(f) \text{ and } h = 0 \text{ on } Z(f)$$

is continuous.

A variant of Proposition 3.4 holds for k-regulous functions, where k is a fixed nonnegative integer.

Lemma 3.5. *Let $f \colon U \to \mathbb{R}$ be a k-regulous function defined on an open semialgebraic subset $U \subseteq \mathbb{R}^n$. Let $g \colon U \setminus Z(f) \to \mathbb{R}$ be a k-regulous function. Then for any integer N large enough, the function $h \colon U \to \mathbb{R}$ defined by*

$$h = f^N g \text{ on } U \setminus Z(f) \text{ and } h = 0 \text{ on } Z(f)$$

is k-regulous.

Proof. Both f and g are continuous semialgebraic functions; see [6, Proposition 2.2.2] or Corollary 2.3 (i) and Remark 2.6. Hence, by Proposition 3.4, h is continuous if N is large enough. Furthermore, h admits a rational representation. Therefore, the proof is complete if $k = 0$.

Suppose that $k \geq 1$ and let $\alpha = (\alpha_1, \dots, \alpha_n)$ be an n-tuple of nonnegative integers with $|\alpha| := \alpha_1 + \dots + \alpha_n \leq k$. If s is a sufficiently large integer (depending only on k), then Proposition 3.4 implies that the function

$$f^s \frac{\partial^{|\alpha|} g}{\partial x_1^{\alpha_1} \dots \partial x_n^{\alpha_n}},$$

defined on $U \setminus Z(f)$, extends by zero through the zero locus $Z(f)$ to a continuous function on U. By the product rule, $f^{s+k} g$ extends by zero to a $\mathcal{C}^k$ function on U that is k-flat on $Z(f)$ (that is, the partial derivatives of order at most k are all equal to 0 at every point in $Z(f)$). □

We need one more observation to prove Theorem 1.4.

Lemma 3.6. *Let I be an ideal of the ring $\mathcal{R}^k(\mathbb{R}^n)$. Then there exists a k-regulous function $g \in I$ with $Z(g) = Z(I)$.*

Proof. If functions $f_1, \dots, f_l$ belong to $\mathcal{R}^k(\mathbb{R}^n)$, then

$$Z(f_1, \dots, f_l) = Z(f_1^2 + \dots + f_l^2)$$

is a C-closed subset of $\mathbb{R}^n$ by Corollary 2.3 (ii) and Remark 2.6.

It suffices to prove that $Z(g_1, \dots, g_r) = Z(I)$ for some k-regulous functions $g_1, \dots, g_r$ in I. Suppose that this is not the case. Then we can find an infinite sequence $\{g_i\}$ in I so that

$$Z(g_1) \supsetneq Z(g_1, g_2) \supsetneq Z(g_1, g_2, g_3) \supsetneq \dots$$

We obtain a contradiction since the sets in this chain are C-closed, and the C-topology is Noetherian. □

Proof of Theorem 1.4. Suppose that a function $f \in \mathcal{R}^k(\mathbb{R}^n)$ vanishes on $Z(I)$. By Lemma 3.6, there exists a function $g \in I$ with $Z(g) = Z(I)$. The function

$$\frac{1}{g} \colon \mathbb{R}^n \setminus Z(f) \to \mathbb{R}$$

is k-regulous. It follows from Lemma 3.5 that if N is a sufficiently large integer, then the function $h \colon \mathbb{R}^n \to \mathbb{R}$ defined by

$$h = \frac{f^N}{g} \text{ on } \mathbb{R}^n \setminus Z(f) \text{ and } h = 0 \text{ on } Z(f)$$

is k-regulous. Hence $f^N = gh$ belongs to I. □

4. Algebra versus geometry

Throughout this section k stands for a fixed nonnegative integer. Following [12, Théorème 6.4], we give a detailed description of relationships between ideals of the ring $\mathcal{R}^k(\mathbb{R}^n)$ and C-closed subsets of $\mathbb{R}^n$.

Theorem 4.1. *For a subset A of $\mathbb{R}^n$, the following conditions are equivalent:*

(a) *A is C-closed in $\mathbb{R}^n$.*
(b) *$A = Z(f)$ for some function f in $\mathcal{R}^k(\mathbb{R}^n)$.*
(c) *$A = Z(I)$ for some ideal I of $\mathcal{R}^k(\mathbb{R}^n)$.*

Proof. Clearly (b) $\Rightarrow$ (c), while (c) $\Rightarrow$ (b) follows from Lemma 3.6. Moreover, (b) $\Rightarrow$ (a) in view of Corollary 2.3 (ii) and Remark 2.6. It remains to prove (a) $\Rightarrow$ (b).

Suppose that (a) holds and let $Y \subseteq \mathbb{R}^n$ be the Zariski closure of A. We use induction on $\dim Y$ to prove that (b) holds. The case $\dim Y = 0$ is obvious since then Y is a finite set.

Assume that $\dim Y \geq 1$. By Lemma 3.1, $Y^{\mathrm{sm}} \subset A$ and hence

$$Y = A \cup Z, \text{ where } Z = \mathrm{Sing}(Y).$$

Let φ and ψ be polynomial functions on $\mathbb{R}^n$ with $Z(\varphi) = Y$ and $Z(\psi) = Z$. Consider the functions

$$\frac{1}{\varphi^2} \colon \mathbb{R}^n \setminus Y = (\mathbb{R}^n \setminus A) \setminus (Z \setminus (A \cap Z)) \to \mathbb{R}, \psi|_{\mathbb{R}^n \setminus A} \colon \mathbb{R}^n \setminus A \to \mathbb{R}.$$

Since $Z(\psi|_{\mathbb{R}^n \setminus A}) = Z \setminus (A \cap Z)$, it follows from Lemma 3.5 that for a sufficiently large integer N, the function $\alpha \colon \mathbb{R}^n \setminus A \to \mathbb{R}$ defined by

$$\alpha = \frac{\psi^{2N}}{\varphi^2} \text{ on } \mathbb{R}^n \setminus Y \text{ and } \alpha = 0 \text{ on } Z \setminus (A \cap Z)$$

is k-regulous.

We claim that the function $\beta \colon \mathbb{R}^n \setminus (A \cap Z) \to \mathbb{R}$ defined by

$$\beta = \frac{\varphi^2}{\varphi^2 + \psi^{2N}} \text{ on } \mathbb{R}^n \setminus Z \text{ and } \beta = 1 \text{ on } Z \setminus (A \cap Z)$$

is k-regulous. This is reduced to showing that β is of class $\mathcal{C}^k$. On $\mathbb{R}^n \setminus A$, we have $\beta = 1/(1 + \alpha)$ since

$$(\mathbb{R}^n \setminus A) \cap (\mathbb{R}^n \setminus Z) = (\mathbb{R}^n \setminus Y).$$

Clearly, β is regular on $\mathbb{R}^n \setminus Z$. Since

$$\mathbb{R}^n \setminus (A \cap Z) = (\mathbb{R}^n \setminus A) \cup (\mathbb{R}^n \setminus Z),$$

we conclude that β is of class $\mathcal{C}^k$, as required.

By construction,

$$Z(\beta) = Y \setminus Z = A \setminus (A \cap Z).$$

Since $\dim Z < \dim Y$, it follows from the induction hypothesis that there exists a k-regulous function $\gamma\colon \mathbb{R}^n \to \mathbb{R}$ with $Z(\gamma) = A \cap Z$. In view of Lemma 3.5 once again, if M is a sufficiently large integer, then the function $f\colon \mathbb{R}^n \to \mathbb{R}$ defined by

$$f = \gamma^M \beta \text{ on } \mathbb{R}^n \setminus (A \cap Z) \text{ and } f = 0 \text{ on } A \cap Z$$

is k-regulous. Evidently, $Z(f) = A$, hence (b) holds. □

Given a subset $A \subseteq \mathbb{R}^n$, we denote by $J_{\mathcal{R}^k}(A)$ the ideal of $\mathcal{R}^k(\mathbb{R}^n)$ consisting of all functions vanishing on A,

$$J_{\mathcal{R}^k}(A) := \{f \in \mathcal{R}^k(\mathbb{R}^n)\colon f(x) = 0 \text{ for all } x \in A\}.$$

As usual, the radical of an ideal I of $\mathcal{R}^k(\mathbb{R}^n)$ will be denoted by $\mathrm{Rad}(I)$.

The following is an equivalent formulation of Theorem 1.4.

Theorem 4.2. *Let I be an ideal of the ring $\mathcal{R}^k(\mathbb{R}^n)$. Then*

$$J_{\mathcal{R}^k}(Z(I)) = \mathrm{Rad}(I).$$

A straightforward consequence of Theorems 4.1 and 4.2 is the following.

Corollary 4.3. *The assignment*

$$\mathbb{R}^n \supseteq A \mapsto J_{\mathcal{R}^k}(A) \subseteq \mathcal{R}^k(\mathbb{R}^n)$$

gives rise to one-to-one correspondences:

(i) *between the C-closed subsets of $\mathbb{R}^n$ and the radical ideals of $\mathcal{R}^k(\mathbb{R}^n)$;*
(ii) *between the irreducible C-closed subsets of $\mathbb{R}^n$ and the prime ideals of $\mathcal{R}^k(\mathbb{R}^n)$;*
(iii) *between the points of $\mathbb{R}^n$ and the maximal ideals of $\mathcal{R}^k(\mathbb{R}^n)$.*

For a function f in $\mathcal{R}^k(\mathbb{R}^n)$, we let $\mathrm{Rad}(f)$ denote the radical of the principal ideal generated by f.

The following is a direct consequence of Theorem 4.2.

Corollary 4.4. *Let f and g be functions in the ring $\mathcal{R}^k(\mathbb{R}^n)$. Then* $\mathrm{Rad}(f) = \mathrm{Rad}(g)$ *if and only if $Z(f) = Z(g)$.*

The radical of an arbitrary ideal can be described as follows.

Corollary 4.5. *Let I be an ideal of the ring $\mathcal{R}^k(\mathbb{R}^n)$. Then*

$$\operatorname{Rad}(I) = \operatorname{Rad}(f)$$

for some $f \in I$. Furthermore, the equality holds if and only if $Z(I) = Z(f)$.

Proof. By Theorem 4.1, $Z(I) = Z(f)$ for some $f \in I$. The proof is complete in view of Theorem 4.2. □

As usual, for f in $\mathcal{R}^k(\mathbb{R}^n)$, we let $\mathcal{R}^k(\mathbb{R}^n)_f$ denote the localization of the ring $\mathcal{R}^k(\mathbb{R}^n)$ with respect to the multiplicatively closed subset

$$\{1\} \cup \{f^m \colon m = 1, 2, \ldots\}.$$

In particular, $\mathcal{R}^k(\mathbb{R}^n)_f$ is the zero ring if f is identically equal to 0. By convention, $\mathcal{R}^k(\emptyset)$ is also the zero ring. Lemma 3.5 allows us to give a geometric description of $\mathcal{R}^k(\mathbb{R}^n)_f$, which will be useful in Section 5.

Corollary 4.6. *Let f be a function in $\mathcal{R}^k(\mathbb{R}^n)$. Then the restriction homomorphism*

$$\mathcal{R}^k(\mathbb{R}^n) \to \mathcal{R}^k(\mathbb{R}^n \setminus Z(f)), h \mapsto h|_{(\mathbb{R}^n \setminus Z(f))}$$

induces an isomorphism between the localization $\mathcal{R}^k(\mathbb{R}^n)_f$ and $\mathcal{R}^k(\mathbb{R}^n \setminus Z(f))$.

Proof. By Lemma 3.5, the induced homomorphism

$$\mathcal{R}^k(\mathbb{R}^n)_f \to \mathcal{R}^k(\mathbb{R}^n \setminus Z(f))$$

is surjective.

If an element $g/f^m \in \mathcal{R}^k(\mathbb{R}^n)_f$ is sent to $0 \in \mathcal{R}^k(\mathbb{R}^n \setminus Z(f))$, then $fg = 0$ in $\mathcal{R}^k(\mathbb{R}^n)$. Hence $g/f^m = 0$, which means that the induced homomorphism is injective. □

5. Regulous sheaves

In this section we present result on *regulous sheaves*, which were introduced and investigated in [12].

Notation/Convention 5.1. Throughout this section, $\mathbb{R}^n$ is endowed with the C-topology and k stands for a nonnegative integer. Thus, a subset $A \subseteq \mathbb{R}^n$ is closed if and only if $A = Z(f)$ for some $f \in \mathcal{R}^k(\mathbb{R}^n)$ (Theorem 4.1). We set $X(f) := \mathbb{R}^n \setminus Z(f)$.

For every open subset U of $\mathbb{R}^n$, the ring $\mathcal{R}^k(U)$ is defined (Definition 1.1). The assignment

$$\mathcal{R}^k: U \mapsto \mathcal{R}^k(U)$$

is a sheaf of rings on $\mathbb{R}^n$, and $(\mathbb{R}^n, \mathcal{R}^k)$ is a locally ringed space.

There is a close connection between the ringed space $(\mathbb{R}^n, \mathcal{R}^k)$ and the affine scheme $\operatorname{Spec}(\mathcal{R}^k(\mathbb{R}^n))$. We first describe relationships between the underlying topological spaces.

By definition, a subset $V \subseteq \operatorname{Spec}(\mathcal{R}^k(\mathbb{R}^n))$ is closed if and only if $V = V(F)$ for some collection $F \subseteq \mathcal{R}^k(\mathbb{R}^n)$, where

$$V(F) := \{\mathfrak{p} \in \operatorname{Spec}(\mathcal{R}^k(\mathbb{R}^n)) : F \subseteq \mathfrak{p}\}.$$

If I is the ideal generated by F, then $V(F) = V(\operatorname{Rad}(I))$. Hence, by Corollary 4.5, $V(F) = V(f)$ for some $f \in I$; here $V(f) = V(\{f\})$. Consequently, each open subset of $\operatorname{Spec}(\mathcal{R}^k(\mathbb{R}^n))$ is of the form

$$D(f) := \operatorname{Spec}(\mathcal{R}^k(\mathbb{R}^n)) \setminus V(f)$$

for some $f \in \mathcal{R}^k(\mathbb{R}^n)$. In particular, each open subset of $\operatorname{Spec}(\mathcal{R}^k(\mathbb{R}^n))$ is affine.

Define a map

$$\iota: \mathbb{R}^n \to \operatorname{Spec}(\mathcal{R}^k(\mathbb{R}^n))$$

by $\iota(x) = \mathfrak{m}_x$ for all $x \in \mathbb{R}^n$, where

$$\mathfrak{m}_x := \{f \in \mathcal{R}^k(\mathbb{R}^n) : f(x) = 0\}$$

($\mathfrak{m}_x = J_{\mathcal{R}^k}(\{x\})$ with notation as in Section 4).

Theorem 5.2. *The map ι is a topological embedding of $\mathbb{R}^n$ onto the subspace $\operatorname{Max}(\mathcal{R}^k(\mathbb{R}^n))$ of $\operatorname{Spec}(\mathcal{R}^k(\mathbb{R}^n))$ consisting of the maximal ideals of $\mathcal{R}^k(\mathbb{R}^n)$. Furthermore:*

(1) $Z(f) = \iota^{-1}(V(f))$ *for every* $f \in \mathcal{R}^k(\mathbb{R}^n)$.

(2) *For each closed subset $Z \subseteq \mathbb{R}^n$ there exists a unique closed subset $\tilde{Z} \subseteq \operatorname{Spec}(\mathcal{R}^k(\mathbb{R}^n))$ such that $Z = \iota^{-1}(\tilde{Z})$.*

(3) $X(f) = \iota^{-1}(D(f))$ *for every* $f \in \mathcal{R}^k(\mathbb{R}^n)$.

(4) *For each open subset $U \subseteq \mathbb{R}^n$ there exists a unique open subset $\tilde{U} \subseteq \operatorname{Spec}(\mathcal{R}^k(\mathbb{R}^n))$ such that $U = \iota^{-1}(\tilde{U})$.*

Proof. By Corollary 4.3, ι induces a bijection of $\mathbb{R}^n$ onto $\operatorname{Max}(\mathcal{R}^k(\mathbb{R}^n))$. It follows immediately that (1) holds.

Suppose that $Z \subseteq \mathbb{R}^n$ and $V \subseteq \mathrm{Spec}(\mathcal{R}^k(\mathbb{R}^n))$ are closed subsets with $Z = \iota^{-1}(V)$. Then $Z = Z(f)$ and $V = V(g)$, which proves (2).

Statements (3) and (4) follow from (1) and (2), respectively.

It is now clear that ι induces a homeomorphism between $\mathbb{R}^n$ and $\mathrm{Max}(\mathcal{R}^k(\mathbb{R}^n))$. □

In view of Proposition 1.3, the affine scheme $\mathrm{Spec}(\mathcal{R}^k(\mathbb{R}^n))$ is not Noetherian if $n \geq 2$. However, the following holds.

Corollary 5.3. *The topological space* $\mathrm{Spec}(\mathcal{R}^k(\mathbb{R}^n))$ *is Noetherian.*

Proof. This follows from Theorem 5.2 since $\mathbb{R}^n$ with the C-topology is a Noetherian topological space in view of Proposition 3.2. □

Next we study sheaves on the spaces under consideration. For any function $f \in \mathcal{R}^k(\mathbb{R}^n)$, we identify the rings $\mathcal{R}^k(\mathbb{R}^n)_f$ and $\mathcal{R}^k(X(f))$ via the canonical isomorphism described in Corollary 4.6. Denoting by $\tilde{\mathcal{R}}^k$ the structure sheaf on $\mathrm{Spec}(\mathcal{R}^k(\mathbb{R}^n))$, we get

$$\iota_* \mathcal{R}^k = \tilde{\mathcal{R}}^k \text{ and } \mathcal{R}^k = \iota^{-1}(\tilde{\mathcal{R}^k})$$

by Theorem 5.2. Furthermore, the category of sheaves of Abelian groups (resp. sheaves of $\mathcal{R}^k$-modules) on $\mathbb{R}^n$ is equivalent to the category of sheaves of Abelian groups (resp. sheaves of $\tilde{\mathcal{R}}^k$-modules) on $\mathrm{Spec}(\mathcal{R}^k(\mathbb{R}^n))$. The equivalence is effected by the direct image functor $\mathcal{R} \mapsto \iota_* \mathcal{F}$, whose inverse is the inverse image functor $\mathcal{G} \mapsto \iota^{-1}\mathcal{G}$. Via these equivalences, quasi-coherent sheaves of $\mathcal{R}^k$-modules on $\mathbb{R}^n$ correspond to quasi-coherent sheaves of $\tilde{\mathcal{R}}^k$-modules on $\mathrm{Spec}(\mathcal{R}^k(\mathbb{R}^n))$.

As a reminder, a *quasi-coherent sheaf* on a ringed space $(X, \mathcal{O}_X)$ is a sheaf $\mathcal{F}$ of $\mathcal{O}_X$-modules which has a local presentation, that is, every point in X has an open neighborhood U in which there is an exact sequence

$$\mathcal{O}_X^{\oplus K}|_U \to \mathcal{O}_X^{\oplus L}|_U \to \mathcal{F}|_U \to 0$$

for some (possibly infinite) sets K and L.

By a *k-regulous sheaf* on $\mathbb{R}^n$ we mean a sheaf of $\mathcal{R}^k$-modules. Our goal is to establish basic properties of quasi-coherent k-regulous sheaves.

For clarity of exposition, it is convenient to recall Cartan's theorem A for affine schemes [17, Theorem 7.16, Corollary 7.17]. For a commutative ring R, consider the affine scheme $Y = \mathrm{Spec}(R)$ with structure sheaf $\mathcal{O}_Y$. Any R-module M determines a quasi-coherent sheaf $\tilde{M}$ of $\mathcal{O}_Y$-modules on Y. The functor $M \mapsto \tilde{M}$ gives an equivalence of categories between the category of R-modules and the category of quasi-coherent sheaves of $\mathcal{O}_Y$-modules on Y. Its inverse is the global section functor $\mathcal{G} \mapsto \mathcal{G}(Y)$. In particular, every quasi-coherent sheaf of $\mathcal{O}_Y$-modules on Y is generated by its global sections.

Returning to our main topic, for any $\mathcal{R}^k(\mathbb{R}^n)$-module M, we define a presheaf $\widehat{M}$ of $\mathcal{R}^k$-modules on $\mathbb{R}^n$ by

$$\widehat{M}(U) := M \otimes_{\mathcal{R}^k(\mathbb{R}^n)} \mathcal{R}^k(U)$$

for every open subset $U \subseteq \mathbb{R}^n$. Since $U = X(f)$ for some $f \in \mathcal{R}^k(\mathbb{R}^n)$, we get

$$\widehat{M}(U) = \widehat{M}(X(f)) = M \otimes_{\mathcal{R}^k(\mathbb{R}^n)} \mathcal{R}^k(\mathbb{R}^n)_f = \tilde{M}(D(f)),$$

where the last equality is the canonical identification. Hence $\widehat{M}$ is actually a sheaf and

$$\widehat{M} = \iota^{-1}\tilde{M}.$$

We immediately get the following variant of Cartan's theorem A.

Theorem 5.4. *The functor $M \mapsto \widehat{M}$ gives an equivalence of categories between the category of $\mathcal{R}^k(\mathbb{R}^n)$-modules and the category of quasi-coherent k-regulous sheaves on $\mathbb{R}^n$. Its inverse is the global section functor $\mathcal{F} \mapsto \mathcal{F}(\mathbb{R}^n)$. In particular, every quasi-coherent k-regulous sheaf on $\mathbb{R}^n$ is generated by its global sections.*

According to Cartan's theorem B for affine schemes, if $\mathcal{G}$ is quasi-coherent sheaf of $\mathcal{O}_Y$-modules on $Y = \operatorname{Spec}(R)$, then $H^i(Y, \mathcal{G}) = 0$ for all $i \geq 1$ [18, Théorème 1.3.1].

The equivalence of the categories of sheaves on $\mathbb{R}^n$ and on $\operatorname{Spec}(\mathcal{R}^k(\mathbb{R}^n))$ via the functors ι_* and ι^{-1} yields the following variant of Cartan's theorem B.

Theorem 5.5. *If $\mathcal{F}$ is a quasi-coherent k-regulous sheaf on $\mathbb{R}^n$, then*

$$H^i(\mathbb{R}^n, \mathcal{F}) = 0 \textit{ for all } i \geq 1.$$

One should keep in mind that coherent algebraic sheaves on $\mathbb{R}^n$ have pathological properties; in particular, Cartan's theorems A and B fail for such sheaves; see [6, Example 12.1.5], [8, Theorem 1], [23]. It is therefore rather surprising that Cartan's theorems A and B hold for quasi-coherent k-regulous sheaves on $\mathbb{R}^n$.

6. Regulous varieties

In this section, k is a fixed nonnegative integer. All subsets of $\mathbb{R}^n$ will be endowed with the C-topology, which is induced by the C-topology on $\mathbb{R}^n$.

Let $A \subseteq \mathbb{R}^n$ be a closed subset and let $f\colon U \to \mathbb{R}$ be a function defined on an open subset $U \subseteq A$.

Definition 6.1. We say that f is *k-regulous* if for every point $p \in U$ there exists a k-regulous function $g_p\colon W_p \to \mathbb{R}$, defined on an open neighborhood $W_p \subseteq \mathbb{R}^n$ of p, such that $f|_{U\cap W_p} = g|_{U\cap W_p}$ (for $k = 0$ this is compatible with Definition 2.4; see Proposition 7.3 in the next section).

The collection $\mathcal{R}^k_A(U)$ of all k-regulous functions on U forms a ring (commutative with 1). Clearly, the assignment

$$\mathcal{R}^k_A\colon U \mapsto \mathcal{R}^k_A(U)$$

is a sheaf of rings on A, and $(A, \mathcal{R}^k_A)$ is a locally ringed space.

Definition 6.2. A locally ringed space $(X, \mathcal{R}^k_X)$, where $\mathcal{R}^k_X$ is a sheaf of functions on X, is called an *affine k-regulous variety* if it is isomorphic to $(A, \mathcal{R}^k_A)$ for some A as above. More generally, $(X, \mathcal{R}^k_X)$ is called a *k-regulous variety* if there exists an open cover $\{U_1, \ldots, U_r\}$ of X such that each $(U_i, \mathcal{R}^k_X|_{U_i})$ is an affine k-regulous variety. A morphism of k-regulous varieties, called a *k-regulous map*, is a morphism of locally ringed spaces.

The product of k-regulous varieties is defined by the same procedure as the product of algebraic varieties.

Any affine real algebraic variety can be regarded in the obvious way as an affine k-regulous variety. Also, any (abstract) real algebraic variety can be regarded as a k-regulous variety.

Proposition 6.3. *Let $(X, \mathcal{R}^k_X)$ be an affine k-regulous variety and let $U \subseteq X$ be an open subset. Then the locally ringed space $(U, \mathcal{R}^k_X|_U)$ is an affine k-regulous variety.*

Proof. We regard X as a closed subset of $\mathbb{R}^n$. By Theorem 4.1, there exists a k-regulous function $f\colon \mathbb{R}^n \to \mathbb{R}$ with $Z(f) = X \setminus U$. Clearly, $g\colon \mathbb{R}^n \times \mathbb{R} \to \mathbb{R}$ defined by $g(x, y) = f(x)y - 1$ is a k-regulous function, and

$$Y := Z(g) \cap (X \times \mathbb{R})$$

is a closed subset of $\mathbb{R}^n \times \mathbb{R}$. Then the map

$$U \to Y, x \mapsto \left(x, \frac{1}{f(x)}\right)$$

is a k-regulous isomorphism, which completes the proof. □

Proposition 6.4. *Let $A \subseteq \mathbb{R}^n$ be a closed subset and let $f\colon A \to \mathbb{R}$ be a k-regulous function. Then there exists a k-regulous function $g\colon \mathbb{R}^n \to \mathbb{R}$ such that $g|_A = f$.*

Proof. Consider the sheaf $\mathcal{R}^k$ on $\mathbb{R}^n$ and let $\mathcal{J} \subseteq \mathcal{R}^k$ be the sheaf of ideals of functions vanishing on A. Clearly, the k-regulous functions on A can be identified with the sections on $\mathbb{R}^n$ of the quotient sheaf $\mathcal{R}^k/\mathcal{J}$. Note that $\mathcal{J}$ is a quasi-coherent k-regulous sheaf on $\mathbb{R}^n$. Hence, in view of Theorem 5.5, the canonical short exact sequence

$$0 \to \mathcal{J} \to \mathcal{R}^k \to \mathcal{R}^k/\mathcal{J} \to 0$$

induces a short exact sequence of global sections. This completes the proof. □

A *k-regulous sheaf* on a k-regulous variety $(X, \mathcal{R}^k_X)$ is a sheaf of $\mathcal{R}^k_X$-modules. Moreover, Cartan's theorems A and B also hold for quasi-coherent k-regulous sheaves on affine k-regulous varieties.

Remark 6.5. In the remainder of the paper, we consider only 0-regulous functions. It is not known whether the results presented in Sections 7, 8 and 9 have some analogs in the k-regulous setting with $k \geq 1$. The notion of regulous vector bundle discussed in Section 10 can easily be generalized to that of k-regulous vector bundle. Moreover, Theorems 10.4 and 10.5 have suitable counterparts for k-regulous vector bundles. However, it is not clear whether deeper results on vector bundles, contained in [28, 30, 34, 35, 38], hold also in the k-regulous framework with $k \geq 1$.

7. Extending regulous functions

Throughout this section, we consider regulous functions in the sense of Definition 2.4.

The main result of Kollár and Nowak [22, Theorem 10] can be stated as follows.

Theorem 7.1. *Let X be an affine real algebraic variety, $Y \subseteq X$ an algebraic subvariety, and $f\colon Y \to \mathbb{R}$ a regulous function. Then there exists a regulous function $g\colon X \to \mathbb{R}$ such that $g|_Y = f$.*

It is not hard to generalize Theorem 7.1 as follows.

Proposition 7.2. *Let X be an affine real algebraic variety, $A \subseteq X$ a constructible E-closed subset, and $f\colon A \to \mathbb{R}$ a regulous function. Then there exists a regulous function $g\colon X \to \mathbb{R}$ such that $g|_A = f$.*

Proof. We may assume without loss of generality that $X = \mathbb{R}^n$. Let $Y \subseteq \mathbb{R}^n$ be the Zariski closure of A. We use induction on $\dim Y$. The case $\dim Y = 0$ is obvious since then Y is a finite set.

Suppose that $\dim Y \geq 1$. By Proposition 3.1, $Y^{\mathrm{sm}} \subseteq A$, hence

$$Y = A \cup Z,$$

where $Z = \mathrm{Sing}(Y)$. Since $\dim Z < \dim Y$, the induction hypothesis implies that there exists a regulous function $h\colon \mathbb{R}^n \to \mathbb{R}$ with $h|_{A\cap Z} = f|_{A\cap Z}$. The function $\varphi\colon Y \to \mathbb{R}$ defined by

$$\varphi = f \text{ on } A \text{ and } \varphi = h \text{ on } Z$$

is continuous, hence regulous. By Theorem 7.1, there exists a regulous function $g\colon \mathbb{R}^n \to \mathbb{R}$ with $g|_Y = \varphi$. Evidently, $g|_A = f$. □

Proposition 7.3. *Let X be an affine real algebraic variety and let $B \subseteq A \subseteq X$ be constructible E-closed subsets. For a function $f\colon A \setminus B \to \mathbb{R}$, the following conditions are equivalent:*

(a) *f is regulous.*

(b) *There exists a regulous function $g\colon \mathbb{R}^n \setminus B \to \mathbb{R}$ with $g|_{A\setminus B} = f$.*

(c) *For every point $p \in A \setminus B$ there exist a constructible C-open neighborhood $W_p \subseteq \mathbb{R}^n$ of p and a regulous function $g_p\colon W_p \to \mathbb{R}$ such that $g_p = f$ on $(A \setminus B) \cap W_p$.*

Proof. Suppose that (a) holds. By Theorem 4.1, there exists a regulous function $\varphi\colon \mathbb{R}^n \to \mathbb{R}$ with $Z(\varphi) = B$. According to Proposition 3.4, if N is a sufficiently large integer, then the function $h\colon A \to \mathbb{R}$ defined by

$$h := \varphi^N f \text{ on } A \setminus B \text{ and } h := 0 \text{ on } B$$

is continuous, hence regulous. In view of Proposition 7.2, there exists a regulous function $H\colon \mathbb{R}^n \to \mathbb{R}$ with $H|_A = h$. Therefore (b) holds with $g := H/\varphi^N$ on $\mathbb{R}^n \setminus B$.

Clearly, (b) implies (c). Suppose then that (c) is satisfied. Since $\mathbb{R}^n$ endowed with the C-topology is a Noetherian topological space, we have

$$A \setminus B \subseteq W_{p_1} \cup \ldots \cup W_{p_r}$$

for some points $p_1, \ldots, p_r$ in $A \setminus B$. It follows that (a) holds. □

8. Curve-regulous and arc-regulous functions

All results discussed in this section are due to Kollár, Kucharz and Kurdyka [21]. We proved that regulous functions in the sense of Definition 2.4 can be characterized by restrictions to algebraic curves or algebraic arcs.

Definition 8.1. Let X be an affine real algebraic variety and let $f\colon W \to \mathbb{R}$ be a function defined on some subset W of X.

We say that f is *regulous on algebraic curves* or *curve-regulous* for short if for every irreducible algebraic curve $C \subseteq X$ the restriction $f|_{W\cap C}$ is a regulous function (equivalently, $f|_{W\cap C}$ is continuous and admits a rational representation).

We say that f is *regulous on algebraic arcs* or *arc-regulous* for short if for every irreducible algebraic curve $C \subseteq X$ and every point $x \in W \cap C$ there exists an open neighborhood $U_x \subseteq W$ of x such that the restriction $f|_{U_x\cap C}$ is a regulous function (equivalently, $f|_{U_x\cap C}$ is continuous and admits a rational representation).

If $f\colon W \to \mathbb{R}$ is curve-regulous, then it is also arc-regulous. The converse does not hold for a rather obvious reason. Indeed, let $H \subseteq \mathbb{R}^2$ be the hyperbola defined by $xy - 1 = 0$. Any function on H that is constant on each connected component of H is arc-regulous, but it must be constant to be regulous.

First we state the main results of [21], and afterwards discuss the key ingredient in their proofs. In [21], curve-regulous (resp. arc-regulous) functions are called *curve-rational* (resp. *arc-rational*).

Theorem 8.2. *Let X be an affine real algebraic variety and let $W \subseteq X$ be a subset that is either open or semialgebraic. For a function $f\colon W \to \mathbb{R}$, the following conditions are equivalent:*

(a) *f is regulous.*
(b) *f is curve-regulous.*

The corresponding result for arc-regulous functions takes the following form.

Theorem 8.3. *Let X be an affine real algebraic variety and let $W \subseteq X^{sm}$ be a connected open subset. For a function $f\colon W \to \mathbb{R}$, the following conditions are equivalent:*

(a) *f is regulous.*
(b) *f is arc-regulous.*

In Theorems 8.2 and 8.3, only the implication (b) $\Rightarrow$ (a) requires a proof. One has to show, in particular, the continuity of f but even this alone is not at all obvious.

Theorem 8.4. *Let X be an affine real algebraic variety and let $f\colon W \to \mathbb{R}$ be an arc-regulous function defined on a semialgebraic subset $W \subseteq X$. Then f is continuous and there exists a sequence of semialgebraic sets*

$$W = W_0 \supseteq W_1 \supseteq \ldots \supseteq W_{m+1} = \emptyset,$$

which are closed in W, such that the restriction of f is a regular function on each connected component of $W_i \setminus W_{i+1}$ for $i = 0, \ldots, m$. In particular, f is a semialgebraic function.

In Theorem 2.8, we already established a connection between regulous functions and arc-analytic functions. For arc-regulous functions, the following holds.

Theorem 8.5. *Let X be an affine real algebraic variety and let $f\colon W \to \mathbb{R}$ be an arc-regulous function defined on an open subset $W \subseteq X$. Then f is continuous and arc-analytic.*

Crucial for the proofs of these results is Theorem 8.7 below. First we need some preparation.

Let X be an affine real algebraic variety. A subset $A \subseteq X$ is called a *smooth algebraic arc* if its Zariski closure C is an irreducible algebraic curve, $A \subseteq C^{\mathrm{sm}}$, and A is homeomorphic to $\mathbb{R}$.

Example 8.6. Consider the function $f\colon \mathbb{R}^2 \to \mathbb{R}$ defined by

$$f(x,y) = \begin{cases} \frac{x^8 + y(x^2-y^3)^2}{x^{10}+(x^2-y^3)^2} & \text{for } (x,y) \neq (0,0) \\ 0 & \text{for } (x,y) = (0,0). \end{cases}$$

The restriction of f is a regular function on each smooth algebraic arc in $\mathbb{R}^2$, but f is not continuous on the curve $x^2 - y^3 = 0$.

As this example shows, continuity cannot be detected on smooth algebraic arcs. Nevertheless, the following holds.

Theorem 8.7. *Let X be an affine real algebraic variety and let $f\colon U \to \mathbb{R}$ be a function defined on a connected open subset $U \subseteq X^{sm}$. Assume that the restriction of f is regular on each smooth algebraic arc contained in U. Then there exists a rational function R on X such that $P := U \cap \mathrm{Pole}(R)$ has codimension at least 2 in U and $f|_{U\setminus P} = R|_{U\setminus P}$.*

Sketch of proof. We may assume that X is irreducible and smooth, with $n := \dim X \geq 2$. The proof is divided into three steps and we provide details for only the first two of them.

Step 1. Assume, in addition, that for some nonempty open semialgebraic subset $V \subseteq U$ the restriction $f|_V$ is a semialgebraic function. Then there exist a rational function R on X and a nonempty open subset $U_0 \subseteq V$ such that $U_0 \subseteq X \setminus \mathrm{Pole}(R)$ and $f|_{U_0} = R|_{U_0}$.

We represent X as $X = \mathbb{X}(\mathbb{R})$, where $\mathbb{X}$ is an irreducible smooth quasiprojective algebraic variety defined over $\mathbb{R}$ and $\mathbb{X}(\mathbb{R})$ denotes its set of real points. By definition of semialgebraic function, there exist a nonempty open subset $W \subseteq V$ and an irreducible algebraic hypersurface $\mathbb{Y} \subseteq \mathbb{X} \times \mathbb{A}^1$ such that the graph $f|_W$ is contained in $\mathbb{Y}$. Note that $f|_W$ admits a rational representation if and only if the first projection $\pi_1 \colon \mathbb{Y} \to \mathbb{X}$ is a birational map, that is, of degree 1. Suppose that $\pi_1 \colon \mathbb{Y} \to \mathbb{X}$ has degree $m \geq 2$. Fix an embedding $\mathbb{X} \subseteq \mathbb{P}^N$ and set $k := N - n - 1$. By Bertini's theorem [41, Theorem 3.3.1], for a general k-dimensional plane $\mathbb{L} \subseteq \mathbb{P}^N$ both $\pi_1^{-1}(\mathbb{X} \cap \mathbb{L})$ and $\mathbb{X} \cap \mathbb{L}$ are irreducible algebraic curves defined over $\mathbb{R}$. We can choose such an $\mathbb{L}$ so that $W \cap \mathbb{L}$ contains a smooth algebraic arc A. Clearly, the restriction $\pi_1^{-1}(\mathbb{X} \cap \mathbb{L}) \to \mathbb{X} \cap \mathbb{L}$ of π_1 has degree m, and the graph of $f|_A$ is contained in $\pi_1^{-1}(\mathbb{X} \cap \mathbb{L})$. Consequently, $f|_A$ does not admit a rational representation. Hence $f|_A$ is not a regular function, a contradiction. It follows that $f|_W$ admits a rational representation, which proves Step 1.

Step 2. Suppose that there exist a nonempty open subset $U_0 \subseteq U$ and a rational function R on X such that $U_0 \subseteq X \setminus \mathrm{Pole}(R)$ and $f|_{U_0} = R|_{U_0}$. Then $P := U \cap \mathrm{Pole}(R)$ has codimension at least 2 in U and $f|_{U \setminus P} = R|_{U \setminus P}$.

We first prove that $f|_{U \setminus P} = R|_{U \setminus P}$. Let $\mathcal{A}$ be the set of all smooth algebraic arcs contained in U. One readily checks that each point $p \in U$ has an arbitrarily small open neighborhood $U(p) \subseteq U$ such that any two points of $U(p)$ belong to an arc in $\mathcal{A}$, contained in $U(p)$. Fix a point $p_0 \in U_0$ and let $p \in U \setminus P$ be an arbitrary point. Let $\gamma \colon [0,1] \to U$ be a continuous path with $\gamma(0) = p_0$ and $\gamma(1) = p$. The compact set $\gamma([0,1])$ can be covered by open sets $U(p_0), U(p_1), \ldots, U(p_r)$ such that $U(p_0) \subseteq U_0$, $p_r = p$, and the intersection $U(p_i) \cap U(p_{i+1})$ is nonempty for $i = 0, \ldots, r-1$. By construction, $f|_{U(p_0) \setminus P} = R|_{U(p_0) \setminus P}$. Using induction on i, we get

$f|_{U(p_i)\setminus P} = R|_{U(p_i)\setminus P}$ for $i = 0, \ldots, r$. Hence $f(p) = R(p)$, which proves the equality $f|_{U\setminus P} = R|_{U\setminus P}$.

Now we show that $\operatorname{codim} P \geq 2$. Suppose to the contrary that $\operatorname{codim} P = 1$. Let B be an arc in $\mathcal{A}$ which meets P transversally at a general point. Then $f|_B$ is a regular function with

$$(f|_B)|_{B\setminus P} = f|_{B\setminus P} = R|_{B\setminus P},$$

hence R cannot have a pole along B, a contradiction.

Step 3. In view of Steps 1 and 2, to complete the proof, it suffices to find a nonempty open semialgebraic subset $V \subseteq U$ such that the restriction $f|_V$ is a semialgebraic function. This can be accomplished by combining some subtle Hartogs-like theorems on functions of several real variables due to Błocki [5] and Siciak [45] with the classical result [7, p. 202, Theorem 6]. □

9. Applications to approximation of continuous maps

There are many interesting open problems concerning approximation by regular maps or regulous maps; see [37] for a short recent account. Approximation of continuous maps means approximation in the compact-open topology. Here we only state two results, which are proved in [26]. The reader may consult [24–27, 29, 32–34, 46, 47] for a more extensive treatment. We focus on maps with values in the unit p-sphere

$$\mathbb{S}^p = \{(u_1, \ldots, u_{p+1}) \in \mathbb{R}^{p+1} : u_1^2 + \ldots + u_{p+1}^2 = 1\}.$$

Theorem 9.1. *For any pair (n, p) of positive integers, each continuous map from $\mathbb{S}^n$ into $\mathbb{S}^p$ can be approximated by regulous maps.*

It is an open problem whether or not each continuous map from $\mathbb{S}^n$ into $\mathbb{S}^p$ can be approximated by regular maps.

Theorem 9.2. *Let X be a compact smooth affine real algebraic variety of dimension p. Then each continuous map from X into $\mathbb{S}^p$ can be approximated by regulous maps.*

Recall that each regular map from $\mathbb{S}^1 \times \mathbb{S}^1$ into $\mathbb{S}^2$ is null homotopic [6, Theorem 13.5.1]. Hence, if a continuous map $f : \mathbb{S}^1 \times \mathbb{S}^1 \to \mathbb{S}^2$ is not null homotopic, then it cannot be approximated by regular maps.

In some approximation results, the following characterization of regulous maps is useful; see [24, p. 528] and [12, Théorème 3.11].

Theorem 9.3. *Let $f\colon X \to Y$ be a map between smooth affine real algebraic varieties. Then the following conditions are equivalent:*

(a) *The map f is regulous.*

(b) *There exists a regular map $\pi\colon X' \to X$, which is the composite of a finite sequence of blowups with smooth algebraic nowhere dense centers, such that $f \circ \pi\colon X' \to Y$ is a regular map.*

Proof. Assume that (a) holds and choose a Zariski open dense subset $X^0 \subseteq X$ such that the restriction $f|_{X^0}$ is a regular map. By the resolution of singularities theorem [19, Theorem 3.27], we may assume that Y is a Zariski open subset of a compact smooth affine real algebraic variety Y^*. By the elimination of indeterminacy theorem [19, Theorem 3.26, Corollary 3.18], there exists a regular map $\pi\colon X' \to X$, which is the composite of a finite sequence of blowups with smooth algebraic nowhere dense centers, such that the map

$$(f|_{X^0}) \circ (\pi|_{\pi^{-1}(X^0)})\colon \pi^{-1}(X^0) \to Y \subseteq Y^*$$

is the restriction of a regular map $g\colon X' \to Y^*$. Since f is a continuous map and the set $\pi^{-1}(X^0)$ is E-dense in X', we get $f \circ \pi = g$ and $g(X') \subseteq Y$, hence (b) holds.

If (b) is satisfied, then f is continuous since π is a surjective map that is proper in the E-topology. It follows from Theorem 2.5 that f is a regulous map, the map π being birational. □

10. Application to vector bundles

In my joint paper with K. Kurdyka [34], we introduced and investigated stratified-algebraic vector bundles which occupy an intermediate position between algebraic vector bundles and topological vector bundles; see also [28, 30, 35]. Stratified-algebraic vector bundles can be identified with regulous vector bundles defined in my joint paper with M. Zieliński [38]. Here we only briefly comment on this identification and do not discuss the main results of the cited papers. From the viewpoint of the present article, it is natural to start with regulous vector bundles.

Notation/Convention 10.1. We let $\mathbb{F}$ stand for either $\mathbb{R}, \mathbb{C}$ or $\mathbb{H}$ (the quaternions). We will consider only left $\mathbb{F}$-vector spaces which play a role if $\mathbb{F} = \mathbb{H}$ since the quaternions are noncommutative. When convenient we identify $\mathbb{F}$ with $\mathbb{R}^{d(\mathbb{F})}$, where $d(\mathbb{F}) := \dim_{\mathbb{R}}(\mathbb{F})$. Unless explicitly stated otherwise, we will use constructible topology on regulous varieties.

Let X be an affine regulous variety.

Definition 10.2. A *regulous* $\mathbb{F}$*-vector bundle* on X is a triple $\xi = (E, p, X)$, where E is a regulous variety (not necessarily affine), $p\colon E \to X$ is a regulous map, and the fiber $E_x := p^{-1}(x)$ is an $\mathbb{F}$-vector space for every point $x \in X$. Furthermore, the following local triviality condition holds: For each point $x \in X$, there exist an open neighborhood $U \subseteq X$ of x, an integer $n \geq 0$ (depending on x), and a regulous isomorphism $\varphi\colon p^{-1}(U) \to U \times \mathbb{F}^n$ such that $\varphi(E_y) = \{y\} \times \mathbb{F}^n$ and the restriction $\varphi_y\colon E_y \to \{y\} \times \mathbb{F}^n$ of φ is an $\mathbb{F}$-linear isomorphism for every $y \in U$.

It is sometimes convenient to denote the total space E of ξ by $E(\xi)$.

Definition 10.3. A *regulous morphism* $h\colon \xi \to \eta$ of regulous $\mathbb{F}$-vector bundles ξ and η on X is a regulous map $h\colon E(\xi) \to E(\eta)$ such that $h(E(\xi)_x) \subseteq E(\eta)_x$ and the restriction $h_x\colon E(\xi)_x \to E(\eta)_x$ of h is an $\mathbb{F}$-linear transformation for every point $x \in X$.

Note that regulous $\mathbb{F}$-vector bundles on X (together with regulous morphisms) form a category.

Given an open subset $U \subseteq X$, we denote by $\mathcal{R}^0(U, \mathbb{F})$ the ring of all regulous functions from U into $\mathbb{F}$. The assignment

$$\mathcal{R}^0_X(\mathbb{F})\colon U \mapsto \mathcal{R}^0_X(\mathbb{F})(U) := \mathcal{R}^0(U, \mathbb{F})$$

is a sheaf of rings of $\mathbb{F}$-valued functions on X. Evidently, $\mathcal{R}^0_X(\mathbb{R})$ is just the structure sheaf $\mathcal{R}^0_X$ of X.

One can relate regulous $\mathbb{F}$-vector bundles on X with locally free $\mathcal{R}^0_X(\mathbb{F})$-sheaves as follows.

Let ξ be a regulous $\mathbb{F}$-vector bundle on X. A *regulous section* $s\colon U \to \xi$ of ξ on an open subset $U \subseteq X$ is a regulous map $s\colon U \to E(\xi)$ such that $s(x) \in E(\xi)_x$ for all $x \in U$. The set $\Gamma^0(U, \xi)$ of all regulous sections of ξ on U is an $\mathcal{R}^0(U, \mathbb{F})$-module. The assignment

$$\Gamma^0_X(\xi)\colon U \mapsto \Gamma^0_X(\xi)(U) := \Gamma^0(U, \xi)$$

is an $\mathcal{R}^0_X(\mathbb{F})$-sheaf on X.

To any regulous morphism $h\colon \xi \mapsto \eta$ of regulous $\mathbb{F}$-vector bundles on X we can associate a morphism $\Gamma^0_X(h)\colon \Gamma^0_X(\xi) \mapsto \Gamma^0_X(\eta)$ of $\mathcal{R}^0_X(\mathbb{F})$-sheaves on X defined by

$$\Gamma^0_X(\xi)(U) = \Gamma^0(U, \xi) \to \Gamma^0(U, \eta) = \Gamma^0_X(\eta)(U), s \mapsto h \circ s$$

for every open subset $U \subseteq X$.

Theorem 10.4. *Let X be an affine regulous variety. The functor $\xi \mapsto \Gamma^0_X(\xi)$ is an equivalence of the category of regulous $\mathbb{F}$-vector bundles on X with the category of locally free $\mathcal{R}^0_X(\mathbb{F})$-sheaves of finite rank on X.*

Also the following holds.

Theorem 10.5. *Let X be an affine regulous variety. The functor $\xi \mapsto \Gamma^0(X, \xi)$ is an equivalence of the category of regulous $\mathbb{F}$-vector bundles on X with the category of finitely generated projective $\mathcal{R}^0(X, \mathbb{F})$-modules.*

These two theorems come from [38]. It is worthwhile to recall that a counterpart of Theorem 10.4 holds for pre-algebraic vector bundles but not for algebraic vector bundles on an affine real algebraic variety; see [6, Section 12.1]. In view of Theorem 10.5 and [34, Proposition 3.9], one can identify regulous vector bundles and stratified-algebraic vector bundles. It is also proved in [38] that the total space of any regulous $\mathbb{F}$-vector bundle is an affine regulous variety.

The main focus of [34] and the subsequent papers [28, 30, 35, 38] is on comparison of algebraic, stratified-algebraic or regulous, and topological vector bundles.

Acknowledgements. The author was partially supported by the National Science Center (Poland) under grant number 2014/15/B/ST1/00046.

References

[1] J. Adamus and H. Seyedinejad, *A proof of Kurdyka's conjecture on arc-analytic functions*, Math. Ann. **369** (2017), 387–395.
[2] J. Adamus and H. Seyedinejad, *Extensions of arc-analytic functions*, Math. Ann. **371** (2018), 685–693.
[3] E. Bierstone and P. Milman, *Arc-analytic functions*, Invent. Math. **101** (1990), 411–424.
[4] M. Bilski, W. Kucharz, A. Valette and G. Valette, *Vector bundles and regulous maps*, Math. Z. **275** (2013), 403–418.
[5] Z. Błocki, *Singular sets of separately analytic functions*, Ann. Polon. Math. **56** (1992), 219–225.
[6] J. Bochnak, M. Coste and M.-F. Roy, *Real Algebraic Geometry*, Ergeb. Math. Grenzgeb. Vol 36, Springer (1998).
[7] S. Bochner and W. T. Martin, *Several Complex Variables*, Princeton University Press, Princeton NJ (1948).
[8] M. Coste and M. M. Diop, *Real algebraic 1-cocycles are Nash coboundaries*, Boll. Un. Mat. Ital. A (7) **6** (1992), 249–254.
[9] A. Czarnecki, *Maximal ideals of regulous functions are not finitely generated*, J. Pure Appl. Algebra **223** (2019), 1161–1166.

[10] C. Fefferman and J. Kollár, *Continuous solutions of linear equations, in: From Fourier analysis and number theory to Radon transforms and geometry*, Dev. Math. vol 28, pp. 233–282, Springer (2013).

[11] J. Fernando, G. Fichou, R. Quarez and C. Ueno, *On regulous and regular images of Euclidean spaces*, Quart. J. Math. **69** (2018), 1327–1351.

[12] G. Fichou, J. Huisman, F. Mangotle and J.-P. Monnier, *Fonctions régulues*, J. Reine Angew. Math. **718** (2016), 103–151.

[13] G. Fichou, J.-P. Monnier and R. Quarez, *Continuous functions on the plane regular after one blowing-up*, Math. Z. **285** (2017), 287–323.

[14] G. Fichou, J.-P. Monnier and R. Quarez, *Weak normalization and seminormalization in real algebraic geometry*, arXiv:1706.04467.

[15] G. Fichou, J.-P. Monnier and R. Quarez, *Integral closures in real algebraic geometry*, arXiv:1810.07556.

[16] G. Fichou, J.-P. Monnier and R. Quarez, *Substitution property for the ring of continuous rational functions*, Singularities — Kagoshima 2017, Proceedings of the 5th Franco-Japanese-Vietnamese Symposium on Singularities (this volume), pp. 71–93, World Scientific Publishing (2020).

[17] U. Görtz and T. Wedhorn, *Algebraic Geometry I. Schemes with Examples and Exercises*, Adv. Lect. Math., Vieweg+Teubner, Wiesbaden, 2010.

[18] A. Grothendieck, *Éléments de géométrie algébrique. III. Étude cohomologique des faisceaux cohérents*, I. Publ. Math, IHES **11** (1961), 167 pp.

[19] J. Kollár, *Lectures on Resolution of Singularities*, Princeton University Press, Princeton, NJ (2007).

[20] S. Iitaka, *Algebraic Geometry. An Introduction to Birational Geometry of Algebraic Varieties*, Springer, Berlin (1982).

[21] J. Kollár, W. Kucharz and K. Kurdyka, *Curve-rational functions*, Math. Ann. **370** (2018), 39–69.

[22] J. Kollár and K. Nowak, *Continuous rational functions on real and p-adic varieties*, Math. Z. **279** (2015), 85–97.

[23] T. Kowalczyk, *A version of Cartan's theorem A for coherent sheaves on real affine varieties*, Internat. J. Math. **29**, no. 8 (2018) 185058, 10 pp.

[24] W. Kucharz, *Rational maps in real algebraic geometry*, Adv. Geom. **9** (2009), 517–539.

[25] W. Kucharz, *Regular versus continuous rational maps*, Topology Appl. **160** (2013), 1375–1378.

[26] W. Kucharz, *Approximation by continuous rational maps into spheres*, J. Eur. Math. Soc. (JEMS) **16** (2014), 1555–1569.

[27] W. Kucharz, *Continuous rational maps into the unit 2-sphere*, Arch. Math. (Basel) **102** (2014), 257–261.

[28] W. Kucharz, *Some conjectures on stratified-algebraic vector bundles*, J. Singul. **12** (2015), 92–104.

[29] W. Kucharz, *Continuous rational maps into spheres*, Math. Z. **283** (2016), 1201–1215.

[30] W. Kucharz, *Stratified-algebraic vector bundles of small rank*, Arch. Math. (Basel) **107** (2016), 239–249.

[31] W. Kucharz, *Nash regulous functions*, Ann. Polon. Math. **119** (2017), 275–289.

[32] W. Kucharz, *Piecewise-regular maps*, Math. Ann. **372** (2018), 1545–1574

[33] W. Kucharz and K. Kurdyka, *Some conjectures on continuous rational maps into spheres*, Topology Appl. **208** (2016), 17–29.

[34] W. Kucharz and K. Kurdyka, *Stratified-algebraic vector bundles*, J. Reine Angew. Math. **745** (2018), 105–154.

[35] W. Kucharz and K. Kurdyka, *Comparison of stratified-algebraic and topological K-theory*, arXiv:1511.04238.

[36] W. Kucharz and K. Kurdyka, *Linear equations on real algebraic surfaces*, Manuscripta Math. **154** (2017), 285–296.

[37] W. Kucharz and K. Kurdyka, *From continuous rational to regulous functions*, Proc. Int. Cong. of Math. (2018), Rio de Janeiro, Vol. 2, 719–747.

[38] W. Kucharz and M. Zieliński, *Regulous vector bundles*, Math. Nachr. **291** (2018), 2252–2271.

[39] K. Kurdyka, *Ensembles semi-algébriques symmétriques par arcs*, Math. Ann. **282** (1988), 445–462.

[40] K. Kurdyka, *Analytic arcs and real analytic singularities*, Panoramas et Synthèses vol 51, Soc. Math. France (2017), 79–106.

[41] R. Lazarsfeld, *Positivity in Algebraic Geometry. I. Classical setting: Line Bundles and Linear Series*, Ergeb. der Math. und ihrer Grengeb. Folge 3, vol. 48, Springer (2004).

[42] K. Kurdyka and A. Parusiński, *Arc-symmetric sets and arc-analytic mappings, in: Arc spaces and additive invariants in real algebraic and analytic geometry*, Panor. Synthèses 24, Soc. Math. France, Paris (2007), 33–67.

[43] J.-P. Monnier, *Semi-algebraic geometry with rational continuous functions*, Math. Ann. **372** (2018), 1041–1080.

[44] A. Parusiński, *Topology of injective endomorphisms of real algebraic sets*, Math. Ann. **328** (2004), 353–373.

[45] J. Siciak, *Singular sets of separately analytic functions*, Colloq. Math. **60/61** (1990), 281–290.

[46] M. Zieliński, *Homotopy properties of some real algebraic maps*, Homology Homotopy Appl. **18** (2016), 287–294.

[47] M. Zieliński, *Approximation of maps into spheres by regulous maps*, Arch. Math (Basel) **110** (2017), 29–34.

PART B

Research Articles

Substitution property for the ring of continuous rational functions

Goulwen Fichou
Univ Rennes, CNRS, IRMAR - UMR 6625, F-35000 Rennes, France
E-mail: goulwen.fichou@univ-rennes1.fr

Jean-Philippe Monnier
LUNAM Université, LAREMA, Université d'Angers
E-mail: jean-philippe.monnier@univ-angers.fr

Ronan Quarez
Université de Rennes
Campus de Beaulieu, 35042 Rennes Cedex, France
E-mail: ronan.quarez@univ-rennes1.fr

We study the substitution property for the ring $\mathcal{K}^0(V)$ of continuous rational functions on a real algebraic affine variety V. We show that $\mathcal{K}^0(V)$ satisfies a substitution property along points; moreover, when V is non-singular, it satisfies also a substitution property along Puiseux arcs, which characterizes $\mathcal{K}^0(V)$.

Keywords: continuous rational function, real algebraic variety, substitution property

2010 MSC: 14P99, 11E25, 26C15

1. Introduction

Given a real algebraic variety $V \subset \mathbb{R}^n$, any morphism $\phi : \mathbb{R}[V] \to \mathbb{R}$ can be seen as an evaluation morphism at a certain point $x_0 \in V$, where $\mathbb{R}[V]$ denotes the coordinate ring of V. For any extension ring B of $\mathbb{R}[V]$ a natural question is to ask whether the evaluation morphism does extend uniquely to B. This question is known as the substitution property, when one considers,

We thank Michel Coste for fruitful discussions on the topic.
The authors benefit from the support of the Centre Henri Lebesgue ANR-11-LABX-0020-01.

more generally, an evaluation morphism $\phi : \mathbb{R}[V] \to R$ into a real closed field extension R of $\mathbb{R}$. Such a property gives a lot of information on the real algebra of the ring B namely, one may derive Artin-Lang property, Positivstellensatz, etc.

Natural classes of rings B to test are the different rings of functions considered in real algebraic geometry. The larger class consists in the ring of continuous (with respect to the Euclidean topology) semi-algebraic functions on V. In that situation, the substitution property is known to be true by [12]. A more rigid class is given by the ring of Nash functions on V, namely real analytic and semi-algebraic functions. Here also the substitution property holds true (see [2, 8.5.2]).

In the present paper, we study the case of the ring $\mathcal{K}^0(V)$ of continuous rational functions on V. Continuous rational functions on a real algebraic variety V are those rational functions that admit a continuous extension along their poles. They form an intermediate ring between regular functions (rational functions without real poles) and continuous semi-algebraic functions. This ring has been intensively studied in several recent works ([4], [5], [6], [7], [8], [9], [10], [11]).

The main results of the paper are the following. We first show that $\mathcal{K}^0(V)$ satisfies the substitution property (Theorem 2.13), generalizing a result in [4]. A natural question is then whether this property characterizes the ring of continuous rational functions. However we prove that this is not the case (Proposition 2.14). This motivates the introduction of the concept of evaluation along Puiseux arcs in section 3. More precisely, we take evaluation morphisms of the form $\phi : \mathbb{R}[V] \to \mathbb{R}[[t^{1/\mathbb{N}}]]$, where $\mathbb{R}[[t^{1/\mathbb{N}}]]_{\text{alg}}$ is the ring of Puiseux power series over $\mathbb{R}$ that are algebraic over real polynomials.

Depending upon our variety V is singular or not, we obtain two opposite results. Firstly, if V is non-singular, then $\mathcal{K}^0(V)$ satisfies the substitution property along arcs and moreover this property characterizes $\mathcal{K}^0(V)$ (see Theorem 3.5). Secondly, it exists a singular variety V such that $\mathcal{K}^0(V)$ does not satisfy the substitution property along arcs (see Theorem 4.7).

2. Substitution on points

2.1. *Definition and first examples*

Definition 2.1. Let A be an $\mathbb{R}$-algebra and B be an A-algebra. We say that B has the substitution property over A if, for any real closed field R, any morphism $A \to R$ admits one and only one lifting to B, namely such that the following diagram is commutative :

$$\begin{array}{ccc} A & \to & R \\ \downarrow & \nearrow & \\ B & & \end{array}$$

If the property is valid for the particular real closed field $R = \mathbb{R}$ we say that B satisfies the weak substitution property.

Recall that an ideal I of A is called real if, for every sequence $a_1, \ldots, a_k$ of elements of A, then $a_1^2 + \cdots + a_k^2 \in I$ implies $a_i \in I$ for $i = 1, \ldots, k$. A field F is called real if (0) is a real ideal in F.

Here is the first easy fact :

Proposition 2.2. *Let us assume that A is a domain whose fraction field is real. If B satisfies the substitution property over A, then the morphism $A \to B$ is injective.*

Proof. Let us assume that $0 \neq a \in A$ is sent onto 0 in B. Since the fraction field of A is real, the null ideal (0) in A is real and hence it is the intersection of all the real prime ideal of A. Then, there is a real prime ideal $\mathfrak{p}$ in A such that $a \notin \mathfrak{p}$. Take then for R the real closure of the residual field at $\mathfrak{p}$. Then, a is sent onto a non-zero element in R, which leads to a contradiction since it should be 0 after lifting the evaluation morphism to B. □

In the following, we will mainly consider rings B which are subrings of the fraction field of a domain A. If $f = p/q$ where p and q lie in A, starting from a morphism $\phi : A \to R$, it is easy to define $\phi(f) = \frac{\phi(p)}{\phi(q)}$ as long as $\phi(q) \neq 0$. This elementary observation gives us the substitution property if B is a subring of the ring of regular functions of A. We recall that the ring of regular functions of A is just the ring $(1 + \sum A^2)^{-1}A$.

Hence, in all the following, we will mainly consider rings B which are subrings of the field of rational functions and which contain the ring of regular functions of A. And the problem we will face is to define $\phi(f)$ when $\phi(q) = 0$.

The first important consequence of Definition 2.1, is that one gets a one-to-one correspondence between the real spectrum of B and A. For background about the real spectrum of a ring, we refer to [2]. In few words, let us say that a point α of the real spectrum of the ring A is an equivalence class of morphisms π from A into a real closed field for the equivalence relation generated by the following : the morphism $\pi : A \to R$

is equivalent to $\pi' : A \to R'$, with R and R' real closed fields, if there exists a factorization of π' through a morphism $R \to R'$.

The real spectrum is endowed with two natural topologies : the real spectrum topology and the constructible topology which are defined as follows. Any element $a \in A$ can be evaluated at any $\alpha \in \mathrm{Spec}_r A$ simply by evaluating it at the morphism $\pi : A \to R$ given by α. Then, $\{\alpha \in \mathrm{Spec}_r A \mid a(\alpha) > 0\}$ gives a basis of open neighborhood for the real spectrum topology. Then, the constructible open subsets are those subsets obtained by boolean combination from the open subsets with respect to the real spectrum topology.

To any semi-algebraic subset S of $\mathbb{R}^n$ is canonically associated a constructible subset $\widetilde{S}$ of $\mathrm{Spec}_r \mathbb{R}[x_1, \ldots, x_n]$. This association is compatible with the boolean operations (intersection, union, complementary) and is such that the semi-algebraic subset $S = \{x \in \mathbb{R}^n \mid p(x) > 0\}$ where $p \in \mathbb{R}[x_1, \ldots, x_n]$ is associated to $\widetilde{S} = \{\alpha \in \mathrm{Spec}_r \mathbb{R}[x_1, \ldots, x_n] \mid p(\alpha) > 0\}$.

Now we state the announced correspondence.

Proposition 2.3. *If B satisfies the substitution property over A, then the induced morphism on the real spectrum $\mathrm{Spec}_r B \to \mathrm{Spec}_r A$ is bijective and continuous (with respect to both the real spectrum topology and the constructible topology).*

Proof. By definition, the induced morphism on the real spectrum $\mathrm{Spec}_r B \to \mathrm{Spec}_r A$ is continuous with respect to the real spectrum topology and also for the constructible topology (see [2, 7.17]).

Surjectivity comes immediately from the existence of the factorization in the substitution property.

Injectivity comes from the uniqueness of the factorization in the substitution property. Indeed, let us consider two points $\beta : B \to R$ and $\beta' : B \to R'$ where R, R' are real closed fields and assume that the compositions $\beta \circ \phi$ and $\beta' \circ \phi$ by $\phi : A \to B$ give rise to the same point in $\mathrm{Spec}_r A$. Then there exists a real closed filed R'' which extends both R and R', which implies that $\beta = \beta'$. $\square$

From now on, we will takes for A a ring of polynomials functions $\mathbb{R}[V]$, or one of its localization, where V is an algebraic variety in $\mathbb{R}^n$ in the sense of [2].

There are two well-known classes of functions whose ring satisfies the substitution property : Nash functions and continuous semi-algebraic functions. More precisely, if $A = \mathbb{R}[X_1, \ldots, X_n]$, then one has the substitution property for the ring of Nash functions on $\mathbb{R}^n$ ([2, 8.5.2]) and one may generalize it to any smooth algebraic variety V in $\mathbb{R}^n$. The substitution property is also valid for the ring of semi-algebraic continuous functions on V. This result can be seen as a consequence of the theory of real closed rings developed by N. Schwartz in [12]. See also the work of J. Fernando on the subject [3].

In these two cases, since a morphism $A \to R$ corresponds to the evaluation at some point $(x_1, \ldots, x_n)$ in $V \subset \mathbb{R}^n$, the substitution property says that any evaluation morphism can be uniquely lifted to the ring B.

Beware that a subring of a ring satisfying the substitution property does not necessarily satisfy the substitution property as it is shown by the following example :

Example 2.4. Let $A = \mathbb{R}[x]$ and $B = \mathbb{R}[x, \sqrt{1+x^2}] \simeq \mathbb{R}[x,y]/(y^2 - (1 + x^2))$. Then, the morphism $A \to \mathbb{R}$ which sends x to 0 (the evaluation morphism at the origin) admits two different liftings to B : the morphism which sends y to 1 and the morphism which sends y to -1. Moreover, B is a subring of the ring of semi-algebraic continuous functions on $\mathbb{R}$.

2.2. *Substitution for continuous rational functions*

Let $V \subset \mathbb{R}^n$ be a real algebraic variety. We will mainly be interested in the ring of continuous rational functions defined on V. These functions are rational functions on V which can be seen as functions defined on the complementary of the zero set of their denominators, and which admits a continuous extension to the whole V. In order to avoid pathological cases, we will mainly assume that the variety V is central which means that the set of non-singular points of V is dense in V for the Euclidean topology. Of course, any non-singular variety is central.

Since we require for our rings to be contained in the field of fraction $\mathbb{R}(V)$ of V, one should consider the natural assumption for the variety V to be moreover irreducible. So let us give now the formal definition of continuous rational functions.

Definition 2.5. Let $V \subset \mathbb{R}^n$ be an irreducible central real algebraic variety. A continuous function $f : V \to \mathbb{R}$ is said to be continuous rational if there exists a non-empty Zariski open subset $W \subset V$ such that the restriction $f_{|W}$ of f to W is a regular function on W. We denote by $\mathcal{K}^0(V)$ the ring of continuous rational functions on V.

Remark 2.6. An alternative ring of functions to work with would be the ring of hereditarily continuous rational (also called regulous) functions on a central variety V defined in [5] as those continuous rational functions which remains rational in restriction to any subvariety. When the variety V is smooth, we recover continuous rational functions, but in general it leads to a proper subring. We will consider this ring solely in Theorem 2.13.

Let us start with an elementary observation about rings of functions satisfying the substitution property. If f is a real function defined on a subset of $\mathbb{R}^n$, we denote by $\mathcal{Z}(f)$ the zero set of f.

Proposition 2.7. *Let $A = \mathbb{R}[x_1, \ldots, x_n]$ and B be a sub A-algebra of $\mathbb{R}(x_1, \ldots, x_n)$ satisfying the substitution property over A.*

Let $f = p/q \in B$ with p and q coprimes. Then,

(i) $\mathcal{Z}(q) \subset \mathcal{Z}(p)$,

(ii) $\mathcal{Z}(q)$ *has codimension at least* 2 *in* $\mathbb{R}^n$.

Proof. To show (i), let us just write $p = fq$ and use the substitution property for B over A.

To show (ii), assume there exists an irreducible factor q_1 of q corresponding to a codimension one variety in $\mathbb{R}^n$. Then, the condition $\mathcal{Z}(q_1) \subset \mathcal{Z}(p)$ implies that q_1 divides p since (q_1) is a real prime ideal (cf. the so-called change of signs criterion [2, 4.5.1]), in contradiction with the coprimality of p and q. □

Remark 2.8. Note that in the one-dimensional case, condition (ii) says already that B is a subring of the ring of regular functions.

Note that the previous proposition generalizes to non-singular varieties :

Proposition 2.9. *Let $A = \mathbb{R}[V]$ where V is a non-singular irreducible algebraic variety in $\mathbb{R}^n$. Let B be a sub A-algebra of $\mathbb{R}(V)$ satisfying the substitution property over A.*

Let $f = p/q \in B$ with p and q coprimes. Then,

(i) $\mathcal{Z}(q) \subset \mathcal{Z}(p)$,

(ii) $\mathcal{Z}(q)$ *has codimension at least* 2 *in* V.

Proof. The coordinates ring $\mathbb{R}[V]_{m_x}$ of the variety V localized at any point $x \in V$ is regular and hence an UFD, so that the proof can be carried out similarly to the proof of Proposition 2.7. □

This result can be used when $B = \mathcal{K}^0(V)$ where $V \subset \mathbb{R}^n$ is non-singular (see Theorem 2.12). On the contrary, the proposition cannot be extended to the case V is singular as illustrated by the example of the cuspidal curve.

Example 2.10. Let $V \subset \mathbb{R}^2$ with equation $y^2 = x^3$, and consider the continuous rational function on V given by $f = y/x$ extended by 0 at the origin. Then, the zero set of its denominator x is the origin, therefore of codimension one in V.

We come now to the main result of this section, which gives the substitution property for the ring of continuous rational functions on a non-singular real algebraic variety. Namely, we generalize [4, 5.4] which relies on Łojasiewicz inequality together with the fact that a continuous rational function on $\mathbb{R}^n$ admits a constructible stratification such that it is regular in restriction to its strata.

This last property extends to any non-singular variety:

Theorem 2.11. *Let $V \subset \mathbb{R}^n$ be a non-singular irreducible real algebraic variety and f be a continuous rational function on V. Then, there exists a stratification of V into Zariski locally closed subsets $S_1, \ldots, S_m$ such that the restriction $f|_{S_k}$ of f to S_k is a regular function.*

Proof. The proof is heavily based on the fact that, since V is non-singular, f is hereditarily rational and hence one may use [10, Théorème 4.1]. □

Let us show now the desired substitution property for continuous rational functions:

Theorem 2.12. *Let $V \subset \mathbb{R}^n$ be a non-singular irreducible real algebraic variety. Then, $\mathcal{K}^0(V)$ satisfies the substitution property over $\mathbb{R}[V]$.*

Moreover, the induced morphism $Spec_r\mathcal{K}^0(V) \to Spec_r\mathbb{R}[V]$ is a homeomorphism with respect to the constructible topology.

Proof. We focus on the unicity of the factorization, since the existence comes from the usual evaluation. Let a morphism $\phi : \mathbb{R}[V] = \mathbb{R}[x_1, \ldots, x_n]/I \to R$ and denote also by ϕ one extension to $\mathcal{K}^0(V)$, namely $\phi : \mathcal{K}^0(V) \to R$.

First, let us see what happens if $R = \mathbb{R}$. Then, one may assume, for simplicity, that the origin o is in V and that $\phi(x_1) = \ldots = \phi(x_n) = 0$. Let $f \in \mathcal{K}^0(V)$. Up to considering $f - f(o)$, one may assume that $f(o) = 0$.

Then, by Łojasiewicz property [2, Proposition 2.6.4] applied to $x_1^2 + \ldots + x_n^2$ and f, there exist an integer N and a continuous rational function g in $\mathcal{K}^0(V)$ such that $f^N = (x_1^2 + \ldots + x_n^2)g$. This algebraic identity implies that $\phi(f) = 0$ which concludes the proof for the case $R = \mathbb{R}$.

Let us now consider a general real closed field R and denote by α the point of the real spectrum of $\mathbb{R}[V]$ given by the morphism ϕ. According to

Theorem 2.11, we know that, for any $f \in \mathcal{K}^0(V)$, there is a Zariski locally closed stratification $V = S_0 \cup \ldots \cup S_m$ such that f is regular on each S_i. Let us assume that $\alpha \in \widetilde{S_0}$; if we set

$$S_0 = \{x \in V \mid r(x) = 0, s(x) \neq 0\}$$

where r, s are polynomials, then $\alpha \in \widetilde{S_0}$ just means that $\phi(r) = 0$ and $\phi(s) \neq 0$.

Then, the zero set of r is contained in the zero set of $s \cdot (qf - p)$ where p, q are polynomials such that $f = p/q$ is regular on S_0. Again, by Łojasiewicz property one gets the existence of an integer N and a continuous rational function g such that $(s(qf - p))^N = rg$. Applying ϕ, and since $\phi(s) \neq 0$ and $\phi(q) \neq 0$ (since p/q is regular on S_0), one gets $\phi(f) = \phi(p)/\phi(q)$ which concludes the first part of the proof.

By Proposition 2.3, the map $\varphi : \mathrm{Spec}_r \mathcal{K}^0(V) \to \mathrm{Spec}_r \mathbb{R}[V]$ is continuous and bijective. Let us consider a constructible set T in $\mathrm{Spec}_r \mathcal{K}^0(V)$ which is a boolean combination of sets of the form $S = \{f = 0, g > 0\}$ where $f, g \in \mathcal{K}^0(V)$. Again by Theorem 2.11, one knows that there is a Zariski locally closed stratification $V = S_0 \cup \ldots \cup S_m$ such that f and g are regular on each S_k. Namely, one may write on S_k, $f = p_k/q_k$ and $g = r_k/s_k$ where p_k, q_k, r_r, s_k are polynomials such that q_k and s_k do not vanish on S_k.

Hence, the image of S by φ is then the union of all $\widetilde{S_k} \cap \varphi(S)$ which can be written as the subset $\widetilde{S_k} \cap \{p_k = 0, r_k s_k > 0\}$ in $\mathrm{Spec}_r \mathbb{R}[V]$. Finally, one gets that $\varphi(T)$ is a constructible subset in $\mathrm{Spec}_r \mathbb{R}[V]$. □

The previous proposition says that if an $\mathbb{R}$-algebra homomorphism from $\mathcal{K}^0(V)$ is the evaluation at a given point in restriction to the polynomials, then it is still the evaluation at this point on any continuous rational functions on V.

So far it does not seem very clear how to get the substitution property for the ring of continuous rational functions on a singular variety since one no longer disposes of the decomposition of rational functions as regular functions on strata.

Although one may obtain formally, by taking some quotients in a commutative diagram, that the ring of regulous functions on a singular variety V satisfies the substitution property over $\mathbb{R}[V]$. By definition, the ring of regulous functions on a real algebraic variety $V \subset \mathbb{R}^n$, given by an ideal $I \subset \mathbb{R}[X_1, \ldots, X_n]$, is just the quotient ring $\mathcal{R}^0(V) = \mathcal{K}^0(\mathbb{R}^n)/\mathcal{I}_{\mathcal{K}^0(\mathbb{R}^n)}(V)$ where $\mathcal{I}_{\mathcal{K}^0(\mathbb{R}^n)}(V) = \{f \in \mathcal{K}^0(\mathbb{R}^n) \mid V \subset \mathcal{Z}(f)\}$. Hence, one may easily derive the substitution property for the ring of regulous functions on V. Namely:

Theorem 2.13. *Let $V \subset \mathbb{R}^n$ be a central irreducible real algebraic variety whose coordinate ring is $\mathbb{R}[V] = \mathbb{R}[X_1, \ldots, X_n]/I$. Then $\mathcal{R}^0(V)$ satisfies the substitution property over $\mathbb{R}[V]$.*

Proof. Let $A = \mathbb{R}[X_1, \ldots, X_n]$, $B = \mathcal{R}^0(\mathbb{R}^n)$ and $\mathcal{I} = \mathcal{I}_{\mathcal{K}^0(\mathbb{R}^n)}(V)$.

The following commutative diagram is a consequence of the universal property for quotient rings :

$$\begin{array}{lll} A & \to A/I & \to R \\ \downarrow & \quad \downarrow & \nearrow \\ B & \to B/\mathcal{I} & \\ \downarrow & & \\ \mathbb{R}(V) & & \end{array}$$

Namely, given an evaluation morphism $A/I \to R$ it gives rise to a unique morphism $A \to R$ which sends I onto 0. Hence, by the substitution property for B, one has an unique lifting morphism $B \to R$ which send $\mathcal{I}$ onto 0. Hence, by the universal property for quotient rings, one has a unique factorization $B/\mathcal{I} \to R$. □

2.3. *The substitution property does not characterize continuous rational functions*

We show that the ring of continuous rational functions is not the biggest subring of rational functions that satisfies the substitution property and hence, the substitution property is not a characterization for the ring of continuous rational functions. This result will lead to the notion of substitution along arcs in the next section for which such a characterization will be available.

But for the classical substitution property, already in the case of the plane one has:

Proposition 2.14. *The ring $\mathcal{K}^0(\mathbb{R}^2)$ is not maximal in $\mathbb{R}(x, y)$ to satisfy the substitution property.*

Before entering into the details of the proof, we will state as an intermediary step that there exist bigger rings that satisfy only the *existence* condition of the lifting in the substitution property. Namely,

Proposition 2.15. *The ring $\mathcal{K}^0(\mathbb{R}^2)$ is not maximal in $\mathbb{R}(x, y)$ among rings satisfying the existence condition of the lifting in the substitution property. Moreover, there does not exist a unique maximal ring satisfying the existence condition of the lifting in the substitution property.*

Proof. For the first point, let us show that the ring $\mathcal{K}^0(\mathbb{R}^2)[\frac{x}{x^2+y^2}]$ satisfies also the existence condition of the lifting in the substitution property.

It suffices to show that the evaluation morphism at the origin ϕ : $\mathcal{K}^0(\mathbb{R}^2) \to \mathbb{R}$ can be lifted to $\mathcal{K}^0(\mathbb{R}^2)[\frac{x}{x^2+y^2}]$. The first step in the proof is to show that one may set $\phi\left(\frac{x}{x^2+y^2}\right) = 0$.

Consider the morphism $\mathcal{K}^0(\mathbb{R}^2)[T] \xrightarrow{\psi} \mathbb{R}(x,y)$ defined by $T \mapsto \frac{x}{x^2+y^2}$. One has

$$\mathcal{K}^0(\mathbb{R}^2)[\frac{x}{x^2+y^2}] \simeq \mathcal{K}^0(\mathbb{R}^2)[T]/\mathrm{Ker}\,\psi.$$

Let $P \in \mathrm{Ker}\,\psi$, and write $P = a_0 + \ldots + a_nT^n$ where the a_i's are continuous rational functions on $\mathbb{R}^2$. In $\mathbb{R}(x,y)[T]$, one may factorize P by $T - p/q$ where $p = x, q = x^2 + y^2$. Hence

$$\frac{(qT-p)(b_0 + \ldots + b_{n-1}T^{n-1})}{d} = a_0 + \ldots + a_nT^n$$

where the b_i's and d are polynomials in $\mathbb{R}[x,y]$.

Then,

$$(*)\begin{cases} -b_0p & = a_0d \\ qb_0 - b_1p & = a_1d \\ \ldots & \\ qb_{n-2} - b_{n-1}p & = a_{n-1}d \\ qb_{n-1} & = a_nd \end{cases}$$

Let r be the valuation of d with respect to the prime $p = x$, namely $v_p(d) = r$. Since p and q are coprime, one gets from the last identity in $(*)$ that $v_p(b_{n-1}) \geq r$ since (p) is a real prime ideal. Likewise, $v_p(b_{n-2}) \geq r$ and more generally $v_p(b_i) \geq r$ for $i = 0, \ldots, n-1$. The first identity in $(*)$ implies that we may write $a_0 = h/k$ with $h, k \in \mathbb{R}[x,y]$, $v_p(h) \geq 1$ and $v_p(k) = 0$. It follows that a_0 vanishes on a non-empty Zariski open subset of $\mathcal{Z}(x)$ and thus on the whole $\mathcal{Z}(x)$. Hence $a_0(0) = 0$.

Hence, the setting $\phi\left(\frac{x}{x^2+y^2}\right) = 0$ is compatible with any algebraic relation verified by $f = \frac{p}{q} = \frac{x}{x^2+y^2}$ over $\mathcal{K}^0(\mathbb{R}^2)$. It defines then a lifting of ϕ to $\mathcal{K}^0(\mathbb{R}^2)[\frac{x}{x^2+y^2}]$.

Note that, if one sets $\phi_\alpha\left(\frac{x}{x^2+y^2}\right) = \alpha \in \mathbb{R}$, one may replace f with $g = f - \alpha$, namely replace x with $p = x - \alpha(x^2+y^2)$ which remains real prime. One can then repeat the previous process to get that $a_0(0) = 0$ for any algebraic relation $a_0 + \ldots + a_ng^n = 0$ where the a_i's are continuous rational functions. Since $\phi_\alpha(g) = 0$, again one has a lifting of ϕ_α to $\mathcal{K}^0(\mathbb{R}^2)[\frac{x}{x^2+y^2}]$. Hence, one has infinitely many liftings of ϕ to $\mathcal{K}^0(\mathbb{R}^2)[\frac{x}{x^2+y^2}]$.

Let us show now the second point. Set $f = \frac{x}{x^2+y^2}$ and $g = \frac{y}{x^2+y^2}$. Let B_1 be a ring that contains $\mathcal{K}^0(\mathbb{R}^2)[f]$ and that satisfies the existence in the substitution property. Then g cannot belong to B_1. Indeed, from the identity $xf + yg = 1$ we get, using that ϕ is the evaluation at the origin, that

$$1 = \phi(xf + yg) = \phi(x)\phi(f) + \phi(y)\phi(g) = 0.$$

Likewise, let B_2 be a ring that contains $\mathcal{K}^0(\mathbb{R}^2)[g]$ and that satisfies the existence in the substitution property. As previously, one has $f \notin B_2$.

Then, there does not exist a unique maximal ring satisfying the existence condition of the lifting in the substitution property. □

The proof of Proposition 2.14 is obtained by slightly modifying the above proof.

Proof of Proposition 2.14. Let us consider the ring

$$B = \mathcal{K}^0(\mathbb{R}^2)[\frac{x}{(x^2+y^2)^n}, n \in \mathbb{N}^*].$$

Setting $f_n = \frac{x}{(x^2+y^2)^n}$, one has $f_n = f_{n+1}(x^2 + y^2)$. Let us note that it suffices to consider the evaluation morphism at the origin $A = \mathbb{R}[x, y] \xrightarrow{\phi} \mathbb{R}$, and show that it admits one and only one lifting to B.

If such a lifting (again denoted by ϕ) of the evaluation at the origin exists, it satisfies necessarily $\phi(f_n) = \phi(f_{n+1})\phi((x^2 + y^2))$. Since $\phi(x^2 + y^2) = 0$, it follows that $\phi(f_n) = 0$ for any n. This shows the unicity of the lifting.

To show the existence, let us consider an algebraic relation of the form $P(f_1, \ldots, f_n) = 0$ where $P = a_0 + \sum_{\alpha>0} a_\alpha x^{\alpha_1} \ldots x^{\alpha_n}$ is a polynomial whose coefficients are continuous rational functions on $\mathbb{R}^2$. The relation $P(f_1, \ldots, f_n) = 0$ can be rewritten as

$$a_0 + \sum_{\alpha>0} a_\alpha \frac{x^{\alpha_1+\ldots+\alpha_n}}{(x^2+y^2)^{\alpha_1+2\alpha_2+\ldots+n\alpha_n}} = 0,$$

which implies

$$a_0(x^2 + y^2)^N = -\sum_{\alpha>0} a_\alpha (x^2 + y^2)^{N(\alpha)} x^{\alpha_1+\ldots+\alpha_n}$$

where N and $N(\alpha)$ are integers.

From the previous identity, it follows that a_0 vanishes on $\mathcal{Z}(x)$ and one gets $\phi(a_0) = 0$. Hence $\phi(P(f_1, \ldots, f_n)) = 0$ as desired. □

In summary, the substitution property as defined in Section 2.1 is not strong enough to characterize continuous rational functions. More precisely, the lifting property is not sufficient to imply continuity, even if the evaluation morphism can be taken over a Puiseux series field. For instance, we may consider a rational function $f = p/q$ on the plane such that $p(0) = q(0) = f(0) = 0$, and an algebraic plane curve at the origin parametrized by two Puiseux series $\gamma = (\alpha(t), \beta(t))$ which can also be seen as an element of the real spectrum of $\mathbb{R}[x, y]$. If γ goes to the origin by assumption, it seems not obvious to deduce that $f(\gamma)$ also goes to the origin.

These considerations lead us to consider a new substitution property along (convergent) arcs in the remaining of the paper.

3. Substitution along arcs

The ring of all formal power series over $\mathbb{R}$ in the indeterminate t will be denoted by $\mathbb{R}[[t]]$. The ring of all formal power series over $\mathbb{R}$ in the indeterminate t which are algebraic over $\mathbb{R}[t]$ will be denoted by $\mathbb{R}[[t]]_{\mathrm{alg}}$.

We consider moreover several power series rings which are all subrings of the field of Puiseux power series. The field of all Puiseux power series over $\mathbb{R}$ in the indeterminate t will be denoted by $\mathbb{R}((t^{1/\mathbb{N}}))$, and its valuation ring by $\mathbb{R}[[t^{1/\mathbb{N}}]]$. The subfield of Puiseux series which are algebraic over $\mathbb{R}[t]$ will de denoted by $\mathbb{R}((t^{1/\mathbb{N}}))_{\mathrm{alg}}$, and its valuation ring by $\mathbb{R}[[t^{1/\mathbb{N}}]]_{\mathrm{alg}}$. From a geometric point of view, an element in the field $\mathbb{R}((t^{1/\mathbb{N}}))_{\mathrm{alg}}$ is identified with a continuous semi-algebraic functions germ $(0, +\infty) \to \mathbb{R}$ on the right at the origin, the ring $\mathbb{R}[[t^{1/\mathbb{N}}]]_{\mathrm{alg}}$ consisting of those germs which can be extended continuously at the origin [2, Section 8.1].

We will use frequently in the sequel the following identification between arcs traced on a variety V, and $\mathbb{R}$-algebra morphisms $\mathbb{R}[V] \to \mathbb{R}[[t]]_{\mathrm{alg}}$. More precisely, if $V \subset \mathbb{R}^n$ is an algebraic variety defined by the polynomial equations $f_1 = \ldots = f_r = 0$, an algebraic formal arc $\gamma(t) = (\gamma_1(t), \ldots, \gamma_n(t)) \in (\mathbb{R}[[t]]_{\mathrm{alg}})^n$ traced on the variety V satisfies $f_1(\gamma(t)) = \ldots = f_r(\gamma(t)) = 0$. In other words, an algebraic formal arc on V is given by an $\mathbb{R}$-algebra morphism $\mathbb{R}[V] \to \mathbb{R}[[t]]_{\mathrm{alg}}$.

3.1. *Definition and first properties*

We state a substitution property along algebraic formal arcs and algebraic Puiseux arcs.

Definition 3.1. Let A be an $\mathbb{R}$-algebra and B be an A-algebra. We say that B has the substitution property over A along algebraic formal arcs (respectively algebraic Puiseux arcs) if any morphism $A \to \mathbb{R}[[t]]_{\text{alg}}$ (resp. $\mathbb{R}[[t^{1/\mathbb{N}}]]_{\text{alg}}$) admits one and only one lifting to B.

Again, we mainly consider the case where A is a ring of polynomial functions. One has the substitution property along algebraic Puiseux arcs for the ring of Nash functions and the ring of continuous semi-algebraic functions.

Proposition 3.2. *Let $A = \mathbb{R}[V]$ be the coordinate ring of a non-singular real algebraic variety $V \subset \mathbb{R}^n$. Let B be either the ring of continuous semi-algebraic functions on V or the ring of Nash functions on V. Then B satisfies the substitution property along Puiseux arcs.*

Proof. First of all, note that since all the possible rings B being subrings of the ring of semi-algebraic continuous functions on V, the evaluation along an arc given by $A \to \mathbb{R}[[t^{1/\mathbb{N}}]]_{\text{alg}}$ can always be lifted to B by considering the usual composition of semi-algebraic arcs.

To prove the uniqueness, recall that $\mathbb{R}[[t^{1/\mathbb{N}}]]_{\text{alg}}$ is included into the field of algebraic Puiseux series $\mathbb{R}((t^{1/\mathbb{N}}))$ which is a real closed field. Then, let us use the fact that the ring B satisfies the substitution property on points (as it has been recalled in subsection 2.1). Since the composite morphism $A \to \mathbb{R}[[t^{1/\mathbb{N}}]]_{\text{alg}} \to \mathbb{R}((t^{1/\mathbb{N}}))_{\text{alg}}$ admits a unique lifting to B, so it is the case for a given morphism $A \to \mathbb{R}[[t^{1/\mathbb{N}}]]_{\text{alg}}$. □

In the sequel, we will mainly consider rings B which are subrings of the field of rational functions of an irreducible variety, and also rings containing the ring of regular functions (since those later obviously satisfy the substitution property along arcs).

Proposition 3.3. *If B satisfies the substitution property along algebraic formal or Puiseux arcs over A, then B satisfies also the weak substitution property on points over A.*

Proof. It suffices to compose with the evaluation morphism $\mathbb{R}[[t]]_{\text{alg}} \to \mathbb{R}$ (respectively with $\mathbb{R}[[t^{1/\mathbb{N}}]]_{\text{alg}} \to \mathbb{R}$) of an arc at its origin, by sending t onto 0. □

Let us also mention an example where the weak substitution on points does not imply the substitution along arcs.

Example 3.4. Let $A = (\mathbb{R}[x])_{(x)}$ and $B = (\mathbb{R}[x,y]/(x^2-y^2))_{(x,y)}$. Then, B has the weak substitution property along points over A but does not satisfy the substitution property over A along arcs. Indeed, the arc $x = t$ written on A can be lifted to two different arcs $\gamma_1 = (t,t)$ and $\gamma_2 = (t,-t)$ on B. Moreover, there is only one morphism $A \to \mathbb{R}$: the one which sends x to 0. This morphism admits one and only one lifting to B, namely the morphism given by $x \mapsto 0$ and $y \mapsto 0$.

3.2. *Substitution along arcs for continuous rational functions*

We show that the ring of continuous rational functions on a non-singular algebraic variety satisfies the substitution property along arcs. Moreover, we show that it is the biggest ring contained in the field of rational functions that one can expect to satisfy this property.

In other words, the substitution property along arcs characterizes, on non-singular varieties, continuous rational functions. It enlightens the importance of substitution along arcs for hereditarily rational functions, as it has already be pointed out in the work of [9] for instance.

Theorem 3.5. *Let $A = \mathbb{R}[V]$ be the coordinate ring of a non-singular irreducible real algebraic variety $V \subset \mathbb{R}^n$. Then the ring $\mathcal{K}^0(V)$ of continuous rational functions on V satisfies the substitution property along arcs. Moreover, if B is a subring of the field $\mathbb{R}(V)$ of rational functions on V which satisfies the substitution property along arcs, then $B \subset \mathcal{K}^0(V)$.*

Proof. First of all, the natural lifting given by the evaluation of a continuous semi-algebraic function along a semi-algebraic arc gives the existence property. For the unicity, remark that $\mathcal{K}^0(\mathbb{R}^n)$ satisfies the substitution property on points (Theorem 2.12). Then, any morphism $\gamma : A \to R$, where $R = \mathbb{R}[[t^{1/\mathbb{N}}]]_{\text{alg}}$ or $\mathbb{R}[[t]]_{\text{alg}}$, gives rise to a morphism $A \to \mathbb{R}((t^{1/\mathbb{N}}))_{\text{alg}}$ which can uniquely be lifted to $\mathcal{K}^0(\mathbb{R}^n)$. This shows the unicity for our desired substitution property, namely there cannot exist another lifting than the natural one.

Let us consider now a ring B satisfying the substitution property along arcs and let us show the inclusion $B \subset \mathcal{K}^0(V)$.

Let us first consider the case $R = \mathbb{R}[[t^{1/\mathbb{N}}]]_{\text{alg}}$. One may use the classical fact (see for instance [13, Chap 6, Lemma 4.2]) that a semi-algebraic function f is continuous on S if and only if $f \circ \gamma$ is continuous for all algebraic Puiseux arcs γ supported in S.

In fact, it is possible to adapt a little bit this result for our purpose :

Lemma 3.6. *Let $f : S \to \mathbb{R}$ be a semi-algebraic function and $T \subset S$ a semi-algebraic subset. Assume that $S \setminus T$ is dense in S.*

Then, f is continuous on S if and only if for any $x \in S$, $f \circ \gamma$ is continuous at 0 for any algebraic Puiseux arc γ supported in $(S \setminus T) \cup \{x\}$ and whose endpoint is x.

Proof. We focus on the non obvious implication, and we show that f is continuous on T.

We claim first that, for any $x_0 \in S$, the restriction $f|_{S\setminus T}$ of f to $S\setminus T$ has limit $f(x_0)$ when $x \in S \setminus T$ goes to x_0 i.e f is continuous on $(S \setminus T) \cup \{x_0\}$.

Indeed, if it were not the case, then there would be $\epsilon > 0$ such that the set

$$\{\|x - x_0\| \mid x \in S \setminus T, |f(x) - f(x_0)| \geq \epsilon\}$$

would contain arbitrarily positive elements. Since this set is semi-algebraic, it contains an interval I with endpoint 0. Now, by the curve selection lemma, there is an algebraic Puiseux arc $\gamma : I \to S \setminus T \cup \{x_0\}$ such that $\|\gamma(t) - x_0\| < \epsilon$ and $|f(\gamma(t)) - f(x_0)| \geq \epsilon$. This would contradict the assumption of the lemma.

Now let us assume that f is not continuous at $x_0 \in T$. There exists a sequence $(x_n)_{n\in\mathbb{N}^*}$ of elements in S whose limit is x_0 such that $(f(x_n))_{n\in\mathbb{N}^*}$ is not convergent toward $f(x_0)$. By density of $S\setminus T$ in S, one may construct a sequence $(x'_n)_{n\in\mathbb{N}^*}$ of elements in $S \setminus T$ close to the sequence $(x_n)_{n\in\mathbb{N}^*}$. Using the triangular inequality and the continuity of $f|_{S\setminus T}$, one gets a contradiction.

This concludes the proof. □

Let $f = p/q$ in B and let γ be an algebraic Puiseux arc well-defined at the origin, which is not contained into the polar locus of f, namely such that $q \circ \gamma \neq 0$. Then, $f \circ \gamma = p \circ \gamma / q \circ \gamma$ is well-defined in $\mathbb{R}((t^{1/\mathbb{N}}))_{\text{alg}}$, and necessarily $p \circ \gamma/q \circ \gamma$ has a non-negative valuation. Hence, $f \circ \gamma$ is continuous.

Moreover, using the substitution property at the origin of the arc, one deduces that the limit is $f(\gamma(0))$. In conclusion, $f \in \mathcal{K}^0(V)$ and hence $B \subset \mathcal{K}^0(V)$.

Let us consider now the case $R = \mathbb{R}[[t]]_{\text{alg}}$. Let $f \in B$ and as previously let γ be an algebraic Puiseux arc, well-defined at the origin, which is not contained in the polar locus of f. There exists an integer m such that $\delta(t) = \gamma(t^m) \in (\mathbb{R}[[t]]_{\text{alg}})^n$. Then, since B satisfies the substitution property along arcs for $R = \mathbb{R}[[t]]_{\text{alg}}$, the arc δ uniquely lifts to B and $f \circ \delta$ is continuous and admits $f(\delta(0))$ as a limit. Composing with the continuous semi-algebraic function $t \mapsto \sqrt[m]{t}$ one gets that $f \circ \gamma$ is continuous and admits $f(\gamma(0))$ as a limit too. □

One natural question is what happens if we replace algebraic Puiseux arcs with formal Puiseux series. It happens that the substitution property remains true for continuous rational functions.

Theorem 3.7. *Let $A = \mathbb{R}[V]$ be the coordinate ring of a non-singular irreducible real algebraic variety $V \subset \mathbb{R}^n$. Then the ring $\mathcal{K}^0(V)$ of continuous rational functions on V satisfies the substitution property along arcs in $\mathbb{R}[[t^{1/\mathbb{N}}]]$.*

Moreover, if B is a subring of the field $\mathbb{R}(V)$ of rational functions on V which satisfies the substitution property along arcs in $\mathbb{R}[[t^{1/\mathbb{N}}]]$, then $B \subset \mathcal{K}^0(V)$.

Proof. Let us consider $\gamma : \mathbb{R}[V] \to \mathbb{R}[[t^{1/\mathbb{N}}]]$. One may see γ as a morphism into $\mathbb{R}((t^{1/\mathbb{N}}))$ which is a real closed field. Then, since $\mathcal{K}^0(V)$ satisfies the substitution property over $\mathbb{R}[V]$ (Theorem 2.12), one gets the unicity for a lifting of γ to $\mathcal{K}^0(V)$.

In order to prove the existence of such a lifting, we show that the unique lifting of γ with values in $\mathbb{R}((t^{1/\mathbb{N}}))$ has, in fact, values in $\mathbb{R}[[t^{1/\mathbb{N}}]]$. By the contrary, let us assume that there is a continuous rational function f such that $\gamma(f) \notin \mathbb{R}[[t^{1/\mathbb{N}}]]$. By Theorem 2.11, there is a Zariski locally closed stratification $V = S_1 \cup \ldots \cup S_t$ such that f is regular with restriction to each stratum. One derives a constructible stratification of the real spectrum of $\mathbb{R}[V]$, namely $\widetilde{V} = \widetilde{S_1} \cup \ldots \cup \widetilde{S_t}$. Moreover, the morphism γ defines a point of $\widetilde{V}$ that we assume to lie in $\widetilde{S_1}$. Since f is regular by stratum, there are two polynomials p, q in $\mathbb{R}[V]$ such that $f = p/q$ is regular with restriction to S_1. In particular $\gamma(q) \neq 0$ and we get $\gamma(f) = \frac{\gamma(p)}{\gamma(q)}$.

The condition $\gamma(f) \notin \mathbb{R}[[t^{1/\mathbb{N}}]]$ says that $\operatorname{ord}_t(\gamma(p)) < \operatorname{ord}_t(\gamma(q))$. By the Artin approximation Theorem ([1]), there exists $\delta : \mathbb{R}[V] \to \mathbb{R}[[t^{1/\mathbb{N}}]]_{\text{alg}}$ that approximates γ as close as desired for the (t)-adic topology. More precisely, if $S_1 = \{r = 0, s \neq 0\}$, where r, s are two polynomials, one may choose δ satisfying

$$\begin{aligned}\delta(r) &= 0,\\ \mathrm{ord}_t(\gamma(p)) &= \mathrm{ord}_t(\delta(p)),\\ \mathrm{ord}_t(\gamma(q)) &= \mathrm{ord}_t(\delta(q)),\\ \mathrm{ord}_t(\gamma(s)) &= \mathrm{ord}_t(\delta(s)).\end{aligned}$$

Then, one deduces that $\mathrm{ord}_t(\delta(p)) < \mathrm{ord}_t(\delta(q))$ and $\delta \in \widetilde{S_1}$. Thus, $\delta(f) = \frac{\delta(p)}{\delta(q)}$. By Theorem 3.5, one knows that $\delta(f) \in \mathbb{R}[[t^{1/\mathbb{N}}]]_{\mathrm{alg}}$ and hence $\mathrm{ord}_t(\gamma(p)) \geq \mathrm{ord}_t(\gamma(q))$, a contradiction. This concludes the proof. □

Remark 3.8. The proof of Theorem 3.7 heavily relies on the existence of a stratification, associated with a given continuous rational function, such that the function is regular on the strata. This result can also be generalized (with exactly the same proof) to a singular real algebraic variety if we replace the ring of continuous rational functions by the ring of regulous functions as considered in Theorem 2.13. Moreover, all the previous results that concern non-singular real algebraic varieties can be extended to central real algebraic varieties with isolated singularities since in this case rational continuous functions are still regulous [11, Theorem 2.25]. However, the question to know whether we get the same notion if we replace algebraic Puiseux arcs with formal ones in the definition of the substitution along arcs for general rings (Definition 3.1) seems open.

Another natural question which arises is what can be said about substitution along arcs for continuous rational functions defined on singular varieties. Note that considering Puiseux arcs rather than formal arcs makes a crucial difference:

Example 3.9. Consider the (singular but central) variety $V \subset \mathbb{R}^3$ defined by the equation $y^3 = z^2x^3$. The arc γ given by $x = y = 0$ and $z = t$ is contained in V. Let us now consider the rational function $f = y/x$ on V. This function is continuous on V because of the relation $f^3 = z^2$. Note now that the power series arc $\gamma \in (\mathbb{R}[[t]]_{\mathrm{alg}})^3$ lifts to $f \circ \gamma = t^{2/3}$ which belongs to $\mathbb{R}[[t^{1/\mathbb{N}}]]_{\mathrm{alg}}$ but not to $\mathbb{R}[[t]]_{\mathrm{alg}}$.

4. Substitution property for singular varieties

We give some examples and partial results under some assumptions and perform also a counterexample showing that there is no hope to get the substitution property along arcs for continuous rational functions defined on a general singular variety.

4.1. *First facts*

From now on, let $V \subset \mathbb{R}^n$ be a singular irreducible central algebraic variety. We first state an elementary result that will imply the weak substitution property.

Proposition 4.1. *Let $A = \mathbb{R}[V]$ be the coordinate ring of $V \subset \mathbb{R}^n$ and B an A-algebra which is a subring of the ring of semi-algebraic continuous functions on V. Assume that B satisfies the Łojasiewicz property over A. Namely for any f and g in B such that $\mathcal{Z}(f) \subset \mathcal{Z}(g)$, there are $h \in B$ and an integer n such that $g^n = fh$.*

Then, B satisfies the weak substitution property over points.

Proof. Let $\phi : A \to \mathbb{R}$ be a morphism. Since B is a subring of the ring of semi-algebraic continuous functions, ϕ admits a canonical lifting to B. Hence, one has only to show the unicity. The argument (already met in the proof of Theorem 2.12) is the following.

One may assume, for simplicity, that $\phi(x_1) = \ldots = \phi(x_n) = 0$. Let us assume then, for simplicity, that the origin o of $\mathbb{R}^n$ is in V. Let $f \in B$. Up to considering $f - f(o)$, one may assume that $f(o) = 0$.

Then, by Łojasiewicz property applied to $x_1^2 + \ldots + x_n^2$ and f, there exist an integer N and a semi-algebraic continuous function $g \in B$ such that $f^N = (x_1^2 + \ldots + x_n^2)g$. This algebraic identity implies that $\phi(f) = 0$ which concludes the proof. □

As a consequence, since the ring $\mathcal{K}^0(V)$ has the Łojasiewicz property, one gets :

Proposition 4.2. *The ring $\mathcal{K}^0(V)$ of continuous rational functions on V satisfies the weak substitution property on points.*

One can get also the substitution along arcs but under some additional hypothesis.

Proposition 4.3. *Let $\pi : \widetilde{V} \to V$ be a birational regular morphism with $\widetilde{V}$ non-singular, such that π is a homeomorphism for the Euclidean topology. Then $\mathcal{K}^0(V)$ is isomorphic to $\mathcal{K}^0(\widetilde{V})$ and satisfies the substitution property along arcs.*

Proof. The canonical morphism $\mathcal{K}^0(V) \to \mathcal{K}^0(\widetilde{V})$, given by the composition by π, is an isomorphism whose inverse isomorphism is given by the composition by the continuous rational map π^{-1} (the composition is still continuous rational because π is birational).

We know, by Theorem 3.5, that $\mathcal{K}^0(\widetilde{V})$ satisfies the substitution property along arcs, since $\widetilde{V}$ is non-singular. Hence, the substitution property along arcs holds also for $\mathcal{K}^0(V)$. Indeed, one has to show that any arc $\gamma : \mathbb{R}[V] \to R$, where $R = \mathbb{R}[[t^{1/\mathbb{N}}]]_{\text{alg}}$ or $\mathbb{R}[[t]]_{\text{alg}}$, admits a unique lifting to $\mathbb{R}[\widetilde{V}]$. One may compose γ with the evaluation morphism $e_u : R \to \mathbb{R}$ sending t onto $u \in [0, \epsilon[$. Since $\widetilde{V} \to V$ is a bijection, the morphism $e_u \circ \gamma$ admits a unique lifting to $\mathbb{R}[\widetilde{V}]$ for any $u \in [0, \epsilon[$. This shows that γ has a unique lifting to $\mathbb{R}[\widetilde{V}]$. □

One may illustrate this result with the following example.

Example 4.4. Let us consider the surface with equation $x^3 = (1 + z^2)y^3$, for which continuous rational functions and regulous functions differ [5]. The blowing-up $\widetilde{V} \to V$ along the z-axis is a homeomorphism and hence the ring $\mathcal{K}^0(V)$ does satisfy the substitution property along arcs. Note moreover that for the continuous rational function x/y, the denominator has a zero set of codimension only one (to be compared with the smooth case given by Proposition 2.9).

One may generalize a bit the previous proposition in a version which will be useful in the next subsection.

Proposition 4.5. *Let $W \subset \mathbb{R}^m$ be an irreducible real algebraic variety, $\pi : W \to V$ a surjective regular birational morphism and assume V is central. Then, $\mathcal{K}^0(V)$ is isomorphic to the ring $\mathcal{K}^0_\pi(W)$ of continuous rational functions on W which are constant on the fibers of π.*

Proof. Let us consider the morphism $\pi^* : \mathcal{K}^0(V) \to \mathcal{K}^0_\pi(W)$ defined by $f \mapsto f \circ \pi$. This morphism is well-defined since, being a composition of continuous functions, the function $f \circ \pi$ is also continuous, and also constant on the fibers of π by construction. It is moreover rational since π is birational. And the morphism π^* is injective since π is surjective.

Finally, π^* is surjective since any continuous function on W constant on the fibers of π goes down to a continuous function on V. By birationality of π, this new function on V remains rational if the function on W was so. □

Remark 4.6. Note that if V is a central variety and $\pi : W \to V$ is a blowing-up, then π is surjective and one may use Proposition 4.5.

4.2. *A counterexample*

The aim of this last subsection is to show that the ring $\mathcal{K}^0(V)$ does not necessarily satisfy the substitution property along arcs if V is singular.

Theorem 4.7. *Let V the central hypersurface in $\mathbb{R}^4$ given by the equation $x^8 = (z_1^2 + z_2^2)y^8$. The ring $\mathcal{K}^0(V)$ of continuous rational functions on V does not satisfy the substitution property along arcs.*

Proof. We are going to prove that the unicity in the substitution fails. First note that the variety V is central, irreducible and singular along the plane of coordinates (z_1, z_2) and also along the y-axis.

Let us remark that the rational function x^2/y^2, which is well-defined outside the singular plane, does extend by continuity on the whole V to the function $\sqrt[4]{z_1^2 + z_2^2}$, and hence x^2/y^2 defines a function g in $\mathcal{K}^0(V)$. Note that it is not the case for the function x/y. Indeed, for all $(x, y, 1, 0) \in V$, one has $\left(\frac{x}{y}\right)^8 = 1$ and hence $\frac{x}{y} \to \pm 1$ as $(x, y) \to (0, 0)$. Since x and y can take positive and negative values on V, $\frac{x}{y}$ does not admit a limit as (x, y) goes to $(0, 0)$.

The blowing-up $\pi : W \to V$ of V along the plane of coordinates (z_1, z_2) can be seen in the chart given by the coordinates

$$x = uv,\ y = v,\ z_1 = w_1,\ z_2 = w_2$$

as the variety with equation $u^8 = w_1^2 + w_2^2$. Hence W is the product of a surface S in $\mathbb{R}^3$, defined by the same equation as the one of W, with a line (the v-axis). The surface S has an isolated singularity at the origin, hence the singular locus of W consists of the v-axis.

Since V is a central variety, the blowing-up π is surjective (see Remark 4.6) and, using Proposition 4.5, one may identify $\mathcal{K}^0(V)$ with the ring $\mathcal{K}^0_\pi(W)$ of continuous rational functions on W which are constant on the fibers of π.

We are going to exhibit an algebraic Puiseux arc contained in the variety V which can be lifted to the ring $\mathcal{K}^0(V)$ in two different ways. Let us denote by $\gamma : [0, \epsilon[\to V$ the polynomial arc defined by $\gamma(t) = (0, 0, 0, t)$. The arc γ is contained in the singular plane of V. Moreover, it can be lifted to W into exactly two different algebraic Puiseux arcs given by

$$\tilde{\gamma}_\pm(t) = (\pm\sqrt[4]{t}, 0, 0, t).$$

Lemma 4.8. *Let $f \in \mathcal{R}_0(V)$. The semi-algebraic function $t \mapsto f \circ \gamma(t)$ is the composition of a one-variable rational function F, regular at any point* *in* $[0,+\infty)$*, with the square root function, i.e.* $(f\circ\gamma)(t)=\frac{U(\sqrt{t})}{V(\sqrt{t})}$ *where* U *and* V *are one-variable polynomials such that* $V(0)\neq 0$.

Proof. It suffices to show the result for the function $f\circ\pi$ composed with a lifting of the arc γ on W (for instance $\tilde{\gamma}_+$) since $f\circ\gamma=f\circ\pi\circ\tilde{\gamma}_+$.

We want to restrict the function $f\circ\pi$, which is continuous rational on W, to the image of $\tilde{\gamma}_+$. The Zariski closure of the image of $\tilde{\gamma}_+$ is the nonsingular irreducible curve C defined by $u^4=w_2$ and $v=w_1=0$ in $S\times\{0\}\subset W$. Note that C is not contained in the singular locus of W. Therefore the restriction of $f\circ\pi$ to C is still rational by [5, Proposition 8], and of course continuous. In particular, it is a regular function on C because C is a nonsingular curve.

Parametrizing C with $\theta\mapsto(\theta,\theta^4)$, one gets the existence of a regular function G on $\mathbb{R}$ such that $f\circ\pi(\theta,0,0,\theta^4)=G(\theta)$. Moreover, this regular function is even since $(\theta,0,0,\theta^4)\in W$ and $(-\theta,0,0,\theta^4)\in W$ lie over the same point $(0,0,0,\theta^4)\in V$ and $f\in\mathcal{K}^0_\pi(W)$. Hence, we claim there is a one-variable rational function F, regular at any point in $[0,+\infty)$, such that $G(\theta)=F(\theta^2)$. Indeed, we may write $G(\theta)=R(\theta)/(1+Q(\theta))$ with R,Q some real one-variable polynomials and such that Q is a sum of squares of one-variable polynomials. Since $G(\theta)=(R(\theta)(1+Q(-\theta)))/((1+Q(\theta))(1+Q(-\theta)))$ and $((1+Q(\theta))(1+Q(-\theta)))$ have even parity then it follows that $(R(\theta)(1+Q(-\theta)))$ has also even parity and we get the claim. To end, one sets $\theta^4=t$ and gets

$$(f\circ\gamma)(t)=G(\sqrt[4]{t})=F(\sqrt{t})$$

as desired. □

Lemma 4.9. *Any morphism* $\phi:\mathcal{K}^0(V)\to\mathbb{R}[[t^{1/\mathbb{N}}]]_{\mathrm{alg}}$ *which extends the morphism given by the evaluation along* γ *on polynomial functions, is entirely determined by the image of the continuous rational function* g *(which extends* x^2/y^2 *by continuity on all* V*).*

Proof. According to Lemma 4.8 and since $(g\circ\gamma)(t)=\sqrt{t}$, for any $f\in\mathcal{K}^0(V)$, there is a one-variable rational function F regular at any point in $[0,+\infty)$ such that $f\circ\gamma=F(g\circ\gamma)$. Hence, for any $t\in[0,\epsilon[$, one has $(f-F\circ g)(\gamma(t))=0$. Note also that $F\circ g\in\mathcal{K}^0(V)$. Since the image of γ is locally contained in the zero set of $x^2+y^2+z_1^2$, one has, by Łojasiewicz equality ([2, Théorème 2.6.6]), the existence of an integer N and a semi-algebraic continuous function h such that

$$(f-F\circ g)^N=(x^2+y^2+z_1^2)h.$$

Note that this equality shows also that $h \in \mathcal{K}^0(V)$. Hence, for any given morphism ϕ, one has $\phi((f - F \circ g)^N) = 0$ which gives $\phi(f) = \phi(F \circ g)$. Since F is regular at 0, one may evaluate its denominator at any arc going through the origin, and hence one gets also $\phi(f) = F(\phi(g))$. □

In particular, one deduces that there exist at most two candidates for such morphisms : either ϕ_+ which sends g onto $\sqrt{t}$ or ϕ_- which sends g onto $-\sqrt{t}$. The morphism ϕ_+ is the natural morphism coming from the evaluation of continuous semi-algebraic functions along γ. The next lemma states that the choice ϕ_- is also possible.

Lemma 4.10. *The setting $\phi_-(g) = -\sqrt{t}$ defines a morphism $\phi_- : \mathcal{K}^0(V) \to \mathbb{R}[[t^{1/\mathbb{N}}]]_{\text{alg}}$ which extends the morphism given by γ on $\mathbb{R}[V]$ the ring of polynomial functions over V.*

Proof. One has to check that ϕ_- does respect algebraic relations existing for elements of $\mathcal{K}^0(V)$ over $\mathbb{R}[V]$, the ring of polynomial functions over V. Let $f_1, \ldots, f_n$ in $\mathcal{K}^0(V)$ such that $P(f_1, \ldots, f_n) = 0$ where P is a polynomial with coefficients in $\mathbb{R}[V]$. One has to show that $\phi_-(P(f_1, \ldots, f_n)) = 0$.

According to the end of the proof of the previous lemma, there exists a one-variable rational function F, regular at 0, such that $\phi_+(P(f_1, \ldots, f_n)) = F(\phi_+(g))$ and $\phi_-(P(f_1, \ldots, f_n)) = F(\phi_-(g))$.

Let us write $F = U/V$ where U and V are one-variable polynomials.

Since the natural morphism ϕ_+ is well-defined, one gets from $P(f_1, \ldots, f_n) = 0$ that $\phi_+(P(f_1, \ldots, f_n)) = 0$. Namely, one has $F(\phi_+(g)) = 0$, which means that $U(\phi_+(g)) = 0$.

One may write $U(x) = S(x^2) + xT(x^2)$ where S and T are one-variable polynomials. Then, $U(t) = 0$ for any $t \in [0, \sqrt{\epsilon}[$, and hence $S = T = 0$, i.e. $U = 0$.

Consequently, $F(\phi_-(g)) = 0$ and hence $\phi_-(P(f_1, \ldots, f_n)) = 0$, the desired equality. □

As a consequence the unicity in the substitution property fails. □

References

[1] M. Artin, *On the solutions of analytic equations*, Invent. Math. **5** (1968), 277–291.

[2] J. Bochnak, M. Coste, M.-F. Roy, Real algebraic geometry, Springer, (1998).

[3] J. F. Fernando, *On the substitution theorem for rings of semi-algebraic functions*, J. Inst. Math. Jussieu **14** (2015), 857–894.

[4] G. Fichou, J.-P. Monnier, R. Quarez, *Continuous functions on the plane regular after one blowing-up*, Math. Z. **285** (2017), no. 1-2, 287–323.
[5] J. Kollár, K. Nowak, *Continuous rational functions on real and p-adic varieties*, Math. Z. **279** (2015), no. 1-2, 85–97.
[6] G. Kreisel, Review of Ershov, Zbl. **374**, 02027 (1978).
[7] W. Kucharz, *Rational maps in real algebraic geometry*, Adv. Geom. **9** (2009), no. 4, 517–539.
[8] W. Kucharz, K. Kurdyka, *Stratified-algebraic vector bundles*, J. Reine Angew. Math. **745** (2018), 105–154.
[9] J. Kollár, W. Kucharz, K. Kurdyka, *Curve-rational functions*, Math. Ann. **370** (2018), no. 1-2, 39–69.
[10] G. Fichou, J. Huisman, F. Mangolte, J.-P. Monnier, *Fonctions régulues*, J. Reine Angew. Math. **718** (2016), 103–151.
[11] J.-P. Monnier, *Semi-algebraic geometry with rational continuous functions*, Math. Ann. **372** (2018), no. 3-4, 1041–1080.
[12] N. Schwartz, *Real closed rings*, Habilitationsschrift, München (1984).
[13] L. Van Den Dries, *Tame Topology and O-minimal Structures*, London Math. Soc. Lecture Note 248, Cambridge Univ. Press (1998).

Milnor fibration, A'Campo's divide and Turaev's shadow

Masaharu Ishikawa

Department of Mathematics, Hiyoshi Campus,
Keio University,
4-1-1 Hiyoshi, Kohoku-ku, Yokohama-shi, Kanagawa, 223-8521, Japan
E-mail: ishikawa@keio.jp

Hironobu Naoe

Department of Mathematics,
Chuo University,
1-13-27 Kasuga, Bunkyo-ku, Tokyo, 112-8551, Japan
E-mail: naoe@math.chuo-u.ac.jp

We give a method for constructing a shadowed polyhedron from a divide. The 4-manifold reconstructed from a shadowed polyhedron admits the structure of a Lefschetz fibration if it satisfies a certain property, which is formulated as an *LF-structure* on a shadowed polyhedron. We will show that the shadowed polyhedron constructed from a divide satisfies this property and the Lefschetz fibration of this polyhedron is isomorphic to the Lefschetz fibration of the divide. Furthermore, applying the same technique to certain free divides we will show that the links of those free divides are fibered with positive monodromy.

Keywords: plane curve singularity, Milnor fibration, divide, shadow, Lefschetz fibration

2010 MSC: Primary: 32S55; Secondary: 32S30, 57M20

1. Introduction

A divide is the image of a generic and relative immersion of a finite number of intervals and circles into the unit disk, which was introduced by N. A'Campo in [3, 4] as a generalization of real morsified curves of complex plane curve singularities [1, 2, 19–21]. A link in S^3 is defined from a divide and this link is fibered if the divide is connected. Furthermore, if a divide is a real morsified curve of a complex plane curve singularity then its link is isotopic to the link of the singularity and the fibration is isomorphic to its Milnor fibration. In [23], the first author reformulated the fibration structure of a divide in terms of a Lefschetz fibration and generalized the definition of divides in the unit disk to those in compact orientable surfaces. In this generalized setting, the unit disk bundle in the cotangent bundle* over a compact orientable surface is the total space of the Lefschetz fibration. In the case of Milnor fibration, the total space corresponds to the Milnor ball, a regular fiber corresponds to a Milnor fiber and its boundary corresponds to the link of the singularity.

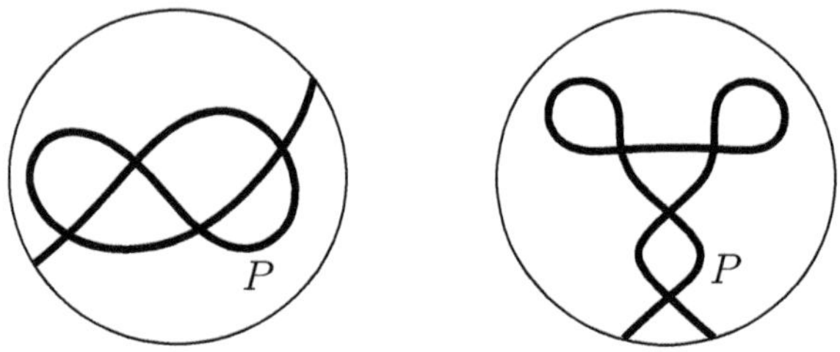

Fig. 1. Divides in the unit disk: The left is a divide of the $(3,5)$-torus knot, which is a real morsified curve of the singularity of $f(x,y)=x^3-y^5$. The right one does not come from a singularity. The link of this divide is 10_{139}.

One may wonder how a Milnor fiber is embedded in the Milnor ball. It is possible to guess the position sensuously, but it is not easy to describe it concretely. In this paper, we use Turaev's shadow [35, 36] to explain how the fiber surface is embedded. Let W be a compact, oriented, smooth 4-manifold with boundary M and L be a link in M. A shadow X of (W,L) is a simple polyhedron obtained from W by collapsing with keeping the link L. Conversely, if the shadow X is given then there exists an assignment $\mathfrak{gl}$ of half integers to regions of X, called a *gleam*, such that the pair (W,L) is recovered from $(X,\mathfrak{gl})$ uniquely. This method is called *Turaev's reconstruction.*

Let P be an admissible divide on a compact orientable surface $\Sigma_{g,n}$ of genus g and with n boundary components. The admissibility condition is needed to have the structure of a Lefschetz fibration in the total space, see Section 2 for the definition of an admissible divide. Now we double the curve of P as follows (see Figure 2):

*The Lefschetz fibration of a divide is constructed in the cotangent bundle of $\Sigma_{g,n}$ rather than the tangent bundle, though it was not carefully observed in [23]. In this paper, according to the original paper of A'Campo, we call an element in the bundle a *tangent vector* though it is a cotangent vector in actuality.

1. double the curve of P;
2. for each endpoint of P, close the corresponding two endpoints of the doubled curve by a small half circle;
3. for each edge of P that is not adjacent to an endpoint, add a crossing between the two edges of the doubled curve parallel to the edge.

The obtained doubled curve is a divide and we denote it by P_2. Note that this doubling method is similar to the one introduced in [15], but here we add a crossing in the middle of each edge of P. Let $(X_P, \mathfrak{gl}_P)$ be a shadowed polyhedron obtained from $\Sigma_{g,n}$ by attaching an annulus along one of the boundary components to each immersed circle of P_2 and assigning a gleam $\mathfrak{gl}_P$ to internal regions as follows:

4. assign $\frac{1}{2}$ to each of the two triangular regions corresponding to an edge of P not adjacent to an endpoint;
5. assign 0 to the bigon corresponding to an endpoint of P;
6. assign -1 to the remaining internal regions.

We call $(X_P, \mathfrak{gl}_P)$ a *shadowed polyhedron of P*.

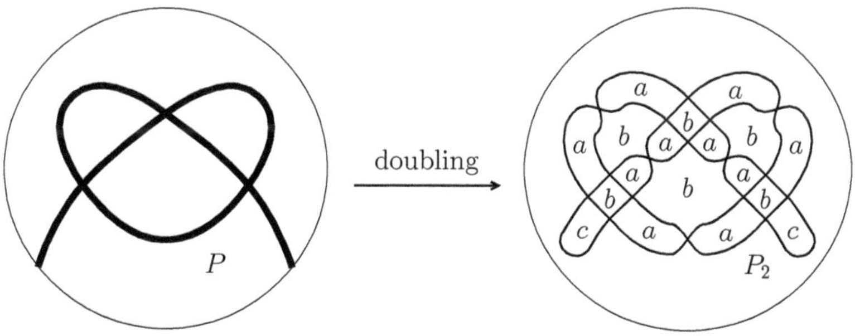

Fig. 2. The figure on the right represents the shadowed polyhedron of the divide on the left. The polyhedron is obtained by attaching an annulus along the doubled curve. The gleams of the internal regions labeled a, b, c are $\frac{1}{2}$, -1, 0, respectively.

In a specific case, a shadowed polyhedron has the structure of a Lefschetz fibration canonically.

Definition 1.1. Let $(X, \mathfrak{gl})$ be a shadowed polyhedron. If there exist a sub-polyhedron X' of X and ordered disk regions $D_1, \dots, D_n \subset X'$ such that

(i) X collapses onto X', so that the gleam $\mathfrak{gl}'$ of X' is induced from $\mathfrak{gl}$,

(ii) ∂D_i and ∂D_j $(i \neq j)$ intersect only at true vertices,

(iii) $X' \setminus (D_1 \cup \cdots \cup D_n)$ is homeomorphic to a compact, orientable surface Σ,

(iv) there exists an orientation on Σ such that the gleam given as the sums of local contributions around crossing points of ∂D_i and ∂D_j with $i < j$ shown in Figure 3 coincides with the gleam $\mathfrak{gl}'$ on each internal region of X' contained in Σ, and

(v) for each $i = 1, \ldots, n$, the gleam $\mathfrak{gl}'$ of the region D_i is -1,

then the tuple $\mathcal{X} = (X'; D_1, \ldots, D_n)$ is called an *LF-structure* on $(X, \mathfrak{gl})$.

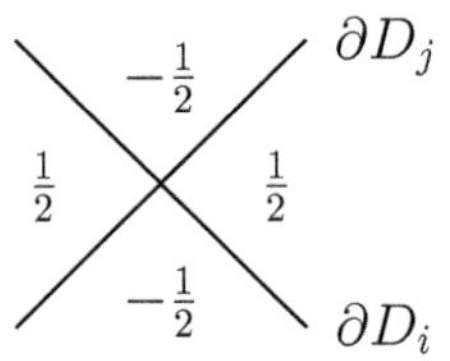

Fig. 3. Local contribution around a crossing of ∂D_i and ∂D_j with $i < j$.

If a shadowed polyhedron $(X, \mathfrak{gl})$ has an LF-structure then the corresponding 4-manifold $W(X, \mathfrak{gl})$ has the structure of a Lefschetz fibration whose regular fiber is Σ and singular points correspond to the internal regions $D_1, \ldots, D_n$. Conversely, a Lefschetz fibration is constructed from $\Sigma \times D^2$, where D^2 is a 2-disk, by attaching 2-handles along disjoint simple closed curves on fibers over ∂D^2 with surface framing minus 1. Hence the polyhedron X obtained from Σ by attaching the cores of the 2-handles with gleam -1 is a shadow of the total space of the Lefschetz fibration and we may assign a suitable gleam $\mathfrak{gl}$ to the internal regions on Σ such that $(X, \mathfrak{gl})$ has an LF-structure. It will be proved in Lemma 3.5 that the shadowed polyhedron $(X_P, \mathfrak{gl}_P)$ of an admissible divide P has an LF-structure $\mathcal{X}_P$.

The main theorem of this paper is the following.

Theorem 1.2. *Let $(X_P, \mathfrak{gl}_P)$ be the shadowed polyhedron of an admissible divide P. Then the Lefschetz fibration of $\mathcal{X}_P$ on $(X_P, \mathfrak{gl}_P)$ coincides with that of P.*

The fiber surface of P is the surface embedded in X_P and bounded by $\partial X_P \setminus \partial \Sigma_{g,n}$. An advantage of X_P is that we can see both of the fiber surface and the surface $\Sigma_{g,n}$ in the polyhedron. In the case of the Milnor fibration, the latter corresponds to the real plane in the Milnor ball. By recovering the total space according to the gleam $\mathfrak{gl}_P$, we may understand precisely how the fiber surface is embedded in the Milnor ball with respect to this real plane.

The detection of the structure of a Lefschetz fibration by an LF-structure can also be used for certain free divides. A free divide is a divide whose endpoints are not necessary on the boundary of the unit disk. It was introduced by Gibson and the first author in [16], where they defined links associated with free divides and studied their properties. In a special case, we can show that a free divide has an LF-structure and has a structure of a Lefschetz fibration. An endpoint of a free divide is called a *free endpoint* if the region adjacent to the endpoint is bounded by the curve of the free divide.

Theorem 1.3. *Let Q be a free divide in the unit disk D consisting of one immersed interval and with one free endpoint. Starting at the free endpoint along Q, let c be the double point of Q met first. Assume either*

(1) *c is on the boundary of the region adjacent to the boundary of D, or*
(2) *the immersed arc on Q connecting c and the non-free endpoint passes exactly one double point.*

Then the link of Q is fibered and the fibration is obtained as the boundary of a Lefschetz fibration. In particular, its monodromy is positive.

Here a monodromy is said to be *positive* if it is represented as a product of right-handed Dehn twists.

Among the free divides listed in [16], for example, the links of 3_{17} and 3_{18} are not fibered. Actually, they do not satisfy the assumption in Theorem 1.3. In the list, there are two knots, up to 10 crossings, that are represented by free divides with one free endpoint, satisfy the conditions in Theorem 1.3 and neither closed positive braids nor links of divides, which are 10_{154} and 10_{161}.

Corollary 1.4. *The fibered knots 10_{154} and 10_{161} are obtained as the boundaries of Lefschetz fibrations. In particular, their monodromies are positive.*

As we mentioned, the link of a divide has the structure of a Lefschetz fibration. A closed positive braid also has this property, which follows from the "anthology" in [32] and the fact that it can be constructed by successive Murasugi-sum's of torus links of type $(2, k)$. One can see that the fiber surfaces of 10_{154} and 10_{161} are obtained by plumbing positive Hopf bands. Hence Corollary 1.4 also follows from "anthology" and plumbings.

The relation between divides and shadows was suggested by Professor Norbert A'Campo when the first author was a student in Universität Basel though he, the first author, could not catch the point at that time. The authors would like to thank him for introducing them to these two interesting topics. They are also grateful to Burak Özbağci for telling us the orientation issue of the bundle in [23], Mikami Hirasawa for telling us about Hopf plumbings of 10_{154} and 10_{161}, and Seiichi Kamada and Yuya Koda for precious comments. The first author is supported by JSPS KAKENHI Grant Number JP16K05140 and Keio University Academic Development Funds for Individual Research. The second author is supported by JSPS KAKENHI Grant Number JP18H05827. This work is supported by JSPS KAKENHI Grant Number JP17H06128.

2. Preliminaries

In this paper, ∂A means the boundary of a topological space A and, for topological spaces A and B with $A \subset B$, $\mathrm{Nbd}(A; B)$ means a small compact neighborhood of A in B.

2.1. *A'Campo's divide*

Let $\Sigma_{g,n}$ be a compact, orientable, smooth surface of genus g and with n boundary components with an arbitrary Riemannian metric, where $g, n \geq 0$.

Definition 2.1. A *divide* P in $\Sigma_{g,n}$ is the image of a generic and relative immersion of a finite number of copies of the unit interval or the unit circle into $\Sigma_{g,n}$. The generic condition is the following:

- the image has neither self-tangent points nor triple points;
- an immersed interval intersects $\partial\Sigma_{g,n}$ at the endpoints transversely;
- an immersed circle does not intersect $\partial\Sigma_{g,n}$.

If $\Sigma_{g,n}$ is closed then we set $N(\Sigma_{g,n}) = \mathrm{Nbd}(\Sigma_{g,n}; T(\Sigma_{g,n}))$, where $T(\Sigma_{g,n})$ is the total space of the tangent bundle of $\Sigma_{g,n}$.

If $\Sigma_{g,n}$ has boundary, we define $N(\Sigma_{g,n})$ as follows: Set $A = \mathrm{Nbd}(\partial\Sigma_{g,n}; \Sigma_{g,n})$ and $B = \Sigma_{g,n} \setminus A$. First thicken B in $T(\Sigma_{g,n})$ as

$$\hat{B} := \{(x, u) \in T(\Sigma_{g,n}) \mid x \in B, u \in T_x(\Sigma_{g,n}), \|u\| \leq \varepsilon\},$$

where $T_x(\Sigma_{g,n})$ is the tangent space to $\Sigma_{g,n}$ at x and $\varepsilon > 0$. Next, set $\alpha := \partial A \setminus \partial\Sigma_{g,n}$, which is the boundary of the annuli A not contained in $\partial\Sigma_{g,n}$, and choose a compact tubular neighborhood $\mathrm{Nbd}(\alpha; T(\Sigma_{g,n}))$ of α suitably such that the boundary of $\mathrm{Nbd}(\alpha; T(\Sigma_{g,n})) \cup \hat{B}$ becomes a smooth 3-manifold. Then we define $N(\Sigma_{g,n}) = \mathrm{Nbd}(\alpha; T(\Sigma_{g,n})) \cup \hat{B}$. Note that $\partial N(\Sigma_{g,n})$ is diffeomorphic to a connected sum of $2g+n-1$ copies of $S^2 \times S^1$ if $n \geq 1$. In particular, it is S^3 if $g = 0$ and $n = 1$.

Definition 2.2. The *link* of a divide P in $\Sigma_{g,n}$ is the set $L(P)$ defined by

$$L(P) := \{(x, u) \in \partial N(\Sigma_{g,n}) \mid x \in P,\ \ u \in T_x(P)\},$$

where $T_x(P)$ is the set of tangent vectors to P at x.

To be precise, as mentioned in the footnote of the second page, we need to reverse the orientation of $N(\Sigma_{g,n})$, or equivalently, need to replace the tangent bundle of $\Sigma_{g,n}$ in the above construction with the cotangent bundle.

Each connected component of $\Sigma_{g,n} \setminus P$ is called a *region* of P. If a region of P is bounded by P then it is called an *inside region*, and otherwise it is called an *outside region*.

Definition 2.3. A divide P in $\Sigma_{g,n}$ is *admissible* if it satisfies the following:

- P is connected;
- each inside region of P is simply-connected;
- each outside region of P is either simply-connected or an annulus such that one boundary component is a component of $\partial\Sigma_{g,n}$ and the other is contained in P;
- each component of $\partial\Sigma_{g,n}$ either does not intersect P or intersects P at an even number of points transversely;
- each circle component of P intersects the other components of P at an even number of points transversely.

In the case where $g = 0$ and $n = 1$ (i.e., $\Sigma_{g,n}$ is a disk), a divide P is admissible if and only if it is connected. In [4], A'Campo proved that if P in $\Sigma_{0,1}$ is connected then $L(P)$ is fibered with positive monodromy. The admissibility condition was introduced in [23] to inherit this fiberedness property to the general setting.

Theorem 2.4 (Ishikawa [23]). *If a divide P is admissible then $L(P)$ is a fibered link in $\partial N(\Sigma_{g,n})$ with positive monodromy.*

The fibration of the fibered link $L(P)$ is obtained as the boundary of a Lefschetz fibration. We here explain how the Lefschetz fibration is obtained briefly. See [23] more precise explanation.

Let $f_P : \Sigma_{g,n} \to \mathbb{R}$ be a Morse function on $\Sigma_{g,n}$ such that $f_P^{-1}(0) = P$ and each inside region of P has exactly one singular point of f_P. The existence of such a Morse function is guaranteed by the admissibility condition. Define a map $F_P : T(\Sigma_{g,n}) \to \mathbb{C}$ by

$$F_P(x,u) = f_P(x) + idf_P(x)(u) - \frac{1}{2}\chi(x)H_{f_P}(x)(u,u),$$

where $i = \sqrt{-1}$, $x \in \Sigma_{g,n}$, $u \in T_x(\Sigma_{g,n})$, H_{f_P} is the Hessian of f_P and $\chi(x)$ is a bump function which is 0 outside small neighborhoods of double points of P and 1 on smaller neighborhoods.

Now let D_η be the disk in $\mathbb{C}$ centered at the origin with sufficiently small radius $\eta > 0$ and restrict F_P to $F_P^{-1}(D_\eta) \cap N(\Sigma_{g,n})$. This is a Lefschetz fibration with only one singular fiber. Note that the number of Morse singularities on the singular fiber is same as the number of double points of P. Let $R_1, \ldots, R_m$ be the inside regions of P and R_i' be the closure of $R_i \backslash F_P^{-1}(D_\eta)$ for $i = 1, \ldots, m$. The total space $N(\Sigma_{g,n})$ can be recovered, up to isotopy, from $F_P^{-1}(D_\eta) \cap N(\Sigma_{g,n})$ by attaching $R_1' \times [0,1], \ldots, R_m' \times [0,1]$ along the simple closed curves $\partial(F_P^{-1}(D_\eta)) \cap (R_1' \cup \cdots \cup R_m')$. We can check directly that the framings of these attachings are those of the fiber surface of the Lefschetz fibration minus 1. Thus the Lefschetz fibration on $F_P^{-1}(D_\eta) \cap N(\Sigma_{g,n})$ extends to $R_1' \times [0,1], \ldots, R_m' \times [0,1]$ after these attachings. We call this fibration the *Lefschetz fibration of the admissible divide P*.

We here note known studies related to divides. A divide is defined first in the unit disk by A'Campo [3, 4]. In this case, the link of a divide is defined in the unit sphere S^3. He then proved that if a divide is connected then the link is fibered, if the divide is a real morsified curve of a complex plane curve singularity then its fibration is isomorphic to the Milnor fibration, and if a divide consists of only immersed intervals then the unknotting number of the link is equal to the number of double points. Furthermore, in [5], he proved that there are many links of divides that are hyperbolic. The link-types of the links of divides had been studied by Couture-Perron [13], Hirasawa [22], Goda-Hirasawa-Yamada [17] and Kawamura [25]. Some mysterious relation between divides and exceptional surgeries had been studied by Yamada [37, 38]. Recently, Fomin-Pylyavskyy-Shustin studied real morsified curves and divides using quivers [14] and Özbağci used divides on compact surfaces for constructing specific Lefschetz fibrations and open book decompositions [33].

2.2. *Turaev's shadow*

If each point of a compact space X has a neighborhood homeomorphic to one of (i)-(v) in Figure 4, then X is called a *simple polyhedron*. The set of points of type (ii), (iii) and (v) is called the *singular set* of X and denoted by $\mathrm{Sing}(X)$. A point of type (iii) is a *true vertex*, and each connected component of $\mathrm{Sing}(X)$ with true vertices removed is called a *triple line*. Each connected component of $X \setminus \mathrm{Sing}(X)$ is called a *region*. Hence a region consists of points of type (i) or (iv). A region is called an *internal region* if it contains no points of type (iv), and a *boundary region* otherwise. The *boundary* of X, denoted by ∂X, is defined as the set of points of type (iv) and (v).

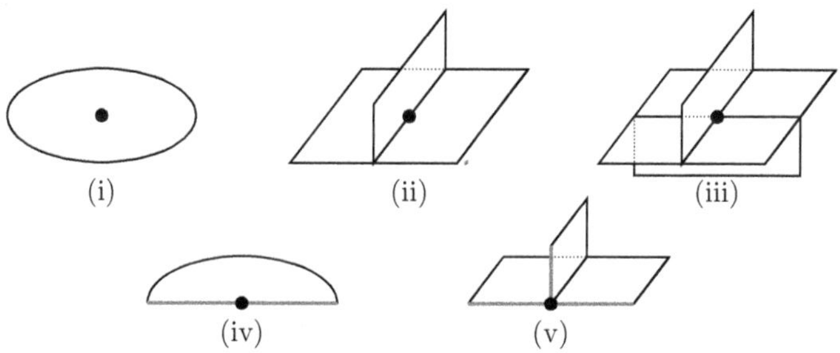

Fig. 4. The local models of a simple polyhedron.

Definition 2.5. Let W be a 4-manifold with boundary and $X \subset W$ be a simple polyhedron that is proper and locally flat in W. If W collapses onto X after giving some triangulation to (W, X), then the polyhedron X is called a *shadow* of W.

Here X is said to be proper in W if $X \cap \partial W = \partial X$ and locally flat in W if there is a local chart (U, φ) around each point of X such that $\varphi(U \cap X)$ is contained in $\mathbb{R}^3 \subset \mathbb{R}^4 = \varphi(U)$. It is easy to see that any handlebody consisting of 0-, 1- and 2-handles admits a shadow [8, 36].

For any simple polyhedron X, one can define the $\mathbb{Z}_2$-*gleam* on each internal region. Let R be an internal region, and $i : F \to X$ be a continuous map extended from the inclusion of R, where F is a compact surface whose interior is homeomorphic to R. Note that the restriction $i|_{\mathrm{Int}(F)}$ coincides with the inclusion of R, and that $i(\partial F) \subset \mathrm{Sing}(X)$. We now see that there exists a local homeomorphism $\tilde{i} : \tilde{F} \to X$ such that its image is a neighborhood of $i(F)$ in X, where $\tilde{F}$ is a simple polyhedron obtained from F by attaching an annulus or a Möbius strip along its core circle to each boundary component of F. Note that $\tilde{F}$ is determined up to homeomorphism from the topology of X. Here the $\mathbb{Z}_2$-gleam $\mathfrak{gl}_2(R)$ of R is defined to be 0 if the number of the attached Möbius strips is even, and 1 otherwise.

Definition 2.6. A *gleam* on a simple polyhedron X is a coloring for all the internal regions of X suth that each value $\mathfrak{gl}(R)$ on an internal region R satisfies $\mathfrak{gl}(R) - \frac{1}{2}\mathfrak{gl}_2(R) \in \mathbb{Z}$. We call a pair $(X, \mathfrak{gl})$ a *shadowed polyhedron.*

Theorem 2.7 (Turaev [36]).

(1) *There exists a canonical way to construct a* 4*-manifold* W *from a given shadowed polyhedron* $(X, \mathfrak{gl})$ *such that* X *is a shadow of* W*. This construction provides a smooth structure on* W *uniquely.*
(2) *For a* 4*-manifold* W *admitting a shadow* X*, there exists a gleam* $\mathfrak{gl}$ *on* X *such that* W *is diffeomorphic to the* 4*-manifold constructed from the shadowed polyhedron* $(X, \mathfrak{gl})$ *according to the way of* (1).

The construction in (1) is called Turaev's reconstruction. A gleam plays a role as a framing coefficient to attach a 2-handle in the original proof of Turaev's reconstruction. It is also regarded as a generalized Euler number of an embedded surface in a 4-manifold. In the case where a 4-manifold is a D^2-bundle over a surface F, the 4-manifold has a shadow F and the Euler number of F coincides with the gleam coming from the above theorem.

As we mentioned in Introduction, if a shadowed polyhedron $(X, \mathfrak{gl})$ has an LF-structure then $W(X, \mathfrak{gl})$ admits the structure of a Lefschetz fibration.

As far as we know, the first paper that relates shadows and singularity theory is the paper of Costantino and Thurston [12], where the Stein factorization of a stable map from a 3-manifold to $\mathbb{R}^2$ is regarded as a shadow after a small perturbation if necessary. In [24], Koda and the first author focused on this relation and studied a relationship between the minimal number of true vertices of shadows and the minimal number of specific singular fibers of stable maps. Shadows are used in the study of quantum invariants by various authors, see for instance [6, 18, 34–36]. In particular Carrega and Martelli constructed a shadow containing a given ribbon surface in D^4 and studied the Jones polynomial of a ribbon link [7]. Concerning studies of

4-manifolds, Costantino studied almost complex structures and Stein structures of 4-manifolds with shadow representatives [9, 11], and the second author studied shadow representatives of corks, which yield exotic pairs of 4-manifolds [29, 31]. A study of classification of 4-manifolds according to the numbers of vertices of shadows is now in progress, see [10, 26–28, 30].

3. From divide to shadow

We first show how to get a shadowed polyhedron of an oriented divide and then prove Theorem 1.2. An oriented divide was introduced in [15] by Gibson and the first author to determine the link-type of the link of a divide.

Definition 3.1. An *oriented divide* $\vec{P}$ in $\Sigma_{g,n}$ is the image of a generic immersion of oriented circles into $\Sigma_{g,n}$.

Definition 3.2. The *link* of an oriented divide $\vec{P}$ in $\Sigma_{g,n}$ is the set $L(\vec{P})$ defined by

$$L(\vec{P}) := \{(x, u) \in \partial N(\Sigma_{g,n}) \mid x \in \vec{P},\ \ u \in T_x^+(\vec{P})\},$$

where $T_x^+(P)$ is the set of tangent vectors to $\vec{P}$ at x in the same direction as $\vec{P}$.

Note that for any oriented link in $\partial N(S_{g,n})$ there exists an oriented divide $\vec{P}$ such that the link is isotopic to $L(\vec{P})$, see [15].

Let $q_1, \ldots, q_\ell$ be the images of circles of $\vec{P}$. For each point $x \in \vec{P}$, let $I(x)$ denote the segment in $N(\Sigma_{g,n})$ consisting of the point x and the points corresponding to the tangent vectors to $\vec{P}$ at x in the same direction as $\vec{P}$. For each $i = 1, \ldots, \ell$, the union $\bigcup_{x \in q_i} I(x)$ is an annulus one of whose boundary component lies on $\Sigma_{g,n}$ and the other lies on $\partial N(\Sigma_{g,n})$. We denote it by $R(q_i)$. Then the union of $\Sigma_{g,n}$ and $R(q_i)$, $i = 1, \ldots, \ell$, constitutes a simple polyhedron embedded in $N(\Sigma_{g,n})$. We denote this polyhedron by $X_{\vec{P}}$. Note that the internal regions of $X_{\vec{P}}$ correspond to the inside regions of $\vec{P}$ on $\Sigma_{g,n}$.

Next we assign a gleam to $X_{\vec{P}}$. For each inside region R of $\vec{P}$, we define a local contribution to the gleam at each double point of $\vec{P}$ on ∂R as shown in Figure 5. In the figure, the curve is a part of $\vec{P}$ along which the annuli $R(q_i)$'s are attached. The gleam of R is given as the sum of the local contributions minus the Euler characteristic of the region. We denote this gleam by $\mathfrak{gl}_{\vec{P}}$.

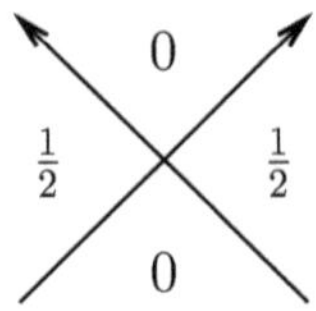

Fig. 5. Local contribution around a double point of an oriented divide.

Lemma 3.3. *The pair $(N(\Sigma_{g,n}), X_{\vec{P}})$ is obtained from the shadowed polyhedron $(X_{\vec{P}}, \mathfrak{gl}_{\vec{P}})$ by Turaev's reconstruction.*

Proof. The vector field on the left in Figure 6 represents the annulus regions of $X_{\vec{P}}$ attached along $\vec{P}$. We may isotope these annulus regions in $N(\Sigma_{g,n})$ relatively to the position represented by the vector field on the right. We denote it by $v_{\partial R}$. The two vectors at the crossing are both horizontal, which means that the corresponding polyhedron is locally embedded in $\mathbb{R}^3$. Hence we can regard $X_{\vec{P}}$ as a shadow of $N(\Sigma_{g,n})$. Let R be an internal region of $X_{\vec{P}}$. It is sufficient to check that the gleam of R determined from the above embedding of $X_{\vec{P}}$ into $N(\Sigma_{g,n})$ coincides with $\mathfrak{gl}_{\vec{P}}(R)$. Let $H = \mathrm{Nbd}(\partial R; N(\Sigma_{g,n}))$ and $\overline{R} = R \setminus \mathrm{Int}(H)$. Note that $\mathrm{Int}(\overline{R})$ is homeomorphic to R. There exists an annulus or a Möbius strip, denoted by A, along $\partial\overline{R}$ in ∂H according to $v_{\partial R}$ as shown in Figure 7. Let $v_{\partial\overline{R}}$ be a non-zero vector field along $\partial\overline{R}$ consisting of vectors tangent to $\partial\overline{R}$ and B be the annulus along $\partial\overline{R}$ in ∂H that is associated with $v_{\partial\overline{R}}$. After suitable perturbation, we may assume that B intersects A transversely finite times only near true vertices. By careful verification of orientation, we may conclude that the local contribution to the gleam near the true vertex is given as in Figure 5.

The obstruction to extend $v_{\partial\overline{R}}$ on the whole $\overline{R}$ is $-\chi(\overline{R}) = -\chi(R)$, which coincides with the self-intersection number of $\overline{R}$ in $N(\Sigma_{g,n})$. Therefore $\mathfrak{gl}_{\vec{P}}(R)$ is given as the sum of the local contributions minus $\chi(R)$. □

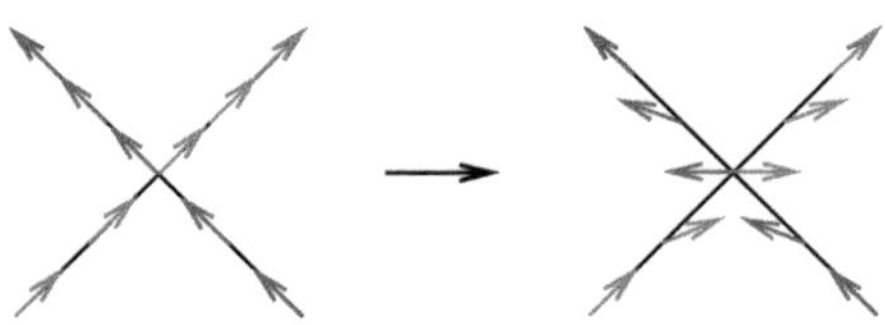

Fig. 6. The vector field that represents the framing of thickening.

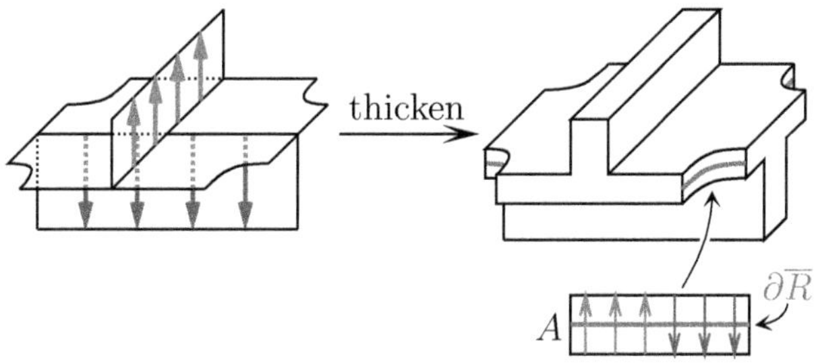

Fig. 7. The neighborhood of a true vertex in a slice $\mathbb{R}^3$ and a part of A.

Remark 3.4. The internal regions R of $X_{\vec{P}}$ lie on $\Sigma_{g,n}$. In the case of divides in the unit disk, in particular the case of real morsified curves, these internal regions lie on the real plane $\mathbb{R}^2 \subset \mathbb{C}^2$.

Lemma 3.5. *The shadowed polyhedron* $(X_P, \mathfrak{gl}_P)$ *of an admissible divide* P *has an LF-structure.*

Proof. The polyhedron X_P is obtained from P by doubling it to the divide P_2 and attaching annuli along P_2 as explained in Introduction. Let X'_P be the polyhedron obtained from X_P by removing the regions adjacent to $\partial\Sigma_{g,n}$ by collapsing from $\partial\Sigma_{g,n}$. We may obtain a smooth surface Σ from X'_P by removing internal regions corresponding to the singularities of f_P. Since P is admissible, the inside regions of P admit a checkerboard coloring with colors, say, black and write. To each edge of P, we assign the orientation induced from the orientation of the write region adjacent to that edge. Two triangular regions of P_2 correspond to the edge and we define that the orientation of the first triangular region is positive and the second one is negative. We can see that the orientations of all triangular regions are consistent, that is, Σ is orientable.

To prove the lemma, it is enough to show that the gleam on each internal region of X'_P coincides with the one determined by the conditions (iv) and (v) in the definition of LF-structure. Note that a bigon with gleam 0 in Step 5 of the doubling (cf. the regions labeled c in Figure 2) is not an internal region of X'_P. Hence we don't need to check its gleam. Let $R_1, \ldots, R_m$ be the internal regions of X'_P corresponding to maxima, saddles and minima of f_P. We order these regions such that $R_1, \ldots, R_{m_1}$ are maxima, $R_{m_1+1}, \ldots, R_{m_2}$ are saddles and $R_{m_2+1}, \ldots, R_m$ are minima. The gleams of these regions are -1, which coincide with the condition (v).

Now we check the coincidence of the gleams on the remaining internal regions, which are the triangular regions of P_2 on Σ. There are two choices of the orientation of Σ, either the one shown on the top in Figure 8 or the opposite one. We fix the orientation shown in the figure.

First we check the local contribution around a crossing point adjacent to a region of a maximum and a region of a minimum. As shown on the bottom in Figure 8, the local contribution given according to Figure 3 is $-\frac{1}{2}$, that is, the local contribution to each of the triangular regions on the top in Figure 8 is $-\frac{1}{2}$.

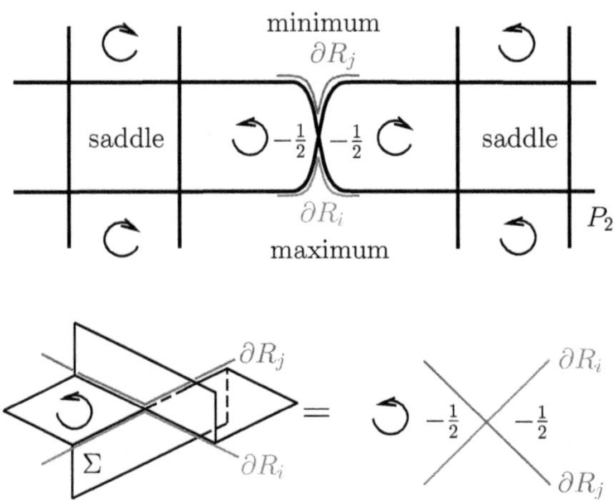

Fig. 8. Local contribution around a crossing point adjacent to a maximum and a minimum.

Next, we check the local contribution around a crossing point adjacent to a region of a maximum and a region of a saddle, and also around a crossing point adjacent to a region of a saddle and a region of a minimum. As shown in Figure 9, the local contribution to each of the triangular regions on the top in Figure 9 is $\frac{1}{2}$.

Summing up these contributions, we see that the gleam on each triangular region of P_2 on Σ is $-\frac{1}{2}+\frac{1}{2}+\frac{1}{2}=\frac{1}{2}$, which coincides with the gleam $\mathfrak{gl}_P$ of that region. This completes the proof. □

Let $\mathcal{X}_P$ denote the LF-structure on $(X_P, \mathfrak{gl}_P)$ determined in the proof of Lemma 3.5.

Proof of Theorem 1.2. Let P be an admissible divide on $\Sigma_{g,n}$, $q_1, \ldots, q_\ell$

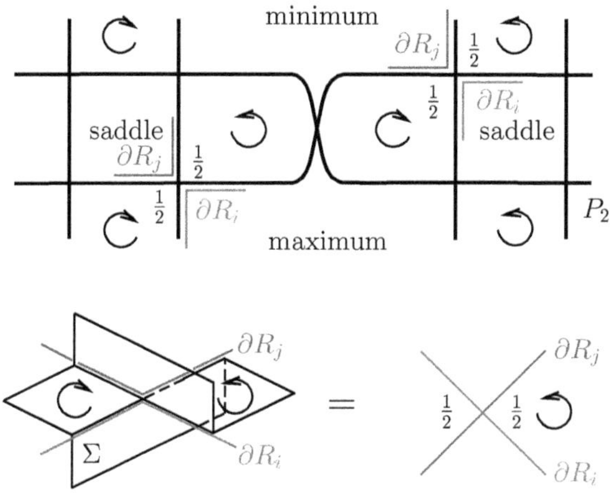

Fig. 9. Local contribution around crossing points adjacent to saddles.

be the images of immersed intervals and circles of P, $R_1, \ldots, R_m$ be the inside regions of P and $c_1, \ldots, c_\delta$ be the double points of P. Assign an orientation to each q_i. For each point $x \in q_i$, let $I_+(x)$ (resp. $I_-(x)$) denote the segment in $N(\Sigma_{g,n})$ consisting of the point x and the points corresponding to the vectors tangent to $\overrightarrow{P}$ at x in the same (resp. opposite) direction to the orientation of q_i. Each of $\bigcup_{x \in q_i} I_+(x)$ and $\bigcup_{x \in q_i} I_-(x)$ is an annulus one of whose boundary component lies on $\Sigma_{g,n}$ and the other lies on $\partial N(\Sigma_{g,n})$. We denote $\bigcup_{x \in q_i} I_+(x)$ and $\bigcup_{x \in q_i} I_-(x)$ by $R_+(q_i)$ and $R_-(q_i)$, respectively. The union of $\Sigma_{g,n}$, $R_+(q_i)$'s and $R_-(q_i)$'s for $i = 1, \ldots, \ell$ is a non-simple polyhedron embedded in $N(\Sigma_{g,n})$, which we denote by $\hat{X}_P$. The singular fiber $F_P^{-1}(0) \cap N(\Sigma_{g,n})$ is isotopic to $\bigcup_{i=1}^{\ell}(R_+(q_i) \cup R_-(q_i))$. As explained in Section 2.1, $N(\Sigma_{g,n})$ is recovered from $F_P^{-1}(D_\eta) \cap N(\Sigma_{g,n})$ by attaching 2-handles corresponding to the inside regions of P.

Next we perturb $\hat{X}_P$ in $N(\Sigma_{g,n})$ so that it becomes simple. The regions $R_+(q_i)$ and $R_-(q_i)$ can be represented by vector fields based on q_i as shown on the left in Figure 10, which we denote by $v_+(q_i)$ and $v_-(q_i)$, respectively. A deformation of $R_+(q_i)$ and $R_-(q_i)$ in $N(\Sigma_{g,n})$ can be represented by a deformation of the base curve q_i and an isotopy of these vector fields. We perturb $\hat{X}_P$ such that the base curve becomes the doubled curve P_2 and the vector fields such that they are based on P_2 and tangent to P_2 in the same direction as $\overrightarrow{P}_2$, see on the right in Figure 10. Here the orientation of $\overrightarrow{P}_2$ is the one induced from the orientation of triangular internal regions given in the first paragraph of the proof of Lemma 3.5. The obtained polyhedron $X_{\vec{P}_2}$ is embedded in $N(\Sigma_{g,n})$ and the embedding is represented by the gleam $\mathfrak{gl}_{\vec{P}_2}$ of the oriented divide $\vec{P}_2$. The shadowed polyhedron $(X_{\vec{P}_2}, \mathfrak{gl}_{\vec{P}_2})$ is nothing but $(X_P, \mathfrak{gl}_P)$ by definition.

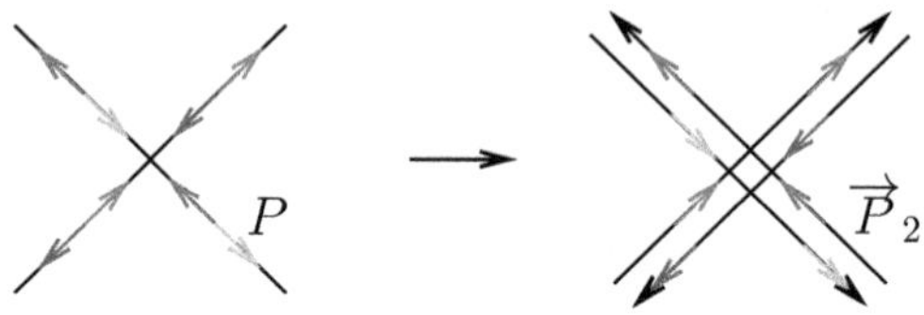

Fig. 10. A deformation of $R_+(q_i)$ and $R_-(q_i)$.

Let Σ denote the surface obtained from X_P by removing all regions contained in the inside regions of P and those containing a double point of P. Since the singular fiber $F_P^{-1}(0) \cap N(\Sigma_{g,n})$ is isotopic to the surface Σ outside small neighborhoods $\mathrm{Nbd}(c_k; N(\Sigma_{g,n}))$, the nearby fiber $F_P^{-1}(t) \cap N(\Sigma_{g,n})$, $0 < |t| \ll 1$, is also isotopic to Σ outside $\mathrm{Nbd}(c_k; N(\Sigma_{g,n}))$'s. In $\mathrm{Nbd}(c_k; N(\Sigma_{g,n}))$, $F_P^{-1}(t) \cap \mathrm{Nbd}(c_k; N(\Sigma_{g,n}))$ and $\Sigma \cap \mathrm{Nbd}(c_k; N(\Sigma_{g,n}))$ are annuli in the 4-ball $\mathrm{Nbd}(c_k; N(\Sigma_{g,n}))$. Furthermore, since $F_P^{-1}(t) \cap \partial\mathrm{Nbd}(c_k; N(\Sigma_{g,n}))$ and $\Sigma \cap \partial\mathrm{Nbd}(c_k; N(\Sigma_{g,n}))$ are isotopic as oriented links in $\partial\mathrm{Nbd}(c_k; N(\Sigma_{g,n}))$, these annuli are isotopic in $\mathrm{Nbd}(c_k; N(\Sigma_{g,n}))$. Hence $F_P^{-1}(t) \cap N(\Sigma_{g,n})$ and Σ are isotopic.

It had been shown in Lemma 3.5 that $(X_P, \mathfrak{gl}_P)$ has the LF-structure $\mathcal{X}_P$. Moreover, in the proof of Lemma 3.5, the order of the internal regions for the definition of LF-structure is maxima, saddles and minima, which is the order of the right-handed Dehn twists of the monodromy of the fibration of the divide P. Both of the Lefschetz fibrations of $\mathcal{X}_P$ on $(X_P, \mathfrak{gl}_P)$ and P are obtained from $F_P^{-1}(D_\eta) \cap N(\Sigma_{g,n}) \cong \mathrm{Nbd}(\Sigma; N(\Sigma_{g,n}))$ by attaching 2-handles corresponding to the inside regions of P along the same vanishing cycles with fiber surface framing minus 1 and with the same order. Thus the two Lefschetz fibrations are isomorphic. This completes the proof. □

4. Lefschetz fibrations of certain free divides

A free divide is a divide in the unit disk whose endpoints are not necessary on the boundary of the disk. Let D denote the unit disk.

Definition 4.1. A *free divide* Q is the image of a generic immersion of intervals and circles into D.

In this paper, we only study free divides consisting of one immersed interval. We further assume that one of the endpoints lies on ∂D and the other is not adjacent to the outside region, which is called a free divide *with one free endpoint.*

Definition 4.2. Let Q be a free divide in D consisting of one immersed interval and with one free endpoint. The *link* of Q is defined to be the link of an oriented divide obtained from Q by doubling it according to the same rule as explained in Introduction.

Remark that though there are two choices for the orientation of the doubled curve Q_2 of Q, the link-type of $L(Q)$ does not depend on this choice since they are isotopic by π-rotation of the fibers of the bundle $\partial N(D) \cap \hat{B} \to B \subset D$. If one immersed interval has two free endpoints then we need to introduce signs to these endpoints to define its link, see [16]. We also remark that we only consider a free divide not in $\Sigma_{g,n}$ but in the unit disk. This is because we do not know the admissibility condition for free divides.

Proof of Theorem 1.3. Let Q be a free divide in the assertion. We first prove case (1). There are two edges adjacent to c and the outside region, one of which is also adjacent to the region whose boundary contains the free endpoint. We denote it by e. To make a shadowed polyhedron of Q, we use the following doubling method:

1. double the curve of Q;
2. for each endpoint of Q, close the corresponding two endpoints of the double curve by a small half circle;
3. for each edge of Q that is neither adjacent to an endpoint nor the edge e, add a crossing between the two edges of the doubled curve parallel the edge.

The doubled curve near the edge e becomes as shown on the left in Figure 11 or its mirror image. We prove the assertion in the former case. The latter case can also be proved by the same argument.

Let $\overrightarrow{Q}$ be an oriented divide obtained from Q by applying this double method, deforming the curve near the free endpoint as shown in Figure 11 and assigning any orientation to the doubled curve. Note that the link-type of the link $L(\overrightarrow{Q})$ of $\overrightarrow{Q}$ does not depend on the choice of the assigned orientation. Obviously, $L(\overrightarrow{Q})$ and $L(Q)$ are isotopic. Hence it is enough to show that $L(\overrightarrow{Q})$ satisfies the properties in the assertion. Let $X'_{\overrightarrow{Q}}$ be the shadowed polyhedron obtained from the shadowed polyhedron of $\overrightarrow{Q}$ by removing the boundary region adjacent to ∂D. We may obtain the surface Σ for an LF-structure from $X'_{\overrightarrow{Q}}$ by removing suitable internal regions as we did for divides in the proof of Lemma 3.5. We set the orientation of Σ as shown on the right in Figure 11. Regarding the vanishing cycle about the bigon in the figure as a saddle, we set the order of the regions $X'_{\overrightarrow{Q}} \setminus \Sigma$ by the order of maxima, saddles and minima as for divides. This order satisfies the condition (iv) of LF-structure. Thus the assertion in case (1) is proved.

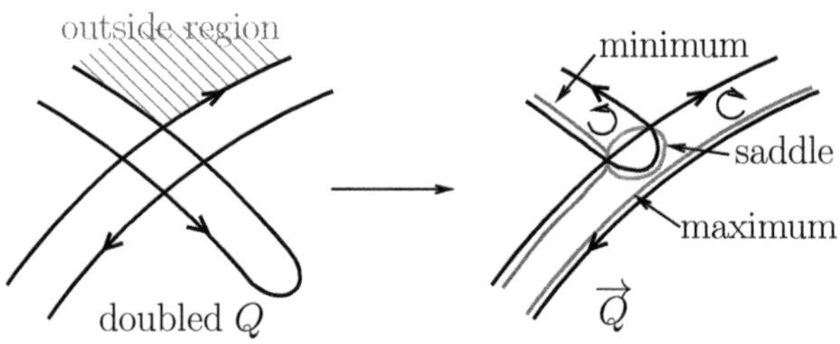

Fig. 11. A deformation of the doubled curve near the free endpoint in case (1).

Next we prove case (2). Let c' be the double point of Q connected to its non-free endpoint by a single edge and e be the edge connecting c and c'. There are two regions adjacent to e, one of which is adjacent to the free endpoint and we denote the other one by R. Let e' be the edge of Q adjacent to c' and lying between R and the outside region. We apply the doubling method to Q as in case (1) with modification that, for each of the edges e and e', we do not add a crossing between the two edges of the doubled curve parallel to the edge. The doubled curve near the edge e becomes as shown on the left in Figure 12 or its mirror image. We prove the assertion in the former case. The latter case can also be proved by the same argument. As in case (1), we define $\overrightarrow{Q}$ as in Figure 12, make $X'_{\overrightarrow{Q}}$, define Σ and fix its orientation. Let C_1, C_2, C_3 be the vanishing cycles shown on the right in Figure 12 and R_1, R_2, R_3 be the internal regions of $X'_{\overrightarrow{Q}}$ corresponding to these cycles. We regard C_1 and C_3 as maxima and set the order of regions $X'_{\overrightarrow{Q}} \setminus \Sigma$ except R_2 by the order of maxima, saddles and minima as in case (1). We then set the order of R_2 as $R_1 < R_2 < R_3$. This order satisfies the condition (iv) of LF-structure and the proof completes. □

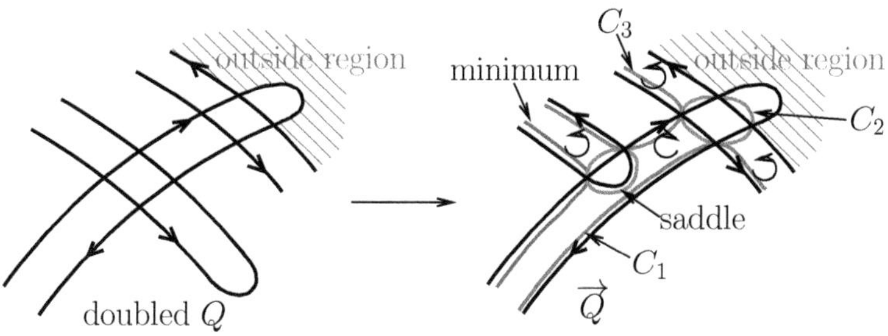

Fig. 12. The doubled curve around the arc connecting c and the non-free endpoint in case (2).

We conclude the paper with one example.

Example 4.3. Let Q be a free divide shown on the left in Figure 13. The shadowed polyhedron is given on the right. The union of the regions with marks of orientations and the annuli attached along $\vec{Q}$ is the surface Σ of the LF-structure, which is a regular fiber of the Lefschetz fibration. Let C_1, C_2, C_3, C_4 be the four vanishing cycles along which the regions R_1, R_2, R_3, R_4 with gleam -1 are attached. The monodromy of the fibered link $L(Q)$ is the product of right-handed Dehn twists $\sigma_1, \sigma_2, \sigma_3, \sigma_4$ along C_1, C_2, C_3, C_4 in this order. The monodromy matrix is given as

$$M_{\sigma_4}M_{\sigma_3}M_{\sigma_2}M_{\sigma_1} = \begin{pmatrix} 1 & 0 & 0 & 0 \\ 0 & 1 & 0 & 0 \\ 0 & 0 & 1 & 0 \\ -1 & 1 & 1 & 1 \end{pmatrix} \begin{pmatrix} 1 & 0 & 0 & 0 \\ 0 & 1 & 0 & 0 \\ 0 & 0 & 1 & -1 \\ 0 & 0 & 0 & 1 \end{pmatrix} \begin{pmatrix} 1 & 0 & 0 & 0 \\ 1 & 1 & 0 & -1 \\ 0 & 0 & 1 & 0 \\ 0 & 0 & 0 & 1 \end{pmatrix} \begin{pmatrix} 1 & -1 & 0 & 1 \\ 0 & 1 & 0 & 0 \\ 0 & 0 & 1 & 0 \\ 0 & 0 & 0 & 1 \end{pmatrix}$$

$$= \begin{pmatrix} 1 & -1 & 0 & 1 \\ 1 & 0 & 0 & 0 \\ 0 & 0 & 1 & -1 \\ 0 & 1 & 1 & -1 \end{pmatrix},$$

where M_{σ_i} is the monodromy matrix of σ_i. The characteristic polynomial of this matrix is $t^4 - t^3 + t^2 - t + 1$, which is the Alexander polynomial of the $(2, 5)$-torus knot. It is known in [16] that the link of this free divide is a $(2, 5)$-torus knot. Actually, we can check that it is a positive $(2, 5)$-torus knot. We can also check it by describing a Kirby diagram of the shadowed polyhedron.

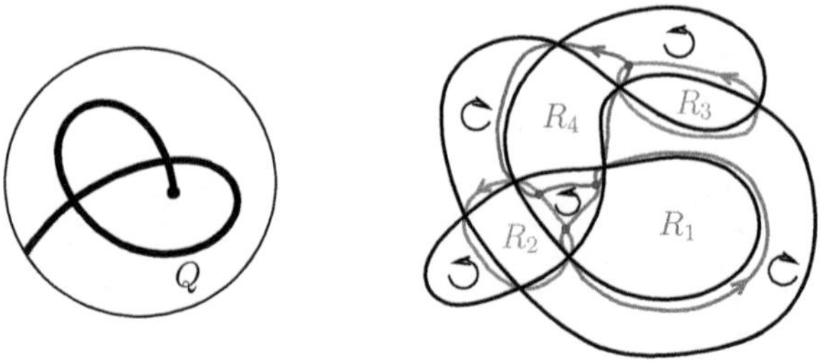

Fig. 13. An example of a free divide and its shadowed polyhedron.

References

[1] N. A'Campo, *Le groupe de monodromie du déploiement des singularités isolées de courbes planes I*, Math. Ann. **213** (1975), 1–32.

[2] N. A'Campo, *Le groupe de monodromie du déploiement des singularités isolées de courbes planes II*, Actes du Congrès International des Mathematiciens, Vancouver, 1974, pp. 395–404.

[3] N. A'Campo, *Real deformations and complex topology of plane curve singularities*, Ann. de la Faculté des Sciences de Toulouse **8** (1999), no. 1, 5–23.

[4] N. A'Campo, *Generic immersions of curves, knots, monodromy and gordian number*, Publ. Math. de l'I.H.E.S. **88** (1998), 151–169.

[5] N. A'Campo, *Planar trees, slalom curves and hyperbolic knots*, Publ. Math. de l'I.H.E.S. **88** (1998), 171–180.

[6] U. Burri, *For a fixed Turaev shadow Jones-Vassiliev invariants depend polynomially on the gleams*, Comment. Math. Helv. **72** (1997), 110–127.

[7] A. Carrega, B. Martelli, *Shadows, ribbon surfaces, and quantum invariants*, Quantum Topol. **8** (2017), no. 2, 249–294.

[8] F. Costantino, Shadows and branched shadows of 3 and 4-manifolds, Scuola Normale Superiore, Edizioni della Normale, Pisa, Italy, 2005.

[9] F. Costantino, *Stein domains and branched shadows of 4-manifolds*, Geom. Dedicata **121** (2006), 89–111.

[10] F. Costantino, *Complexity of 4-manifolds*, Experiment. Math. **15** (2006), no. 2, 237–249.

[11] F. Costantino, *Branched shadows and complex structures on 4-manifolds*, J. Knot Theory Ramifications **17** (2008), no. 11, 1429–1454.

[12] F. Costantino, D. Thurston, *3-manifolds efficiently bound 4-manifolds*, J. Topol. **1** (2008), no. 3, 703–745.

[13] O. Couture, B. Perron, *Representative braids for links associated to plane immersed curves*, J. Knot Theory Ramifications **9** (2000), 1–30.

[14] S. Fomin, P. Pylyavskyy, E. Shustin, *Morsifications and mutations*, preprint, arXiv:1711.10598 [math.GT]

[15] W. Gibson, M. Ishikawa, *Links of oriented divides and fibrations in link exteriors*, Osaka J. Math. **39** (2002), 681–703.
[16] W. Gibson, M. Ishikawa, *Links and gordian numbers associated with generic immersions of intervals*, Topology Appl. **123** (2002), 609–636.
[17] H. Goda, M. Hirasawa, Y. Yamada, *Lissajous curves as A'Campo divides, torus knots and their fiber surfaces*, Tokyo J. Math. **25** (2002), 485–491.
[18] M. N. Goussarov, *Interdependent modifications of links and invariants of finite degree*, Topology **37** (1998), 595–602.
[19] S. M. Gusein-Zade, *Intersection matrices for certain singularities of functions of two variables*, Funct. Anal. Appl. **8** (1974), 10–13.
[20] S. M. Gusein-Zade, *Dynkin diagrams of singularities of functions of two variables*, Funct. Anal. Appl. **8** (1974), 295–300.
[21] S. M. Gusein-Zade, *The monodromy groups of isolated singularities of hypersurfaces*, Russian Math. Surveys **32** (1977), 23–69.
[22] M. Hirasawa, *Visualization of A'Campo's fibered links and unknotting operation*, Proceedings of the First Joint Japan-Mexico Meeting in Topology (Morelia, 1999), Topology Appl. **121** (2002), no. 1-2, 287–304.
[23] M. Ishikawa, *Tangent circle bundles admit positive open book decompositions along arbitrary links*, Topology **43** (2004), 215–232.
[24] M. Ishikawa, Y. Koda, *Stable maps and branched shadows of 3-manifolds*, Math. Ann. **367** (2017), no. 3-4, 1819–1863.
[25] T. Kawamura, *Quasipositivity of links of divides and free divides*, Topology Appl. **125** (2002), no. 1, 111–123.
[26] Y. Koda, B. Martelli, H. Naoe, *Four-manifolds with shadow-complexity one*, preprint, arXiv:1803.06713 [math.GT]
[27] Y. Koda, H. Naoe, *Shadows of acyclic 4-manifolds with sphere boundary*, to appear in Algebr. Geom. Topol., arXiv:1905.00809 [math.GT]
[28] B. Martelli, *Four-manifolds with shadow-complexity zero*, Int. Math. Res. Not. **2011**, no. 6, 1268–1351.
[29] H. Naoe, *Mazur manifolds and corks with small shadow complexities*, Osaka J. Math. **55** (2018), no. 3, 479–498.
[30] H. Naoe, *Shadows of 4-manifolds with complexity zero and polyhedral collapsing*, Proc. Amer. Math. Soc. **145** (2017), no. 10, 4561–4572.
[31] H. Naoe, *Corks with large shadow-complexity and exotic 4-manifolds*, Experiment. Math., https://doi.org/10.1080/10586458.2018.1514332, in press.
[32] W. Neumann, L. Rudolph, *Unfoldings in Knot Theory*, Math. Ann. **278** (1987), 409–439.
[33] B. Ozbagci, *Genus one Lefschetz fibrations on disk cotangent bundles of surfaces*, preprint, arXiv:1803.01649 [math.GT]
[34] A. Shumakovitch, *Shadow formula for the Vassiliev invariant of degree two*, Topology **36** (1997), no. 2, 449–469.
[35] V. G. Turaev, *Shadow links and face models of statistical mechanics*, J. Differential Geom. **36** (1992), no. 1, 35–74.
[36] V. G. Turaev, *Quantum invariants of knots and 3-manifolds*, De Gruyter Studies in Mathematics, vol 18, Walter de Gruyter & Co., Berlin, 1994.
[37] Y. Yamada, *Finite Dehn surgery along A'Campo's divide knots*, Singularity

theory and its applications, Adv. Stud. Pure Math., 43, Math. Soc. Japan, Tokyo, 2006, pp. 573–583.

[38] Y. Yamada, *Lens space surgeries as A'Campo's divide knots*, Algebr. Geom. Topol. **9** (2009), no. 1, 397–428.

On $(2, p)$ quasi torus curves and weak Zariski pairs

Masayuki Kawashima

Department of Information Science,
Okayama University of Science,
1-1 Ridai-cho, Kitaku, Okayama 700-0005, Japan
E-mail: kawashima@mis.ous.ac.jp

In this paper, we construct pairs of plane curves such that they have the same degree and the same singularities. One is a quasi torus curve of type $(p, 2)$ and the other is a torus curve of type $(p, 2)$. For the construction, we use Nagata transformations. If p is even, their complements in $\mathbb{P}^2$ have different topologies. Thus they are a weak Zariski pair.

Keywords: quasi torus curve, weak Zariski pair

2010 MSC: 14H20,14H30,14H45

1. Introduction

Let $\mathcal{M}_d$ be the set of reduced affine plane curves of degree d. Let p and q be positive integers such that $p > q \geq 2$. We say that an affine plane curve $C = \{f = 0\} \in \mathcal{M}_d$ is *a torus curve of type* (p, q) if the defining polynomial f of C is written as

$$f(x, y) = f_a(x, y)^p + f_b(x, y)^q, \quad a, b \geq 1, \quad \deg f_j = j, \quad j = a, b.$$

We also say that the above expression is *a torus decomposition of* C. We denote by $\mathcal{T}_d(p, q)$ the subset of $\mathcal{M}_d$ which consists of torus curves of type (p, q). We also put $\mathcal{T}_d^*(p, q) := \{\{f_a^p + f_b^q = 0\} \in \mathcal{T}_d(p, q) \mid ap = bq = d\}$. Note that if d is a common multiple of a and b, then we can take $a = d/p$ and $b = d/q$ and hence $\mathcal{T}_{pq}(p, q)$ is not empty. If d is not a common multiple of a and b, $\{f_a^p + f_b^q = 0\} \in \mathcal{T}_d(p, q)$ and $\max\{pa, qb\} > d$, then the higher terms of $f_a^p + f_b^q$ must be canceled out. This phenomenon is related with *non-reduced degenerations of plane curves* which are considered by M. Oka in [8].

Torus curves are well-known class of plane curves and studied by many mathematicians. In particular, $(3, 2)$ torus curves of sextics are studied deeply. There are works on the classification of the singularities ([9, 11]), the determination of the topological invariants such as the fundamental groups ([10]) and the Alexander polynomials ([6]) and degenerations of sextics ([8]).

In this paper, we also consider another class of plane curves which are called *quasi torus curves of type* (p, q) (cf. [2, 5]).

Definition 1. We say that $C = \{f = 0\} \in \mathcal{M}_d$ is *a quasi torus curve of type* (p, q) if there exist three polynomials f_a, f_b and f_c such that they satisfy the following condition:

$$f_c(x, y)^{pq} f(x, y) = f_a(x, y)^p + f_b(x, y)^q \qquad (*)$$

where $a, b, c \geq 1$ and $\deg f_j = j$. In the above affine equations, we also assume that any two polynomials of f, f_a, f_b and f_c are coprime. We say that such a decomposition $(*)$ is *a quasi torus decomposition of* C. We denote by $\mathcal{QT}_d(p, q)$ the subset of $\mathcal{M}_d$ which consists of (p, q) quasi torus curves of degree d and we put $\mathcal{QT}_d^{(c)}(p, q) \subset \mathcal{QT}_d(p, q)$:

$$\mathcal{QT}_d^{(c)}(p, q) := \{C \in \mathcal{QT}_d(p, q) \mid d + cpq = pa \text{ or } bq\}.$$

H. Tokunaga studied $(2, 3)$ torus decompositions of sextics via Albanese images of 6-fold cyclic multiple planes and the structures of Mordell-Weil groups of elliptic surfaces. He also gave a criterion to check if a sextic is of $(2, 3)$ torus type ([12, 13]). We call this criterion "Tokunaga's criterion" ([7]). In [4], we also studied $(2, 3)$ torus decompositions for some quartics using line degenerated torus curves ([8]). For these quartics, there is a unique $(2, 3)$ torus decomposition except for three cuspidal quartics. A three cuspidal quartic $Q = \{q = 0\}$ has at least two torus decompositions and one quasi torus decomposition of type $(2, 3)$. Their decompositions are given as follows:

$$\begin{aligned} q(x, y) &= f_1(x, y)^3 + f_2(x, y)^2, \quad \deg f_i = i, \ i = 1, 2 \\ q(x, y) &= g_2(x, y)^3 + g_3(x, y)^2, \quad \deg g_i = i, \ i = 2, 3 \\ h_1(x, y)^6 q(x, y) &= h_3(x, y)^3 + h_5(x, y)^2, \quad \deg h_i = i, \ i = 1, 3, 5. \end{aligned}$$

These equations show that the three cuspidal quartic Q is contained both in $\mathcal{QT}_4(3, 2)$ and in $\mathcal{T}_4(3, 2)$. Thus a given plane curve may have several decompositions. For general (p, q), there are not many known examples of (p, q) quasi torus curves.

We are interested in the following fundamental problems:

– The existence of plane curves in $\mathcal{QT}_d(p, q)$ for given a pair (p, q).

– The existence of plane curves in $\mathcal{QT}_d(p,q) \cap \mathcal{T}_d(p,q)$ for some (p,q).
– The topologies of curves in $\mathcal{QT}_d(p,q)$.
In this paper, we consider only the case $q = 2$. First we will show the following:

Proposition 1. *For any positive integer $p > 2$, there exists a $(p,2)$ quasi torus curve $C = \{f = 0\} \in \mathcal{QT}_d^{(1)}(p,2)$ which has the following quasi torus decomposition:*

$$f_1(x,z)^{2p} f(x,z) = \begin{cases} f_{p+2}(x,z)^p + f_{\frac{(p+2)p}{2}}(x,z)^2 & \text{if } p \text{ is even,} \\ f_{p+1}(x,z)^p + f_{\frac{(p+1)p}{2}}(x,z)^2 & \text{if } p \text{ is odd} \end{cases}$$

where $d = p^2$ if p is even and $d = p(p-1)$ if p is odd and $\deg f_j = j$.

Singularities of C in Proposition 1 have many possibilities (See Remark 3 in §4). In our construction of $(p,2)$ quasi torus curves, we use an n-fold Nagata transformation which is also called elementary transformations and p-gonal curves of $(p,2)$ torus type in a Hirzebruch surface of degree $2k$ (See §2, 3).

Using Proposition 1, we construct a pair of plane curves C and D such that they have the same degree and the same singularities but C has a quasi torus decomposition of type $(p,2)$ and D has a torus decomposition of type $(p,2)$. In statements, we use notation of singularities:

$$B_{a,b} : u^a + v^b = 0, \quad B_{a,b} \circ B_{c,d} : (u^a + v^b)(u^c + v^d) = 0.$$

Theorem 1. *Let p be a positive integer which is greater than 2. We put d is p^2 if p is even and $p(p-1)$ if p is odd. Then there exist two plane curves $C = \{f = 0\}$ and $D = \{g = 0\} \in \mathcal{M}_d$ and they satisfy the following properties:*

(1) $C \in \mathcal{QT}_d^{(1)}(p,2)$ *and* $D \in \mathcal{T}_d^*(p,2)$.

(2) $$\operatorname{Sing} C = \operatorname{Sing} D = \begin{cases} \left\{ B_{2p,2}, \frac{1}{2}(p^2+2p-4)B_{p,2}, S_1 \right\} & \text{if } p \text{ is even,} \\ \left\{ B_{2p,2}, \frac{1}{2}(p^2+p-4)B_{p,2}, S_2 \right\} & \text{if } p \text{ is odd} \end{cases}$$

where $S_1 = B_{4,2} \circ B_{(p+2)(p-2),(p+1)(p-2)}$ *and* $S_2 = B_{(p+1)(p-2),p(p-2)}$.

(3) *The defining polynomials f and g have the following decompositions:*

$$f_1(x,z)^{2p} f(x,z) = \begin{cases} f_{p+2}(x,z)^p + f_{\frac{p(p+2)}{2}}(x,z)^2 & \text{if } p \text{ is even,} \\ f_{p+1}(x,z)^p + f_{\frac{p(p+1)}{2}}(x,z)^2 & \text{if } p \text{ is odd,} \end{cases}$$

and

$$g(x,z)=\begin{cases} g_p(x,z)^p+g_{\frac{p^2}{2}}(x,z)^2 & \text{if } p \text{ is even,}\\ g_{p-1}(x,z)^p+g_{\frac{p(p-1)}{2}}(x,z)^2 & \text{if } p \text{ is odd.}\end{cases}$$

For the curve $C=\{f=0\}$ in Theorem 1, we have the following properties:

Proposition 2. *Under the notations of Theorem 1, we have:*

(1) *If $p>3$, then $C\in \mathcal{QT}_d^{(1)}(p,2)$ is not contained in $\mathcal{T}_d^*(p,2)$.*
(2) *If $p=3$, then $C=\{f=0\}$ is contained in both $\mathcal{QT}_6^{(1)}(3,2)$ and $\mathcal{T}_6^*(3,2)$. Hence f has at least two decompositions:*

$$f_1(x,z)^6f(x,z)=f_4(x,z)^3+f_6(x,z)^2$$
$$f(x,z)=g_2(x,z)^3+g_3(x,z)^2.$$

Note that for the case $p=3$ in Proposition 2, the singularities of C is $\{4A_2,A_5,E_6\}$. In [9], M. Oka and D. T. Pho showed that the moduli space of sextics of $(2,3)$ torus type which have the singularities $\{4A_2,A_5,E_6\}$ is irreducible and the generic Alexander polynomial $\Delta_C(t)$ of C is $(t^2-t+1)^2$ (cf. [6]).

Corollary 1. *Under the notations of Theorem 1, we have:*

(1) *If $p>3$, then $D\in\mathcal{T}_d^*(p,2)$ is not contained in $\mathcal{QT}_d^{(1)}(p,2)$.*
(2) *If $p=3$, then C and D are in the same moduli space.*

Next we consider the topologies of plane curves C and D which are obtained by Theorem 1.

Definition 2. A pair of plane curves (C_1,C_2) is called *a weak Zariski pair* if they have the same degree and singularities, but the complement $\mathbb{P}^2\setminus C_1$ is not homeomorphic to the complement $\mathbb{P}^2\setminus C_2$.

Under the assumption p is even, we construct pairs (C,D) of plane curves in Theorem 1 such that

- $C\in\mathcal{QT}_{p^2}^{(1)}(p,2)\setminus\mathcal{T}_{p^2}^*(p,2)$ and $D\in\mathcal{T}_{p^2}^*(p,2)\setminus\mathcal{QT}_{p^2}^{(1)}(p,2)$.
- $\operatorname{Sing} C=\operatorname{Sing} D$
$=\{B_{2p,2},(\frac{p^2}{2}+p-2)B_{p,2},B_{4,2}\circ B_{(p+2)(p-2),(p+1)(p-2)}\}$.

Proposition 3. *Suppose that p is even and we put $p = 2s$. Then C and D are reducible curves and they are factorized as the following:*

$$C = C_1 \cup C_2 \qquad D = D_1 \cup D_2$$

where $\deg C_1 = 2s(s-1)$, $\deg C_2 = 2s(s+1)$ *and* $\deg D_1 = \deg D_2 = 2s^2$. *In particular, the first homologies of their complements in $\mathbb{P}^2$ are different. That is, the pair (C, D) is a weak Zariski pair.*

2. Preliminaries

2.1. *Notations*

Let $p : \Sigma_d \to \mathbb{P}^1$ be a Hirzebruch surface of degree d and let $\Delta_{\infty,d}$ be the exceptional section with $\Delta_{\infty,d}^2 = -d$. Let (X_0, X_1, X_2) and (Y_0, Y_1) be homogeneous coordinates of $\mathbb{P}^2$ and $\mathbb{P}^1$ respectively. Using these coordinates, Σ_d is defined as

$$\Sigma_d := \{((X_0, X_1, X_2), (Y_0, Y_1)) \in \mathbb{P}^2 \times \mathbb{P}^1 \mid X_1 Y_1^d = X_2 Y_0^d\}$$

and $p : \Sigma_d \to \mathbb{P}^1$ is the canonical projection. There are four affine spaces which cover Σ_d. We use only two affine spaces W_d^1, $W_d^2 \subset \Sigma_d$ with coordinates (y_d, τ_d) and (z_d, τ_d) respectively where

$$y_d = X_2/X_0, \;\; z_d = X_0/X_2, \;\; \tau_s = Y_0/Y_1$$

and they are glued by the relation $y_d z_d = 1$. Note that the exceptional section $\Delta_{\infty,d}$ is defined as $\{y_d = 0\}$ in W_d^1 and we denote the their origin by $O_d^i = (0,0) \in W_d^i$ for $i = 1, 2$. We denote the fiber over $\tau_d = 0$ in Σ_d by F_∞ and the affine line $F_\infty^\circ := F_\infty \setminus \Delta_{\infty,d}$.

2.2. *p-gonal curves*

Let $B \subset \Sigma_d$ be a reduced curve such that $\Delta_{\infty,d} \not\subset B$. We call B *a generalized p-gonal curve* if B intersects with a generic fiber at p points. A generalized p-gonal curve B is called a *p-gonal curve* if $B \cap \Delta_{\infty,d} = \emptyset$.

Let f_i be the defining equation of $B \cap W_d^i$ with the equality $f_1(y_d, \tau_d) = y_d^p f_2(z_d, \tau_d)$ on $W_d^1 \cap W_d^2$. Using the affine coordinates (z_d, τ_d), the local equation $f_2(z_d, \tau_d)$ is written as

$$f_2(z_d, \tau_d) = \sum_{i=0}^{p} b_i(\tau_d) z_d^i, \qquad \deg b_i(\tau_d) = d(p-i).$$

2.3. *Nagata transformations*

Let P be a fixed point in $\Sigma_d \setminus \Delta_{\infty,d}$ and let F be the fiber which passes through P. *A (negative) Nagata transformation* $N : \Sigma_d \dashrightarrow \Sigma_{d-1}$ is a birational transformation which consists of the blowing-up at P and the blowing-down on the strict transform of F. An *n-fold* Nagata transformation is a composition of n-times Nagata transformations.

We express an n-fold Nagata transformation $N^n : \Sigma_d \dashrightarrow \Sigma_{d-n}$ on the fiber $F = p^{-1}(0)$ using local coordinates $(z_d, \tau_d) \in W_d^2$ and $(z_{d-n}, \tau_{d-n}) \in W_{d-n}^2$. Let $P_d, P_{d-1}, \dots, P_{d-n+1}$ be blowing-up points and we put $P_{d-i} := (\alpha_i, 0) \in W_{d-i}^2$ for $i = 0, \dots, n-1$. Assume that the first center is $(\alpha_0, 0) = (0, 0)$. Computing compositions, we have the following:

$$N^n(z_d, \tau_d) = (z_{d-n}, \tau_{d-n}) = \left(\frac{z_d - \varphi(\tau_d)}{\tau_d^n}, \tau_d \right) \quad \text{where} \quad \varphi(\tau_d) = \sum_{i=1}^{n-1} \alpha_i \tau_d^i.$$

Let B be a p-gonal curve in Σ_d and then the image $B' := N^n(B)$ is a generalized p-gonal curve in Σ_{d-n}. More precisely, B' is defined as

$$(\star) \qquad \begin{aligned} B' : \ f_2'(z_{d-n}, \tau_{d-n}) &= \frac{1}{\tau_{d-n}^M} f_2(z_{d-n}\tau_{d-n}^n + \varphi(\tau_{d-n}), \tau_{d-n}) = 0, \\ M :&= \sum_{i=0}^{n-1} m(N^i(B), P_{d-i}) \end{aligned}$$

where $m(N^i(B), P_{d-i})$ is the multiplicity of $N^i(B)$ at P_{d-i}.

2.4. *Singularities of $N^n(B)$*

Let F_d^2 be the fiber which is defined as $\{\tau_d = 0\} \subset W_d^2$. Suppose that $B \subset \Sigma_d$ satisfies the following two conditions:

- B has a $B_{\ell,2}$ singularity at $O_d^2 \in W_d^2$ and its tangent cone does not contain F_d^2.
- B is smooth and intersects transversely at each intersection point with the fiber F_d^2 except for O_d^2.

Under the above conditions, the intersection $B \cap (F_d^2 \setminus \{O_d^2\})$ consists of distinct $p - 2$ points.

Case $n = 1$. The local geometries of $N(B)$ at the intersection point with F_{d-1}^2 are as follows:

- If $\ell = 2$, then $N(B)$ intersects transversely with F_{d-1}^2 at two points.
- If $\ell = 3$, then $N(B)$ is tangent to F_{d-1}^2 with order 2.

- If $\ell > 3$, then $N(B)$ has a $B_{\ell-2,2}$ singularity.

Now we consider the topological type of $N(B)$ at O_d^1. If $p = 3$, then $N(B)$ is smooth and intersects transversely with $\Delta_{\infty,d-1}$ at O_d^1. If p is greater than 3, then $N(B)$ has a $B_{p-2,p-2}$ singularity at O_d^1.

Case $n \geq 2$**.** We take blowing-up centers $\{P_d = O_d^2, P_{d-1}, \ldots, P_{d-n+1}\}$ of N_-^n as follows:

- If $\ell - 2n \geq 0$, then P_{d-i} is the singular point of $N^i(B)$ on F_{d-i}^2 for $i = 1, \ldots, n-1$.
- If $\ell - 2n < 0$ and ℓ is even, $\ell = 2s$ for some s, then P_{d-i} is the singular point of $N^i(B)$ on F_{d-i}^2 for $i = 0, \ldots, s-1$ and P_{d-i} is any point in $F_{d-i}^2 \setminus N^i(B)$ for $i = s, \ldots, n-1$.
- If $\ell - 2n < 0$ and ℓ is odd, $\ell = 2s+1$ for some s, then P_{d-i} is the singular point of $N^i(B)$ on F_{d-i}^2 for $i = 0, \ldots, s-1$ and P_{d-s} is the tangent point of $N^s(B)$ on F_{d-s}^2 and P_{d-i} is any point at $F_{d-i} \setminus N^i(B)$ for $i = s+1, \ldots, n-1$.

Note that $N^n(B)$ has a singularity at O_{d-n}^1. By simple calculations, the singularity of $N^n(B)$ at $O_{d-n}^1 \in \Sigma_{d-n}$ is given as the following:

(1) If $\ell - 2n \geq 0$, then $(N^n(B), O_{d-n}^1) \sim B_{p-2,n(p-2)}$.
(2) If $\ell - 2n < 0$, then $(N^n(B), O_{d-n}^1) \sim B_{2,2n-\ell} \circ B_{p-2,n(p-2)}$.

Remark 1. If $\ell - 2n < 0$, then $N^n(B)$ intersects with $\Delta_{\infty,d-n}$ at O_{d-n}^2 and the intersection multiplicity of $N^n(B)$ and $\Delta_{\infty,d-n}$ at O_{d-n}^1 is $np - \ell$.

2.5. *A p-gonal model of a plane curve*

Let $C \subset \mathbb{P}^2$ be a reduced curve of degree $s > p+1$ and we assume that C has a singular point P with multiplicity $s-p$. Let $\psi : (\Sigma_1, \Delta_{\infty,1}) \to (\mathbb{P}^2, P)$ be a blowing-up at P and then the strict transform $\tilde{C} \subset \Sigma_1$ of C is a generalized p-gonal curve. Taking several blow-ups and blow-downs, we can obtain a p-gonal curve in a Hirzebruch surface. We call such a plane curve *a p-gonal model of C*.

We can regard a p-gonal curve $B = \{f_2 = 0\}$ as a p-gonal model of a plane curve $C = \{f = 0\} \subset \mathbb{P}^2$.

Let $B' = \{f_2' = 0\} \subset \Sigma_1$ be a generalized p-gonal curve such that $B' \cap \Delta_{\infty,1} = \{O_1^1\}$ with $I(B', \Delta_{\infty,1}; O_1^1) = m$. Let $\psi : \Sigma_1 \to \mathbb{P}^2$ be a blowing-up at $(0,1,0)$. We consider the pull-buck ψ^*C of C by ψ.

Assertion 1. *Suppose that* $\psi^*C \sim B' + k\Delta_{\infty,d}$. *Then* $k = m$ *and* $\deg C =$

$p+m$. The defining polynomial f of C is obtained as

$$f(x_0,x_2) := x_0^{p+m} f_2'(1/x_0, x_2/x_0), \quad x_0 = X_0/X_1,\ x_2 = X_2/X_1.$$

Proof. As $B' \cdot \Delta_{\infty,d} = m$ and $\Delta_{\infty,d} \cdot \psi^* C = 0$, we have $B' \sim p\Delta_{\infty,d} + (p+m)F$ and $k = m$ where F is a general fiber in Σ_1. On the other hand, we have $\deg C = F \cdot \psi^* C = F \cdot (B' + m\Delta_{\infty,d}) = p+m$. □

Remark 2. Suppose that B' has a $B_{a_1,m_1} \circ B_{a_2,m_2}$ singularity at O_1^1 such that $m_1 + m_2 = m < a_1 + a_2$. Then C has a $B_{a_1+m_1,m_1} \circ B_{a_2+m_2,m_2}$ singularity.

Next we assume that $B_1 = \{f_2'(z_1,\tau_1) = 0\} \subset W_1^2$ is the image of a p-gonal curve $B = \{f_2(z_1,\tau_1) = 0\} \subset W_d^2$ by a $(d-1)$-fold Nagata transformation over $\tau = 0$. Then $f_2'(z_1,\tau_1)$ is defined as $(\star)$ in §2.3:

$$f_2'(z_1,\tau_1) = \frac{1}{\tau_1^M} f_2(z_1\tau_1^{d-1} + \varphi(\tau_1), \tau_1)$$

where M and $\varphi(\tau_1)$ are as in §2.3. We suppose that $B_1 \cap \Delta_{\infty,d} = \{O_1^1\}$ and the intersection multiplicity $I(B_1, \Delta_{\infty,d}; O_1^1)$ is m. Then, by the above assertion, the defining polynomial f of C is obtained as the following form:

$$f(x_0,x_2) = \frac{x_0^{M+m+p}}{x_2^M} f_2(x_2^{d-1}/x_0^d + \varphi(x_2/x_0), x_2/x_0) \qquad (\sharp)$$

where (x_0,x_2) are the local coordinates of $U_1 \subset \mathbb{P}^2$.

3. p-gonal curves of (p,q) torus type

Let p and q be positive integers such that $p > q \geq 2$. Let B be a p-gonal curve in Σ_{qk} where k is some positive integer. We say that B is *of (p,q) torus type* if there are three sections f_0, f_p and f_q of $\mathcal{O}(\Delta_{\infty,qk})$, $\mathcal{O}(\Delta_{\infty,qk} + pkF)$ and $\mathcal{O}(\Delta_{\infty,qk} + qkF)$, respectively, such that B is the zero set of the section $f_q^p + f_0^{p-q} f_p^q$. In the affine space $(W_{qk}^2, (z_2,\tau_2))$, the defining equation f_2 of B is given as $f_2(z_{qk},\tau_{qk}) = g(z_{qk},\tau_{qk})^p + h(z_{qk},\tau_{qk})^q$ where

$$g(z_{qk},\tau_{qk}) = z_{qk} + b_q(\tau_{qk}), \quad h(z_{qk},\tau_{qk}) = b_{p-q}(\tau_{qk}) z_{qk} + b_p(\tau_{qk})$$

where $\deg b_i(\tau_{qk}) \leq ik$ for $i = p$, q and $p-q$.

3.1. *Singularities of (p,q) torus type*

Let $G := \{g = 0\}$ and $H := \{h = 0\}$ be the corresponding curves of (p,q) torus type $B = \{f_2 = 0\}$ in W_{qk}^2. Let $P \in B$ be a singular point. If P is contained in $G \cap H$, we call P is an *inner singularity*. Otherwise P is called an *outer singularity*. An inner singularity P is written as $(-b_q(s), s)$ for some $s \in \mathbb{C}$ and we put $\Delta_1(\tau_{qk}) := h(-b_q(\tau_{qk}), \tau_{qk})$. Then the multiplicity of $\Delta_1(\tau_{qk})$ at s corresponds to the intersection multiplicity of G and H at P.

It is known that the discriminant of f_2 with respect to z_{qk} is given as

$$\Delta(\tau_{qk}) := c\,\Delta_1(\tau_{qk})^{p(q-1)}\Delta_2(\tau_{qk}), \quad c \in \mathbb{C}^*$$

where

$$\Delta_2(\tau_{qk}) = p^p\Delta_1(\tau_{qk})^{p-q} + (q-p)^{p-q}q^q b_{p-q}(\tau_{qk})^p.$$

Now we study singularities of p-gonal curves of (p, q) torus type. By similar arguments in Lemma 1 in [1], we have the following.

Lemma 1. *Let B be a p-gonal curve of (p, q) torus type in Σ_{qk} and let $P = (-b_q(s), s) \in B$ be an inner singular point with the intersection multiplicity ι. If $s \notin \{\Delta_2 = 0\}$, then B has a $B_{p\iota,q}$ singularity at P.*

4. Proof of Proposition 1

4.1. *Strategy*

Let B be a p-gonal curve of $(p, 2)$ torus type in Σ_{2k} whose defining polynomial f_2 in the affine space $(W_2^{2k}, (z_{2k}, \tau_{2k}))$ is written as $f_2 = g^p + h^2$ where g and h are defined as in §3. By Lemma 1, B has a $B_{\iota p,2}$ singularity at each intersection point of G and H where ι is the intersection multiplicity at the corresponding intersection point.

In this section, we take n as $2k-1$. Let $\pi : \Sigma_{2k} \dashrightarrow \mathbb{P}^2$ be the composition of $N^n : \Sigma_{2k} \dashrightarrow \Sigma_1$ and a blowing-up $\psi : \Sigma_1 \to \mathbb{P}^2$ such that $\psi(\Delta_{\infty,1}) = (0, 1, 0) \in \mathbb{P}^2$.

We consider an affine curve $C = \{f = 0\}$ which is obtained as in §2.5. Using (♯) in §2.5, f is defined as

$$f(x_0, x_2) = \frac{x_0^{M+m+p}}{x_2^M} f_2(x_2^{n-1}/x_0^n + \varphi(x_2/x_0), x_2/x_0). \qquad (\sharp)$$

If $x_0^{M+m+p} f_2$ can be expressed as a polynomial of $(p, 2)$ torus type $k_1^p + k_2^2$ and $M = 2p$, then the above equation (♯) gives a $(p, 2)$ quasi torus decomposition in $\mathbb{C}[x_0, x_2]$:

$$x_2^{2p} f(x_0, x_2) = k_1(x_0, x_2)^p + k_2(x_0, x_2)^2.$$

We will show that the existence of a $(p,2)$ torus curve $B \subset \Sigma_{2k}$ for some integer k which satisfies the above conditions. We divide our proof into two steps:

Step 1. Consider the existence of $B \subset \Sigma_{2k}$ and $N^n : \Sigma_{2k} \dashrightarrow \Sigma_1$ such that the multiplicity M is $2p$.

Step 2. Compute the defining polynomial f of C.

4.2. *Step 1*

Now we take k as $k := [p/2] + 1$. Suppose that $B \subset \Sigma_{2k}$ has a $B_{2p,2}$ singularity as an outer singularity on F^2_{2k}. Let $N^n : \Sigma_{2k} \dashrightarrow \Sigma_1$ be an n-fold Nagata transformation over F^2_{2k} where its centers $\{P_{2k}, \dots, P_1\}$ satisfy:

- If p is odd, then we take the blow-up center P_{d-i} of $N^i(B) \subset \Sigma_{2k-i}$ to be the singular point on F^2_{2k-i}.
- If p is even, then $k = \frac{1}{2}(p+2)$ and $n = p+1$. We take the former p points as the singular point of $N^i(B) \subset \Sigma_{2k-i}$ on F^2_{2k-i} and the last point is any point on $F^2_1 \setminus N^p(B)$.

Then the sum of the multiplicities of M is $2p$. Such a p-gonal curve exists for any p. For example, we take $b_q(\tau_{2k})$, $b_{p-2}(\tau_{2k})$ and $b_p(\tau_{2k})$ as

$$b_2(\tau_{2k}) = 1 + \tau_{2k}^{2k}, \quad b_p(\tau_{2k}) = 1 + \frac{p}{2}\tau_{2k}^{2k} + \tau_{2k}^{2p} + \tau_{2k}^{pk}, \quad b_{p-2}(\tau_{2k}) = \frac{p}{2}.$$

Then $f_2 = g^p - h^2$ has an outer $B_{2p,2}$ singularity at O^2_2. Therefore there exists a $(p,2)$ torus curve B such that it has an outer $B_{2p,2}$ singularity on F_∞.

4.3. *Step 2*

In this step, we compute the defining polynomial f of C.

Suppose that p is even. The image $N^p(B) \subset \Sigma_2$ has no outer singularity. On the other hand, $N^p(B)$ has a singularity at O^1_2 and the type of this singularity is $B_{p-2,\,p(p-2)}$ by (1) in §2.4. Now we take the last Nagata transformation $\Sigma_2 \dashrightarrow \Sigma_1$ whose center is as in Step 1. Then $N^{p+1}(B)$ has a singularity at O^1_1 and then, by (2) in §2.4, the type of the singularity is given as

$$(N^{p+1}(B), O^1_1) \sim B_{2,2} \circ B_{p-2,\,(p+1)(p-2)}.$$

Note that the intersection multiplicity of $N^{p+1}(B)$ and $\Delta_{\infty,1}$ at O^1_1 is $p(p-1)$ by Remark 1.

Finally we take the blowing-down $\psi : (\Sigma_1, \Delta_{\infty,1}) \to (\mathbb{P}^2, (0,1,0))$. Using ($\sharp$), the defining polynomial f of C is obtained as follows:

$$\begin{aligned} f(x_0,x_2) :&= \frac{x_0^{p(p+2)}}{x_2^{2p}} f_2(x_2^{p+1}/x_0^{p+2} + \varphi(x_2/x_0), x_2/x_0) \\ &= \frac{1}{x_2^{2p}} \Bigg(\Big(x_2^{p+1} + x_0^{p+2} c_2(x_2/x_0)\Big)^p \\ &\quad + \Big(b_{p-2}(x_2/x_0) x_0^{(p-2)(\frac{p}{2}+1)} x_2^{2p+1} + x_0^{\frac{p(p+1)}{2}} c_p(x_2/x_0)\Big)^2 \Bigg) \end{aligned}$$

where $c_2 := \varphi + b_2$ and $c_p := b_{p-2}\varphi + b_p$. We put

$$\begin{aligned} f_{p+2}(x_0,x_2) &:= x_2^{p+1} + x_0^{p+2} c_2(x_2/x_0), \\ f_{\frac{(p+2)p}{2}}(x_0,x_2) &:= b_{p-2}(x_2/x_0) x_0^{(p-2)(\frac{p}{2}+1)} x_2^{2p+1} + x_0^{\frac{p(p+1)}{2}} c_p(x_2/x_0). \end{aligned}$$

As $\deg c_2 = p+2$ and $\deg c_p = \frac{(p+2)p}{2}$, f_{p+2} and $f_{\frac{(p+2)p}{2}}$ are also polynomials in $\mathbb{C}[x_0,x_2]$. Hence f is a $(p,2)$ quasi torus curve which satisfies the following relation:

$$x_2^{2p} f(x_0,x_2) = f_{p+2}(x_0,x_2)^p + f_{\frac{(p+2)p}{2}}(x_0,x_2)^2.$$

By the same argument for the case p is odd, we have the assertion of Proposition 1.

Remark 3. Let C be a $(p,2)$ quasi torus curve which is constructed in Proposition 1. The set of singularities of C is given by

$$\operatorname{Sing}(C) = \begin{cases} \{\operatorname{Sing}(B) \setminus \{B_{2p,2}\}, B_{4,2} \circ B_{(p+2)(p-2),\,(p+1)(p-2)}\} & \text{if } p \text{ is even,} \\ \{\operatorname{Sing}(B) \setminus \{B_{2p,2}\}, B_{(p+1)(p-2),\,p(p-2)}\} & \text{if } p \text{ is odd.} \end{cases}$$

4.4. *Proof of Theorem 1*

Let $B = \{f_2 = g^p + h^2 = 0\}$ be a p-gonal curve of $(p,2)$ torus type in Σ_{2k}. We will show the assertion for only the case that p is even. Doing the same argument, we can obtain the assertion for the case that p is odd. To prove the assertion, we take $k := \frac{p}{2} + 1$. Suppose that H is smooth and the intersection multiplicities of G and H are

$$I(G,H;P) = 2, \quad I(G,H;Q_i) = 1$$

where $G \cap H = \{P, Q_1, \ldots, Q_{pk-2}\}$. Moreover we can suppose that B has an outer $B_{2p,2}$ singularity as in the proof of Proposition 1. Hence B has $2B_{2p,2}$ and $(pk-2)B_{p,2}$ singularities.

Now we construct two plane curves $C = \{f = 0\}$ and $D = \{g = 0\}$ such that

- Sing C = Sing D
- $C \in \mathcal{QT}(p, 2; p^2)$ and $D \in \mathcal{T}(p, 2; p^2)$.

Let N_1, $N_2 : \Sigma_{p+2} \dashrightarrow \Sigma_1$ be two $(p+1)$-fold Nagata transformations such that the first center of N_1 (resp. N_2) is the outer $B_{2p,2}$ singularity (resp. the inner $B_{2p,2}$ singularity) and we will assign other centers below.

A $(p, 2)$ quasi torus curve $C = \{f = 0\}$ is obtained as in Proposition 1. Indeed, let $\pi_1 : \Sigma_{p+2} \dashrightarrow \mathbb{P}^2$ be the composition of $N_1 : \Sigma_{p+2} \dashrightarrow \Sigma_1$ and a blowing-up $\psi_1 : \Sigma_1 \to \mathbb{P}^2$. Then the defining polynomial f of the image $C := \pi_1(B)$ satisfies the following quasi toric decomposition:

$$x_2^{2p} f(x_0, x_2) = f_{p+2}(x_0, x_2)^p + f_{\frac{(p+2)p}{2}}(x_0, x_2)^2, \quad \deg f = p^2.$$

By Remark 2, the set of the singularities of C is given as:

$$\{B_{2p,2}, (p^2/2 + p - 2)B_{p,2}, B_{4,2} \circ B_{(p+2)(p-2),(p+1)(p-2)}\}.$$

A $(p, 2)$ torus curve $D = \{g = 0\}$ is obtained as follows. We assume that an inner singularity $B_{2p,2}$ is located at $P = O_2^1$. We decompose N_2 into $N \circ N^p$ where $N^p : \Sigma_{p+2} \dashrightarrow \Sigma_2$ is the p-fold Nagata transformation over $\tau = 0$. By a direct calculation, the defining polynomial $f_p(z_2, \tau_2)$ of $N^p(B)$ in Σ_2 is given as the following:

$$f_p(z_2, \tau_2) = g_p(z_2, \tau_2)^p + h_p(z_2, \tau_2)^2$$
$$g_p(z_2, \tau_2) = z_2\tau_2^{p-2} + b_2^{(p)}(\tau_2), \quad h_p(z_2, \tau) = b_{p-2}(\tau_2)z_2 + b_p^{(p)}(\tau_2)$$

where $\deg b_2^{(p)}(\tau_2) = p$, $\deg b_{p-2}(\tau_2) = \dfrac{1}{2}(p-2)(p+2)$ and $\deg b_p^{(p)}(\tau_2) = \dfrac{p^2}{2}$. Then the intersection of $N^p(B)$ and F_∞° consists of two points and $N^p(B)$ is smooth at these points. We take the Nagata transformation $N : \Sigma_2 \dashrightarrow \Sigma_1$ whose center is not in the intersection points of $N^p(B)$ and F_∞°. By (♯) in §2.5, we have the defining polynomial g of the plane curve $D := \pi_2(B)$ as the following:

$$\begin{aligned} g(x_0, x_2) &:= x_0^{p^2} f_p\left(x_2/x_0^2, x_2/x_0\right) \\ &= \left(x_0^p g_p\left(x_2/x_0^2, x_2/x_0\right)\right)^p + \left(x_0^{\frac{p^2}{2}} h_p\left(x_2/x_0^2, x_2/x_0\right)\right)^2 \\ &= k_p(x_0, x_2)^p + k_{\frac{p^2}{2}}(x_0, x_2)^2 \end{aligned}$$

where k_p and $k_{\frac{p^2}{2}}$ are polynomials of degree p and $p^2/2$ respectively. Thus $D = \{g = 0\}$ is a $(p, 2)$ torus curve which has the same singularities as those of $C = \{f = 0\}$ by the construction. Thus the proof of Theorem 1 is completed.

4.5. *Proof of Proposition 2*

Let $C = \{f = 0\}$ be a $(p, 2)$ quasi torus curve which is constructed in the proof of Theorem 1. First we show that if $p > 3$, then C is not contained in $\mathcal{T}^*(p, 2; d)$ where $d = p^2$ if p is even and $p(p-1)$ if p is odd.

Recall that if p is even, then the set of singularities of C is given as

$$\mathrm{Sing}\,(C) = \{B_{2p,2}, (p^2/2 + p - 2)B_{p,2}, S_1\}$$

where $S_1 := B_{4,2} \circ B_{p^2-4,(p+1)(p-2)}$. Suppose that f has two decompositions as the following forms:

$$\begin{aligned} x_2^{2p} f(x_0, x_2) &= f_{p+2}(x_0, x_2)^p + f_{\frac{p(p+2)}{2}}(x_0, x_2)^2, \\ f(x_0, x_2) &= k_p(x_0, x_2)^p + k_{\frac{p^2}{2}}(x_0, x_2)^2. \end{aligned}$$

Put $C_{p+2} := \{f_{p+2} = 0\}$ and $D_p := \{k_p = 0\}$. Let $O^* = [0, 1, 0]$ and P_i be singular points such that $(C, O^*) \sim S_1$ and $(C, P_i) \sim B_{p,2}$ for $i = 1, \ldots, p^2/2 + p - 2$. All $B_{p,2}$ singularities and S_1 are inner singularities of both decompositions. Hence O^* and P_i must be contained in $C_{p+2} \cap D_p$ for all i. On the other hand, by the form of the defining polynomial, C_{p+2} has the multiplicity $p+1$ at O^* (See §4.3). As the multiplicity of S_1 is $p(p-1)$ at O^*, D_p must have at least the multiplicity $p - 1$ at O^*. Then we have the following inequality by Bézout theorem:

$$(p+2)p \geq (p+1)(p-1) + \frac{p^2}{2} + p - 2 = \frac{3}{2}p^2 + p - 3.$$

Hence p must satisfy the inequality $0 \geq p^2 - 2p - 6$. As p is greater than 3, the above inequality is an obvious contradiction. Thus we have the assertion for the case p is even. Doing the same argument for the case p is odd, we have the assertion of Proposition 2 for the case $p > 3$.

Next we consider the case $p = 3$. For this case, we give an example of two decompositions. In the construction of two decompositions, we use so-called Tokunaga's criterion ([12]). Let g_2, g_3, f_4 and f_6 be polynomials which are defined as:

$$\begin{aligned} g_2(x_0,x_2) &:= a_1\left((75-31s)x_2^2+((48s-112)x_0-16s+48)x_2-8x_0^2\right),\\ g_3(x_0,x_2) &:= a_2\big((229s-1179)x_2^3+((5592-360s)x_0+112s-672)x_2^2\\ &\quad+((-7710-2s)x_0+(64s+2112))x_0x_2+2912x_0^3\\ &\qquad-(1056+32s)x_0^2\big).\\ f_4(x_0,x_2) &:= 3x_2^4+\frac{1}{4}\left((9-s)x_0+12\right)x_2^3+\frac{1}{4}(s-69)x_0^2x_2^2+12x_0^3x_2-3x_0^4,\\ f_6(x_0,x_2) &:= \frac{1}{32}(315-21s)x_2^6+\frac{1}{8}\left((3s-207)x_0+63-3s\right)x_2^5\\ &\quad+\left(\frac{1}{32}(69s-363)x_0-18\right)x_0x_2^4+\left(\frac{1}{4}(345-9s)x_0+9\right)x_0^2x_2^3\\ &\qquad+\frac{1}{4}\left(3s-351\right)x_0^4x_2^2+36x_0^5x_2-6x_0^6 \end{aligned}$$

where

$$s=\sqrt{-3},\quad a_1=\frac{75+31s}{8508},\quad a_2=-\frac{1179+229s}{364}.$$

Then $f := b_1\, g_2^3 + b_2\, g_3^2 \in \mathcal{T}^*(2,3;6)$ has another decomposition:

$$108x_2^6 f(x_0,x_2) = 4f_4(x_0,x_2)^3 + 3f_6(x_0,x_2)^2$$

where $b_1 = \frac{18073}{4096} - \frac{1575}{2048}s$ and $b_2 = -\frac{1033}{24672940032} + \frac{35}{1542058752}s$. Thus we have $C = \{f=0\} \in \mathcal{QT}^{(1)}(2,3,6) \cap \mathcal{T}^*(2,3;6)$.

4.6. *Proof of Proposition 3*

In Proposition 3, we suppose that p is even. We constructed a pair of plane curve (C,D) in Theorem 1 so that

(1) $C \in \mathcal{QT}(p,2;p^2)$ and $D \in \mathcal{T}(p,2;p^2)$.
(2) $\operatorname{Sing} C = \operatorname{Sing} D = \{B_{2p,2}, (\frac{p^2}{2}+p-2)B_{p,2}, B_{4,2}\circ B_{(p+2)(p-2),(p+1)(p-2)}\}$.

We will show that the pair (C,D) is a weak Zariski pair. Put $p=2s$ where s is greater than or equal to 2.

Recall that the decompositions of $C=\{f=0\}$ and $D=\{g=0\}$ are as follows:

$$\begin{aligned} f_1(x_0,x_2)^{4s} f(x_0,x_2) &= f_{2s+2}(x_0,x_2)^{2s} - f_{2s(s+1)}(x_0,x_2)^2,\\ g(x_0,x_2) &= g_{2s}(x_0,x_2)^{2s} - g_{2s^2}(x_0,x_2)^2. \end{aligned}$$

Putting $p_1 := f_{2s+2}^s - f_{2s(s+1)}$, $p_2 := f_{2s+2}^s + f_{2s(s+1)}$, $q_1 := g_{2s}^s - g_{2s^2}$ and $q_2 := g_{2s}^s + g_{2s^2}$, the above decompositions are written as

$$f_1(x_0, x_2)^{4s} f(x_0, x_2) = p_1(x_0, x_2) p_2(x_0, x_2),$$
$$g(x_0, x_2) = q_1(x_0, x_2) q_2(x_0, x_2).$$

By the form of $g(x_0, x_2)$, D is decomposed as $D = D_1 \cup D_2$ and D_i is irreducible for $i = 1, 2$. By the construction, the degrees of D_1 and D_2 are $2s^2 = p^2/2$. For C, there are also two irreducible components C_1 and C_2.

Assertion 2. $\deg C_1 = 2s(s-1)$ *and* $\deg C_2 = 2s(s+1)$.

Proof. Let B be a $2s$-gonal curve in Σ_{2s+2} and let $\pi : \Sigma_{2s+2} \dashrightarrow \mathbb{P}^2$ be a birational map such that B has $2B_{4s,2}$ singularities and $\pi(B) = C$ as in the proof of Theorem 1. As one of these singularities is outer, we assume that O_2^1 is an outer $B_{4s,2}$ singularity. Let $f_{21}(z, \tau) = g(z, \tau)^{2s} - h(z, \tau)^2$ be the defining polynomial of B. Then we can factorize $f_{21}(z, \tau)$ as

$$\begin{aligned} f_{21}(z, \tau) &= (g(z, \tau)^s - h(z, \tau))(g(z, \tau)^s + h(z, \tau)) \\ &= k_1(z, \tau)\, k_2(z, \tau) \end{aligned}$$

where $\deg_z k_i = s$ and $\deg_\tau k_i = 2s(s+1)$ for $i = 1, 2$. As O_2^1 is outer, we may assume that O_2^1 is in $\{k_1 = 0\} \setminus \{k_2 = 0\}$. On the other hand, we have

$$\begin{aligned} x_2^{4s} f(x_0, x_2) &= x_0^{4s(s+1)} f_{21}(x_2^{2s+1}/x_0^{2s+2} + \varphi(x_2/x_0), x_2/x_0) \\ &= \left(x_0^{2s(s+1)} k_1(x_2^{2s+1}/x_0^{2s+2} + \varphi(x_2/x_0), x_2/x_0)\right) \\ &\qquad \times \left(x_0^{2s(s+1)} k_2(x_2^{2s+1}/x_0^{2s+2} + \varphi(x_2/x_0), x_2/x_0)\right). \end{aligned}$$

Note that $x_0^{2s(s+1)} k_1$ and $x_0^{2s(s+1)} k_2$ are polynomials of degree $2s(s+1)$. As $O_2^1 \in \{k_1 = 0\}$ is an A_{4s-1} singular point, x_2^{4s} must divide $x_0^{2s(s+1)} k_1$ by the same argument as in the proof of Step 2 in Proposition 1. Hence we can put $w_1(x_0, x_2) := x_0^{2s(s+1)} k_1 / x_2^{4s}$ as a polynomial in $\mathbb{C}[x_0, x_2]$ and its degree of w_1 is $2s(s+1) - 4s = 2s(s-1)$. Putting $w_2 = x_0^{2s(s+1)} k_2$, we have

$$f(x_0, x_2) = w_1(x_0, x_2) w_2(x_0, x_2), \ \deg w_1 = 2s(s-1), \ \deg w_2 = 2s(s+1).$$

□

In general, let V be a plane curve having r irreducible components $V_1, \ldots, V_r$ with $\deg V_i = d_i$. Then the first homology is $\mathbb{Z}^{r-1} \oplus \mathbb{Z}/(d_1, \ldots, d_r)\mathbb{Z}$ where $(d_1, \ldots, d_r)$ is the greatest common divisor of $d_1, \ldots, d_r$. (cf. Proposition 1.3. [3]).

Now we consider the first homologies of C and D. The degrees of irreducible components C_i and D_j are

$$\deg C_1 = 2s(s-1), \quad \deg C_2 = 2s(s+1), \quad \deg D_1 = \deg D_2 = 2s^2.$$

Note that the greatest common divisors of $2s(s-1)$ and $2s(s+1)$ is not $2s^2$. Hence the first homology of $\mathbb{P}^2 \setminus C$ and $\mathbb{P}^2 \setminus D$ are different. Therefore we have the assertion in Proposition 3.

References

[1] B. Audoubert, T. C. Nguyen, and M. Oka. On Alexander polynomials of torus curves. *J. Math. Soc. Japan*, 57(4):935–957, 2005.

[2] J.-I. Cogolludo-Agustín and A. Libgober. Mordell-Weil groups of elliptic threefolds and the Alexander module of plane curves. *J. Reine Angew. Math.*, 697:15–55, 2014.

[3] A. Dimca. *Singularities and the topology of hypersurfaces.* Universitext. Springer-Verlag, New York, 1992.

[4] M. Kawashima and K. Yoshizaki. On $(2;3)$ torus of decompositions of QL-configurations. *SUT J. Math.*, 46(1):93–117, 2010.

[5] V. S. Kulikov. On plane algebraic curves of positive Albanese dimension. *Izv. Math.*, 59(6):1173–1192, 1995.

[6] M. Oka. Alexander polynomial of sextics. *J. Knot Theory Ramifications*, 12(5):619–636, 2003.

[7] M. Oka. Zariski pairs on sextics 1. *Vietnam J. Math.*, 33(SI):81–92, 2005.

[8] M. Oka. Tangential Alexander polynomials and non-reduced degeneration. In *Singularities in geometry and topology*, pages 669–704. World Sci. Publ., Hackensack, NJ, 2007.

[9] M. Oka and D. Pho. Classification of sextics of torus type. *Tokyo J. Math.*, 25 (2002), no. 2, pages 399–433, 2002.

[10] M. Oka and D. Pho. Fundamental group of sextic of torus type. In A. Libgober and M. Tibar, editors, *Trends in Singularities*, pages 151–180. Birkhäuser, Basel, 2002.

[11] D. T. Pho. Classification of singularities on torus curves of type $(2;3)$. *Kodai Math. J.*, 24(1):259–284, 2001.

[12] H.-o. Tokunaga. $(2,3)$ torus sextics and the Albanese images of 6-fold cyclic multiple planes. *Kodai Math. J.*, 22(2):222–242, 1999.

[13] H.-o. Tokunaga. Irreducible plane curves with the Albanese dimension 2. *Proc. Amer. Math. Soc.*, 127(7):1935–1940, 1999.

Key polynomials in dimension 2

Wael Mahboub

Lebanese International University
Mouseitbeh - PO Box: 146404
Mazraa, Beirut, Lebanon

Mark Spivakovsky

Institut de Mathématiques de Toulouse
UMR 5219 du CNRS, Université Paul Sabatier
118, route de Narbonne,
31062 Toulouse cedex 9, France

Solomon Lefschetz Laboratory (LaSoL) / CNRS UMI 2001
Av. Universidad s/n, Periferica Morelos
62210 Cuernavaca, Mor., Mexico
E-mail: mark.spivakovsky@math.univ-toulouse.fr

Keywords: Valuations, Regular local rings, Key polynomials

1. Introduction

Throughout this paper all the rings considered will be commutative with 1. Let $(R, \mathfrak{m}, k)$ be a regular local ring of dimension 2 and F the field of fractions of R. Consider the poset $(\mathcal{V}, \leq)$ of normalized valuations of F centered at R (see §2).

In [4] C. Favre and M. Jonsson prove that $(\mathcal{V}, \leq)$ has the structure of a parametrized, rooted, non-metric tree when $R = \mathbb{C}[[x, y]]$, the ring of formal power series over the field of complex numbers. The proof of C. Favre and M. Jonsson is based on associating to each valuation in $\mathcal{V}$ a set of key polynomials, a concept introduced by S. MacLane in [9] and [10]. Below we will refer to this set as a **complete set of key polynomials** (see §4 for its definition and properties).

In [5] A. Granja generalizes this result to the case when R is any two-dimensional regular local ring. A. Granja gives a proof based on associating to each valuation in $\mathcal{V}$ a sequence of point blowing ups.

In this paper we give a new proof of A. Granja's result when R is any two-dimensional regular local ring, using appropriate complete sequences of key polynomials, based on the work of M. Vaquié [17] for valuations of arbitrary rank, and the work of F. J. Herrera Govantes, W. Mahboub, M. A. Olalla Acosta and M. Spivakovsky ([6], [7]) for valuations of rank 1 over fields of arbitrary characteristic.

We use the notion of key polynomials introduced in [3] and [14]. We give a simple construction of a complete set of key polynomials associated to a valuation of the field $k(x, y)$ where k is the residue field of R and x, y are independent variables. For explicit constructions of key polynomials on particular cases, see [15], [4], [8].

We start by stating in §2 the basic facts related to valuations needed in this paper. Then we establish, in §3, a natural order-preserving bijection between valuations of F centered at R and valuations of $k(x, y)$ centered at $k[x, y]_{(x,y)}$. This is the content of Theorem 3.12. It consists of describing a one-to-one correspondence between the set of valuations centered at R and the set of simple sequences of local point blowing ups (see Corollary 3.11).

In §4 we give the definition of key polynomials. We state the needed facts about key polynomials and construct a complete set of key polynomials associated to a valuation ν of $k(x, y)$. This is our main tool for the proof of Theorem 6.4. We also define invariants of valuations centered in regular two-dimensional local rings.

Then we consider two comparable valuations, $\mu \leq \nu$, and study the structure of their key polynomials sets and the relation between the invariants of those valuations. This is done in §5. Using this comparison, we prove that the infimum of any two elements of $\mathcal{V}$ exists (Theorem 5.8) and that any increasing sequence in $\mathcal{V}$ has a majorant in $\mathcal{V}$ (Theorem 5.9). We note that a more general version of the latter result — one for rings of arbitrary dimension — is given in Lemma 3.9 (i) of [13].

Finally, in §6 we prove the main theorem of this paper, Theorem 6.4. This Theorem asserts that $\mathcal{V}$ has a tree structure.

We thank the referee for a very careful reading of the paper and for numerous useful comments that helped improve the exposition. We also thank Josnei Novacoski for very useful discussions and for sharing his insights.

2. Basics

Let R be a regular noetherian local ring of dimension 2. Denote by $\mathfrak{m}$ its maximal ideal and let F be the quotient field of R.

A **valuation** of F is a function $\nu : F \longrightarrow \bar{\mathbb{R}} = \mathbb{R} \cup \{\infty\}$ such that for all $f, g \in F$:

(V_1) $\nu(f+g) \geq \inf(\nu(f), \nu(g))$,

(V_2) $\nu(f \cdot g) = \nu(f) + \nu(g)$.

It is an easy exercise to check that if ν is not constant, then axiom (V_2) implies

(V_3) $\nu(1) = 0$.

Let Γ be a totally ordered abelian group. A **Krull valuation** of F is a function

$$\nu : F \longrightarrow \Gamma \cup \{\infty\}$$

satisfying (V_1), (V_2) and (V_3) such that $\nu^{-1}(\infty) = 0$.

If ν is a valuation or a Krull valuation of F, we say that ν is **centered** at R if ν is non-negative on R and strictly positive on $\mathfrak{m}$. We say that ν is **proper** if $\nu(F \setminus \{0\}) \neq \{0\}$ and $\nu(\mathfrak{m}) \neq \{\infty\}$.

If ν and ν' are two valuations of F, then we say that ν and ν' are **equivalent**, and write $\nu \sim \nu'$, if there exists a non-zero real number c such that for all $f \in F$ we have $\nu(f) = c\nu'(f)$.

Let $\mathcal{V} = \{\nu \mid \nu \text{ proper valuation centered at } R\}/\sim$. When working with an element of $\mathcal{V}$, we will tacitly fix a valuation representing it, so in practice we will work with valuations instead of classes of valuations. We will consider only normalized valuations, in the sense that $\nu(\mathfrak{m}) := \inf\{\nu(f) \mid f \in \mathfrak{m}\} = 1$. Indeed, we can represent any element ν of $\mathcal{V}$ by a uniquely determined normalized valuation after multiplying all the values by $\frac{1}{\nu(\mathfrak{m})}$.

For an element $\nu \in \mathcal{V}$ we will denote by Γ_ν the *augmented* value group of ν, that is, $\Gamma_\nu := \nu(F) \subset \bar{\mathbb{R}}$.

If ν is a valuation (resp. a Krull valuation), the set

$$R_\nu := \{f \in F \mid \nu(f) \geq 0\};$$

is a local ring called the **valuation ring associated to** ν, with maximal ideal

$$\mathfrak{m}_\nu := \{f \in F \mid \nu(f) > 0\}.$$

The **rank** of ν, denoted by $\mathrm{rk}(\nu)$, is the Krull dimension of R_ν. In our situation $\mathrm{rk}(\nu)$ is at most 2 by Abhyankar's inequality.

Remark 2.1. We have $\mathrm{rk}(\nu) = 1$ if and only if $\nu(F \setminus \{0\}) \subset \mathbb{R}$ (resp. the group $\nu(F \setminus \{0\})$ can be embedded into the additive group $\mathbb{R}$ of real numbers).

If S is a ring contained in R_ν, the **center** of ν in S is the prime ideal $\mathfrak{m} := \mathfrak{m}_\nu \cap S$. In this situation we also say that ν **is centered at** $\mathfrak{m}$. If $(S, \mathfrak{m})$ is a local domain, we will sometimes say that ν **is centered at** $(S, \mathfrak{m})$.

Definition 2.2. For two local rings $(R_1, \mathfrak{m}_1)$ and $(R_2, \mathfrak{m}_2)$, we say that R_2 **dominates** R_1 if $R_1 \subset R_2$ and $\mathfrak{m}_1 = R_1 \cap \mathfrak{m}_2$. If, in addition, R_1 and R_2 are domains with the same field of fractions, we will say that R_2 **birationally dominates** R_1

Notation. In the above situation we will write $(R_1, \mathfrak{m}_1) < (R_2, \mathfrak{m}_2)$ or simply $R_1 < R_2$.

Remark 2.3. Let $(S, \mathfrak{m})$ be a local domain, contained in F. A valuation ν of F is centered at $\mathfrak{m}$ if and only if we have $(S, \mathfrak{m}) < (R_\nu, \mathfrak{m}_\nu)$.

Remark 2.4.

(1) The valuation ν is uniquely determined by its valuation ring R_ν. For a proof, see [16], Proposition 1.4.

(2) Consider a local domain $(R, \mathfrak{m})$ with field of fractions F. The following conditions are equivalent:

   (a) $(R, \mathfrak{m})$ is of the form $(R_\nu, \mathfrak{m}_\nu)$ for some valuation ν of F

   (b) for every $f \in F$ either $f \in R_\nu$ or $f^{-1} \in R_\nu$ (or both)

   (c) the ring $(R, \mathfrak{m})$ is maximal with respect to the relation of birational domination.

For a proof, see [16], Proposition 1.4 and [2], Chap. 6, §2, n^{0}2, Théorème 1, page 85.

Below we reprove the equivalence (a)$\Longleftrightarrow$(c) in the special case when R is a 2-dimensional regular local ring.

We will use the following partial order on the set of valuations of the field F, centered at R: For two valuations μ and μ' centered at R, we will say that $\mu \leq \mu'$ if $\mu(f) \leq \mu'(f)$ for all $f \in R$.

We denote by $\tilde{\nu}_{\mathfrak{m}}$ the multiplicity valuation, that is, $\tilde{\nu}_{\mathfrak{m}}(f) := \max\{i \mid f \in \mathfrak{m}^i\}$ for all $f \in R$. We note that $\tilde{\nu}_{\mathfrak{m}}$ is the smallest element of $\mathcal{V}$. We say that the **multiplicity** of f at $\mathfrak{m}$ is $\tilde{\nu}_{\mathfrak{m}}(f)$.

If ν is a valuation centered at R then ν determines a Krull valuation ν', centered at R. Furthermore, $\mathrm{rk}(\nu') = 2$ if and only if $\nu^{-1}(\infty) \neq \{0\}$. Indeed, if $\nu^{-1}(\infty) = \{0\}$, then $\nu = \nu'$ is a Krull valuation of rank 1. Otherwise, if $\nu^{-1}(\infty) \neq \{0\}$, then $\nu^{-1}(\infty)$ is a principal prime ideal of R generated by an irreducible element $f \in R$. For each $g \in R - \{0\}$, write $g = f^s h$, where f does not divide h (that is, $\nu(h) < \infty$). Define $\nu'(g) = (s, \nu(h)) \in \mathbb{Z} \oplus \mathbb{R}$. For $G = \frac{g_1}{g_2} \in F - \{0\}$, put $\nu(G) = \nu(g_1) - \nu(g_2)$. It is clear that ν' determines a Krull valuation on F, centered at R.

Conversely, if ν' is a Krull valuation of F centered at R then ν' determines a valuation on R. Indeed, let $\mathfrak{m}_{\nu'}$ be the maximal ideal of $R_{\nu'}$. If $\mathrm{rk}(\nu') = 1$, put $\nu = \nu'$. Otherwise, let Γ_1 be the isolated subgroup of Γ of rank 1 (that is, the smallest non-zero isolated subgroup of Γ). Let P' be the prime ideal of $R_{\nu'}$ associated to Γ_1:

$$P' = \{f \in R_{\nu'} \mid \nu(f) \in \Gamma \setminus \Gamma_1\}.$$

Let $P = P' \cap R$. For each $f \in R$, if $f \in P$, put $\nu(f) = \infty$. Otherwise, put

$$\nu(f) = \nu'(f) \in \Gamma_1.$$

Then ν is a valuation centered at R.

For an element $\beta \in \Gamma_\nu$, let $\mathbf{P}_\beta := \{f \in F \mid \nu(f) \geq \beta\}$, $\mathbf{P}_{\beta+} := \{f \in F \mid \nu(f) > \beta\}$. Let $\mathrm{gr}_\nu F = \bigoplus_{\beta \in \Gamma_1} \frac{\mathbf{P}_\beta}{\mathbf{P}_{\beta+}}$. For an element $\beta \in \Gamma_1$ and an element $f \in F$ such that $\nu(f) = \beta$, we will denote by $\mathrm{in}_\nu f$ the natural image of f in $\frac{\mathbf{P}_\beta}{\mathbf{P}_{\beta+}}$.

3. Valuations and blowing ups

The aim of this section is to describe a natural order-preserving bijection between valuations of F centered at R and valuations of $k(x, y)$ centered at $k[x, y]_{(x,y)}$.

Remark 3.1. Throughout the paper we will commit the following abuse of notation. We will use the letters x, y to denote both the generators of the field $k(x, y)$ over k and a regular system of parameters of R. Since in each case we will specify clearly with which ring we are working, this should cause no confusion.

A **simple** sequence π^* of local point blowings up of *Spec* R is a sequence of the form

$$(R, \mathfrak{m}) \xrightarrow{\pi_1} (R_1, \mathfrak{m}_1) \xrightarrow{\pi_2} \dots \xrightarrow{\pi_i} (R_i, \mathfrak{m}_i) \xrightarrow{\pi_{i+1}} \dots \tag{3.1}$$

where π_i is given by considering the blowing up *Spec* $R_i \xrightarrow{\pi_i^*}$ *Spec* R_{i-1} along $\mathfrak{m}_{i-1}$, picking a point $\xi_i \in \pi_i^{*-1}(\mathfrak{m}_{i-1})$ and putting $R_i := \mathcal{O}_{X_i,\xi_i}$. Let $\Pi(R)$ denote the set of all the simple sequences (finite or infinite) of local point blowings up of *Spec* R. Fix an element $f \in R \setminus \{0\}$. Let μ denote the multiplicity of f at $\mathfrak{m}$. Assume that

$$f \notin (x, y^{\mu+1}) . \tag{3.2}$$

Definition 3.2. A monomial ideal in a regular local ring A, with regular system of parameters $(u_1, \dots, u_s)$ is an ideal in A generated by monomials in $(u_1, \dots, u_s)$.

Let $I(x, y, f)$ denote the smallest *monomial* ideal containing f. (3.2) is equivalent to saying that $y^\mu \in I(x, y, f)$.

Let $e(x, y, f) := \min\left\{ \frac{\alpha}{\mu-\beta} \;\middle|\; x^\alpha y^\beta \in I(x, y, f), \beta < \mu \right\} \in \frac{1}{\mu!}\mathbb{Z} \cup \{\infty\}$, where we adopt the convention that the minimum of the empty set is infinity.

Definition 3.3. The **first characteristic exponent** of f at $\mathfrak{m}$ is the supremum of $e(x, y, f)$, where (x, y) runs over all the regular systems of parameters of R satisfying (3.2).

Fix a real number e. For a real number ξ, let I_ξ denote the monomial ideal of R generated by all the monomials $x^\alpha y^\beta$ such that $\alpha + e\beta \geq \xi$.

Definition 3.4. The **monomial** valuation $\nu_{x,y,e}$ of R, associated to the data (x, y, e) is the valuation defned by $\nu_{x,y,e}(g) = \max\{\xi \in \mathbb{R} \mid g \in I_\xi\}$.

Let R^* denote the set of units in R.

Proposition 3.5. *Let* e_0 *be the first characteristic exponent of* f *at* $\mathfrak{m}$. *Let* $e = e(x, y, f)$.

The following conditions are equivalent:

1. $e < e_0$,
2. e *is an integer, and there exists a regular system of parameters of the form* $(y - ux^e, x)$, *with* u *a unit of* R, *satisfying* $e(x, y, f) < e(x, y - ux^e, f)$.

3. *e is an integer and* $\text{in}_{\nu_{x,y,e}} f$ *is the μ-th power of a linear form in* $\text{in}_{\nu_{x,y,e}} y$ *and* $\text{in}_{\nu_{x,y,e}} x^e$*; more specifically, there exist $c, d \in k$ such that*

$$\text{in}_{\nu_{x,y,e}} f = (\text{in}_{\nu_{x,y,e}} y - c\ \text{in}_{\nu_{x,y,e}} x^e)^\mu. \tag{3.3}$$

Proof. 1) $\Longrightarrow$ 2) Since $e_0 > e$, there exists a change of coordinates

$$x_1 = a_1 x + a_2 y^{l_1}, \tag{3.4}$$

$$y_1 = b_1 y + b_2 x^{l_2} \tag{3.5}$$

with $a_1, a_2, b_1, b_2 \in R^*$ such that

$$e_1 := e(x_1, y_1, f) > e(x, y, f). \tag{3.6}$$

Replacing x by $a_1^{-1}(x_1 - a_2 y^{l_1})$ does not change e, therefore, we may assume that $x_1 = x$. Since $b_1 \in R^*$, we may assume that $b_1 = 1$. Now we will prove that $e = l_2$.

Write

$$f = \sum_{i+je_1 \geq \mu e_1} a_{ij} x^i y_1^j = \sum_{i+je_1 \geq \mu e_1} a_{ij} x^i \left(y + b_2 x^{l_2}\right)^j, \tag{3.7}$$

where $a_{0\mu} \in R^*$. Consider a monomial of the form $a_{ij} x^i y_1^j$ with

$$i + je_1 \geq \mu e_1. \tag{3.8}$$

The element $x^i y_1^j$ belongs to the monomial ideal of R generated by the set

$$\left\{x^{i+sl_2} y^{j-s} \mid s \in \{0, \ldots, j\}\right\}.$$

Let

$$e' = \min\{e_1, l_2\}. \tag{3.9}$$

Let us prove that $e = e'$. Indeed, if i, j satislfy (3.8) then

$$i \geq (\mu - j)e_1. \tag{3.10}$$

If $s \in \{0, \ldots, j\}$ then, since $l_2 \geq e'$, we obtain

$$i + sl_2 \geq (\mu - j)e_1 + sl_2 \geq (\mu - (j - s))e'. \tag{3.11}$$

Thus $e' \leq e$. Combined with (3.6) and (3.9), this proves that

$$e' = l_2 < e_1. \tag{3.12}$$

Combining (3.12) with (3.11), we obtain

$$i + sl_2 \geq (\mu - (j - s))l_2 \tag{3.13}$$

and the inequality is strict unless $i = 0$ and $j = \mu$. Thus

$$\begin{aligned} e &= \min\left\{ \frac{\alpha}{\mu-\beta} \;\middle|\; x^\alpha y^\beta \in I(x,y,f), \beta < \mu \right\} \\ &= \min\left\{ \frac{s l_2}{\mu-(\mu-s)} \;\middle|\; s \in \{1,\dots\mu\} \right\} = l_2. \end{aligned}$$

Therefore, $e \in \mathbb{N}$ and $y_1 = y + ux^e$ with $u \in R^*$ satisfies the conclusion of (2).

2) $\Longrightarrow$ 3) Let $e' = e(x, y - ux^e, f)$ and write $f = (y - ux^e)^\mu + \sum\limits_{\substack{i+je' \geq \mu e' \\ (i,j) \neq (0,\mu)}} a_{ij}x^i(y - ux^e)^j$. To prove 3) it is sufficient to prove that $\nu_{x,y,e}\left(\sum\limits_{\substack{i+je' \geq \mu e' \\ (i,j) \neq (0,\mu)}} a_{ij}x^i(y-ux^e)^j\right) > \mu e$. Now, $\sum\limits_{\substack{i+je' \geq \mu e' \\ (i,j) \neq (0,\mu)}} a_{ij}x^i(y-ux^e)^j$ is contained in the monomial ideal (with respect to (x,y)), generated by monomials of the form $\binom{j}{s}a_{ij}x^iy^{j-s}(ux^e)^s$ with $0 \leq s \leq j$, $i + e'j \geq \mu e'$ and if $j = \mu$ then $i > 0$. Now we have to prove that the quantity $q = i + se + (j - s)e$ is strictly greater than μe. We have $q = i + je$. If $j = \mu$, then $i > 0$ and $q > \mu e$. If $j > \mu$ then $q > \mu e$. If $j < \mu$, then $q = i + je \geq e'(\mu - j) + je = \mu e - \mu e + e'(\mu - j) + je = \mu e + (\mu - j)(e' - e) > \mu e$. This completes the proof of (3).

3) $\Longrightarrow$ 1) Choose $u \in R^*$ such that the natural image of u in k is c. We have

$$f = (y - ux^e)^\mu + \sum_{i,j} a_{ij}x^iy^j,$$

with $\nu_{x,y,e}\left(\sum a_{ij}x^iy^j\right) > \mu e$, that is $i + je > \mu e$ for all the (i,j) appearing in the sum.

Put $y_1 = y - ux^e$. We will prove that $e' = e(x, y_1, f) > e$.

We have $f = y_1^\mu + \sum\limits_{i,j}\sum\limits_{s=0}^{j}\binom{j}{s}a_{ij}x^{i+es}y^{j-s}$ with $i + je > \mu e$ for each (i,j) in the sum.

Now we have $e' \geq \frac{i+es}{\mu-(j-s)} > \frac{(\mu-j)e+es}{\mu-(j-s)} = e$ whenever $(j - s) < \mu$. □

Remark 3.6. Let e denote the first characteristic exponent of f. If R is quasi-excellent, f is reduced and $\mu \geq 2$, we have

$$e \in \frac{1}{\mu!}\mathbb{N}, \tag{3.14}$$

that is, $1 \leq e < \infty$. Since in this paper we work with arbitrary regular two-dimensional local rings and not just the quasi-excellent ones, we will not use this fact in the sequel.

Fix a simple sequence of point blowings up as in (3.1). Let $\mu_i(f)$ and $e_i(f)$ denote, respectively, the multiplicity and the first characteristic exponent of the strict transform of f in R_i.

Lemma 3.7. *At least one of the following conditions holds:*

(1) $(\mu_{i+1}, e_{i+1}) <_{lex} (\mu_i, e_i)$

(2) $e_i(f) = \infty$.

Proof. To simplify the notation, we will consider the case when $i = 0$, so that $R_0 = R$. Assume that $e_i(f) \neq \infty$. Let f_1 be the strict transform of f in R_1. We will follow the notation of ([20], Appendix 5, pages 365–367). Namely, let $\bar{g}$ denote the directional form of the local blowing up π_1 and $\bar{f}$ the natural image of f in $\mathrm{gr}_{\mathfrak{m}} R$.

Let $\mu = \mu_0$ denote the multiplicity of f at $\mathfrak{m}$.

Since $\bar{f}$ is a homogeneous polynomial of degree μ, the greatest power of $\bar{g}$ that could divide $\bar{f}$ is $\bar{g}^\mu$. If $\bar{g}^\mu$ does not divide $\bar{f}$, then by ([20], Appendix 5, page 367, Proposition 2), we have $\mu_1 < \mu$ and (1) of the Lemma holds.

Assume that $\bar{g}^\mu \mid \bar{f}$. Then $\deg \bar{g} = 1$ and there exists a regular system of parameters (x, y) such that $\bar{g} = \bar{y}$ and

$$f = y^\mu + \sum_{i+j>\mu} a_{ij} x^i y^j,$$

where the a_{ij} are units of R. Let $e = e_0$ denote the first characteristic exponent of f and choose (x, y) in such a way that $e = e(x, y, f)$. Write $f = y^\mu + \sum\limits_{\substack{i+je=\mu e \\ (i,j)\neq(0,\mu)}} a_{ij} x^i y^j + \sum\limits_{i+je>\mu e} a_{ij} x^i y^j$. Since $e < \infty$, there exists $(i, j) \neq (0, \mu)$, with $i + je = \mu e$.

Now $f_1 = y_1^\mu + \sum\limits_{\substack{i+je=\mu e \\ (i,j)\neq(0,\mu)}} a_{ij} x_1^{i+j-\mu} y_1^j + \sum\limits_{i+je>\mu e} a_{ij} x_1^{i+j-\mu} y_1^j.$

Note that for each (i, j) with $i + je = \mu e$ we have $(i + j - \mu) + j(e-1) = \mu(e-1)$ and for each (i, j) with $i + je > \mu e$ we have $(i + j - \mu) + j(e - 1) > \mu(e - 1)$.

If $s > e - 1$ then for (i, j) with $i + je = \mu e$ we have $(i + j - \mu) = (\mu - j)(e - 1) < (\mu - j)s$, hence $(i + j - \mu) + js < \mu s$.

If $e - 1 < 1$ then $\mu_1 < \mu$. Otherwise, if $e - 1 \geq 1$, the above considerations prove that $e(x_1, y_1, f_1) = e - 1$.

By Proposition 3.5, $\mathrm{in}_{\nu_{x,y,e}} f$ is not a μ-th power of a linear form in $\mathrm{in}_{\nu_{x,y,e}} y$ and $\mathrm{in}_{\nu_{x,y,e}} x^e$. Hence $\mathrm{in}_{\nu_{x_1,y_1,e-1}} f_1$ is not a μ-th power of a linear form in $\mathrm{in}_{\nu_{x_1,y_1,e-1}} y_1$ and $\mathrm{in}_{\nu_{x_1,y_1,e-1}} x_1^e$. Therefore $e_1 = e(x_1, y_1, f_1) = e - 1 < e$. In all the cases (1) of the Lemma holds. □

Lemma 3.8. *Let π^* be an infinite sequence of local blowings up belonging to $\Pi(R)$. Write π^* as in (3.1). Take an element $f \in R \setminus \{0\}$.*

(1) *If f is a unit in R_{j_0} for some j_0, then f is a unit in R_j for all $j \in \mathbb{N}$.*

(2) *At least one of the following conditions holds:*

  (a) *there exists $i \in \mathbb{N}$ such that $f = x_i^s y_i^t u$ where x_i and y_i are regular parameters of R_i, s and t are natural numbers and u is a unit of R_i*

  (b) *there exists $i_0 \in \mathbb{N}$ such that $e_i(f) = \infty$ for all $i \geq i_0$.*

Proof. 1. This follows directly from the fact that for all $j \in \mathbb{N}$, we have $\mathfrak{m}_j = R_j \cap \mathfrak{m}_{j+1}$.

2. First note that if x_i and y_i are regular parameters of R_i, then either $x_i = x_{i+1} y_{i+1}$ and $y_i = y_{i+1}$ or $x_i = x_{i+1}$ and $y_i = \frac{y_i}{x_i} x_{i+1}$, with $\frac{y_i}{x_i}$ is either a unit in R_{i+1} or equal to y_{i+1}.

Assume that condition (b) does not hold. From Lemma 3.7 we deduce that for each $j' \in \mathbb{N}$ there exists $j > j'$ with $\mu_j < \mu_{j'}$. Hence there exists $j \in \mathbb{N}$ with $f_j \in R^*$. Now by definition of f_j, $f_{j-1} = x^{\mu_{j-1}} f_j$, therefore, using the paragraph above and induction, we get the result. □

For an element $\pi^* \in \Pi(R)$ we denote $\bar{R} = \bigcup_i R_i$. The ring $\bar{R}$ is an integral domain with quotient field F, dominating R and R_i for each $i \in \mathbb{N}$.

Proposition 3.9. *Assume that there exists $f \in R$, satisfying condition (2)(b) in Lemma 3.8. Then there exists a unique Krull valuation ν on F such that R_ν dominates $\bar{R}$.*

Moreover, we have:

1. $\mathrm{rk}\ \nu = 2$.

2. *The set of elements in R satisfying condition (2)(b) in Lemma 3.8 is a prime ideal generated by an irreducible element g.*
3. *Let g_i denote the strict transform of g in R_i. Then ν is the composition of the g-adic valuation of F with the unique rank one Krull valuation, centered in the one-dimensional local rings $\frac{R_i}{(f_i)}$ for each $i \in \mathbb{N}$.*

Proof. Since $\bar{R}$ is an integral domain with field of fractions F, there exists a valuation ring R_ν dominating $\bar{R}$. Let Γ denote the value group of this valuation. Now to prove the uniqueness of ν it is sufficient to prove the conditions (1), (2) and (3).

1. Let f_i denote the strict transform of f in R_i. We have $\mu_i = \mu_{i_0} > 0$ for all $i \geq i_0$. Let (x_{i_0}, y_{i_0}) be a regular system of parameters of R_{i_0} such that $f_{i_0} = y_{i_0}^{\mu_{i_0}} +$ terms of higher degree. Since $\mu_i = \mu_{i_0}$, we have $x_{i_0} = x_{i_0+1}$. Now $f_{i_0+1} = y_{i_0}^{\mu_{i_0}} +$ terms of higher or equal degree. Repeating the same reasoning, we see that for each $i \geq i_0$ we have $x_i = x_{i_0}$ and $f_i = x_{i+1}^{\mu_{i+1}} f_{i+1} = x_{i_0}^{\mu_{i_0}} f_{i+1}$. Thus $f_{i_0} = x_{i_0}^{(i-i_0)\mu_{i_0}} f_i$ for each $i > i_0$. Since $f_i \in R_\nu$ for all i, we have $\nu(f_i) > 0$ for all i. Hence $\nu(f_{i_0}) > i\mu_{i_0}\nu(x_{i_0})$ for all i, so $\nu(f_{i_0})$ cannot belong to a subgroup of Γ of rank 1. Therefore $\mathrm{rk}(\nu) = 2$.
2. Let Γ_1 denote the isolated subgroup of Γ of rank 1 (that is, the unique proper non-trivial subgroup of Γ). Let P' be the prime ideal of R_ν associated to Γ_1. Let $P = P' \cap R$. Then P is a prime ideal of height 1 in R, therefore it is generated by an irreducible element g. Now $f = hg^n$ with $h \notin P$, hence $\nu(h) \in \Gamma_1$. Therefore, by the proof of (1), there exists i such that h_i the strict transform of h in R_i is a unit. Now the strict transform of f in R_i is $h_i g_i^n$. Therefore g must also satisfy condition (2)(b) in Lemma 3.8. An element of R satisfies condition (2)(b) in Lemma 3.8 if and only if it belongs to P.
3. This is a direct consequence of (1) and (2).

□

Proposition 3.10. *Let π^* be an element of $\Pi(R)$ and write π^* as in (3.1). Suppose that R does not contain an element f satisfying condition (2) (b) of Lemma 3.8. The following statements hold.*

(1) *The ring $\bar{R}$ is a valuation ring with field of fractions F, dominating R and R_i for each $i \in \mathbb{N}$.*
(2) *Conversely, let R_μ be a valuation ring with field of fractions F, dominating R and R_i for each $i \in \mathbb{N}$. Then $R_\mu = \bar{R}$.*

In other words, the simple blowing up sequence π^* and the valuation μ determine each other uniquely; they are equivalent sets of data.

Proof. (1) Since $R < R_i < R_j$ for all natural numbers $i \leq j$, $\bar{R}$ is a domain with quotient field F, dominating R and R_i for each $i \in \mathbb{N}$. First, consider the case when the sequence

$$\pi^* : (R, \mathfrak{m}) \xrightarrow{\pi_1} (R_1, \mathfrak{m}_1) \xrightarrow{\pi_2} \dots \xrightarrow{\pi_n} (R_n, \mathfrak{m}_n)$$

is finite. Then $\bar{R} = R_n$. By definition, $\mathfrak{m}_n$ is principal and $(R_n, \mathfrak{m}_n)$ is a discrete valuation ring.

Next, assume that π^* is infinite. To prove that $\bar{R}$ is a valuation ring, consider an element $f \in F^*$, and write $f = \frac{f_1}{f_2}$ where f_1, $f_2 \in R \setminus \{0\}$.

By Lemma 3.8 there exists $i \in \mathbb{N}$ such that $f_1 = x_i^{s_1} y_i^{t_1} u_1$ and $f_2 = x_i^{s_2} y_i^{t_2} u_2$, where x_i and y_i are local parameters in R_i, s_1, t_1, s_2 and t_2 are natural numbers and u_1, u_2 are units in R_i. Hence

$$f = x_i^s y_i^t u, \tag{3.15}$$

where s and t are integers (not necessarily positive) and u is a unit in R_i. If both s and t are non-negative then $f \in R_i \subset \bar{R}$, as desired. If both s and t are non-postive then $\frac{1}{f} \in R_i \subset \bar{R}$, as desired. Otherwise assume, without loss of generality, that $s > 0$ and $t < 0$. Now after another blowing up, we have the following three possibilities:

$$f = x_{i+1}^s y_{i+1}^{s+t} u \quad \text{or} \tag{3.16}$$

$$f = x_{i+1}^{s+t} y_{i+1}^t u \quad \text{or} \tag{3.17}$$

$$f = y_{i+1}^{s+t} v \tag{3.18}$$

where x_{i+1} and y_{i+1} are local parameters in R_{i+1} and, in the last equation, v is a unit in R_{i+1}. If (3.18) holds and $s + t \geq 0$ then $f \in R_{i+1} \subset \bar{R}$. If (3.18) holds and $s + t \leq 0$ then $f^{-1} \in R_{i+1} \subset \bar{R}$. According to Remark 2.4 (2), $\bar{R}$ is a valuation ring. Finally, if either (3.16) or (3.17) holds, we notice that the blowing up π_{i+1} has strictly decreased the quantity $|s| + |t|$. Since this quantity cannot decrease indefinitely, after finitely many steps we will arrive either at (3.15) with s and t of the same sign or at (3.18), thus reducing the problem to one of the previous cases. Note also that if f is of the form (3.15) with s and t of the same sign then the blowing up π_{i+1} brings f to the form (3.18). This completes the proof of (1).

(2) Conversely, let R_μ be a valuation ring such that $R_i < R_\mu$ for all $i \in \mathbb{N}$. Taking the direct limit as i tends to infinity, we obtain $\bar{R} < R_\mu$. Now

part (2) follows from (1) and the implication (a)$\Longrightarrow$(c) of Remark 2.4 (2). However, we give below a direct proof of (2) for the sake of completeness.

If the sequence π^* is finite, its last ring R_n is a discrete valuation ring. Let x_n be a local parameter of R_n. We have $\mu(x_n) > 0$ since μ is centered at $\mathfrak{m}_n$. Now any element f of F^* can be written as $f = x_n^s u$ where $s \in \mathbb{Z}$ and u is a unit of R_n (hence also a unit of R_μ) and therefore $f \in R_n$ if and only if $f \in R_\mu$.

If π^* is infinite, let $f \in F^*$. As in the proof of part (1), there exists i such that $f = x_i^s u$ or $f = y_i^s u$ with x_i and y_i local parameters in R_i, $s \in \mathbb{Z}$ and u is a unit in R_i. Now since π^* is infinite, we have $\nu(x_i) > 0$ and $\nu(y_i) > 0$ and u is also a unit of R_μ since $R_i < R_\mu$. Hence to say that $f \in \bar{R}$ is equivalent to saying that $s \geq 0$, which is equivalent to saying that $\mu(f) \geq 0$. Therefore $\bar{R} = R_\mu$. □

Corollary 3.11. *The set of valuations of F centered at R is in a natural one-to-one correspondence with* $\Pi(R)$.

Proof. For the sake of completeness, we now give an explicit description of the element of $\Pi(R)$, associated to a given valuation μ by the above bijection and vice versa.

Let μ be a valuation centered at R. The center of μ in R is $\xi_0 := \mathfrak{m}$. Consider the point blowing up $\pi_1^* : X_1 \longrightarrow$ *Spec* R along ξ_0. The center of μ in X_1 is the unique point $\xi_1 \in X_1$ whose local ring $\mathcal{O}_{X_1,\xi_1}$ is dominated by R_μ. Put $R_1 := \mathcal{O}_{X_1,\xi_1}$, and let $\mathfrak{m}_1 := \mathfrak{m}_{X_1,\xi_1}$ be its maximal ideal. If $\mathfrak{m}_1$ is principal, stop here. Otherwise, fix $x_1, y_1 \in R_1$ such that $\mathfrak{m}_1 = (x_1, y_1)$. We have $(R, \mathfrak{m}) < (R_1, \mathfrak{m}_1) < (R_\mu, \mathfrak{m}_\mu)$. Now repeat the same procedure with $(R, \mathfrak{m})$ replaced by $(R_1, \mathfrak{m}_1)$. Continuing in this way we obtain the simple sequence $\pi^*(\mu)$ (finite or infinite) of local point blowings up of *Spec* R.

Now we have two cases:

Case 1: The ring R does not contain an element f satisfying condition (2) (b) in Lemma 3.8.

Letting $\bar{R} = \lim\limits_{i\to\infty} R_i$, we have $\bar{R} = R_\mu$.

Conversely, take any element $\pi^* \in \Pi(R)$ and let $\bar{R}$ be as in Proposition 3.10. Let μ to be the valuation on F with valuation ring $\bar{R}$. It is clear that $\pi^*(\mu)$ described above is equal to π^*.

Case 2: The ring R does contain an element f satisfying condition (2) (b) in Lemma 3.8 then by Proposition 3.9 the valuation ν is uniquely determined by π^*. □

Recall that k denotes the residue field of R.

Theorem 3.12. *There is a natural order preserving bijection between valuations of F centered at R and valuations of $k(x,y)$ centered at $k[x,y]_{(x,y)}$.*

Proof. By Proposition 3.10, the set of valuations of F centered at R is in a natural one-to-one correspondence with $\Pi(R)$. Also by Proposition 3.10 applied to $k[x,y]_{(x,y)}$, the set of valuations of $k(x,y)$ centered at $k[x,y]_{(x,y)}$ is in a natural one-to-one correspondence with $\Pi(k[x,y]_{(x,y)})$. Finally, there is a natural one-to-one correspondence between $\Pi(R)$ and $\Pi(k[x,y]_{(x,y)})$. Clearly all those correspondences fit together to give a natural order preserving bijection between valuations of F centered on R and valuations of $k(x,y)$ centered on $k[x,y]_{(x,y)}$. □

4. A Complete Set of Key Polynomials

Let $\tilde{\nu} \in \mathcal{V}$. Fix local coordinates x and y such that $\nu(x) = 1$. Let $K = k(x)$. Let ν be the valuation of $k(x,y)$ corresponding to $\tilde{\nu}$ under the bijection of Theorem 3.12.

The goal of this section is to construct a set of polynomials, complete for ν (the definition is given below). This set will be our main tool for constructing the valuative tree.

4.1. *Definition and Basic Properties of Key Polynomials*

For each strictly positive integer b, we write $\partial_b := \frac{\partial^b}{b!\partial y^b}$, the b-th formal derivative with respect to y.

For each polynomial $P \in K[y]$, let $\epsilon_\nu(P) := \max\limits_{b\in\mathbb{N}} \left\{\frac{\nu(P)-\nu(\partial_b P)}{b}\right\}$,

$$I_\nu(P) := \left\{ b \in \mathbb{N} \;\middle|\; \frac{\nu(P) - \nu(\partial_b P)}{b} = \epsilon_\nu(P) \right\}.$$

and $b_\nu(P) := \min I_\nu(P)$.

Definition 4.1. Let Q be a monic polynomial in $K[y]$, with $\nu(Q) \geq \nu(y)$. We say that Q is an **abstract key polynomial** for ν if for each polynomial f satisfying

$$\epsilon_\nu(f) \geq \epsilon_\nu(Q),$$

we have $\deg(f) \geq \deg(Q)$.

For the rest of the paper, we will say **key polynomial** for **abstract key polynomial**.

For a monic polynomial Q in $K[y]$ and a $g \in K[y]$ we can write g in a unique way as

$$g = \sum_{j=0}^{s} g_j Q^j, \tag{4.1}$$

with all the $g_j \in K[y]$ of degree strictly less than $\deg(Q)$.

Definition 4.2. For every monic polynomial Q and every polynomial g in $K[y]$, we call the expression (4.1) the **Q-expansion** of g. We define $\nu_Q(g) := \min_{0 \leq j \leq s} \nu(g_j Q^j)$. We call ν_Q the **truncation** of ν with respect to Q.

Proposition 4.3 (Proposition 12 of [3]). *If Q is a key polynomial for ν then ν_Q is a valuation.*

The following proposition is a direct consequence of Proposition 19 [3] that states that each key polynomial for ν is ν-irreducible.

Proposition 4.4. *If Q is a key polynomial for ν then Q is irreducible.*

Proposition 4.5. *Every monic linear polynomial Q in $K[y]$ is a key polynomial for ν.*

Proof. For any monic linear polynomial $Q \in K[y]$ and for any $c \in K$, we have

$$\epsilon_\nu(Q) = \nu(Q) > -\infty = \epsilon_\nu(c).$$

□

The first part of the next proposition is Theorem 27 of [3] and the second part is obvious.

Proposition 4.6.

(1) *Let μ be a valuation of $K(y)$ such that $\mu < \nu$, and let Q be a monic polynomial of minimal degree in y such that $\mu(Q) < \nu(Q)$. Then Q is a key polynomial for ν.*
(2) *Furthermore, we have $\mu < \nu_Q \leq \nu$ and the value group Γ_Q of ν_Q is equal to $\Gamma_\mu + \beta\mathbb{Z}$ where Γ_μ is the value group of μ and $\beta = \nu(Q)$.*

Let μ be a valuation of $K(y)$ such that $\mu < \nu$, and let Q be a monic polynomial of minimal degree in y such that $\mu(Q) < \nu(Q)$. Let $\beta \in \bar{\mathbb{R}}$, with $\mu(Q) < \beta$.

We define a new valuation μ' in the following way:

For a polynomial $g \in K[y]$, let $g = \sum\limits_{j=0}^{s} g_j Q^j$ be the Q-expansion of g. Put

$$\mu'(g) := \min_{0 \leq j \leq s} \{\mu(g_j) + j\beta\}.$$

We call μ' the **augmented valuation** constructed from μ, Q, and β, and we denote it by $[\mu,\ Q,\ \beta]$.

For further details on augmented valuations, see [17].

4.2. *A Complete Set of Key Polynomials: the Definition*

Let $\beta_0 = \nu(x) = 1$ and $\beta_1 = \nu(y)$. Let $\Gamma_\nu = \nu(F) \subset \bar{\mathbb{R}}$ denote the augmented value group of ν.

For an element $\beta \in \Gamma_\nu$, let $\mathbf{P}_\beta$ be as defined at the end of §2, but with F replaced by $K(y)$:

$$\mathbf{P}_\beta = \{f \in K(y) \mid \nu(f) \geq \beta\}.$$

Definition 4.7. A **complete set of key polynomials** for ν is a set

$$\mathbf{Q} = \{Q_i\}_{i \in I}$$

where I is a well ordered set, each Q_i is a key polynomial in $K[y]$ for ν, and for each $\beta \in \Gamma_\nu$ the additive group $\mathbf{P}_\beta \cap K[y]$ is generated by products of the form $a \prod\limits_{j=1}^{s} Q_{i_j}^{\gamma_j}$, $a \in K$, such that $\sum\limits_{j=1}^{s} \gamma_j \nu(Q_{i_j}) + \nu(a) \geq \beta$.

In [6] it is proved that every valuation ν admits a complete set $\mathbf{Q} = \{Q_i\}_{i \in I}$ of key polynomials.

Remark 4.8. If $\mathbf{Q} = \{Q_i\}_{i \in I}$ is a complete set of key polynomials for ν, we will always assume that the well ordering of I has the following property: for $i < j$ in I, we have $\nu(Q_i) < \nu(Q_j)$.

4.3. *Basic Structure*

Let μ be a valuation of $K(y)$ with $\mu < \nu$. Suppose that the subset $\Gamma_{\mu+}$ of positive values of $\Gamma_\mu = \mu(K(y))$ is a well ordered set (with the standard order relation in $\bar{\mathbb{R}}$). Note that this assumption is equivalent to saying that $\Gamma_\mu \cong \mathbb{Z}$.

We will use the following notation:

1. Let $d_\mu(\nu)$ be the minimal degree of a monic polynomial f in $K[y]$ satisfying $\mu(f) < \nu(f)$.
2. Put
$$\Phi_\mu(\nu) := \left\{Q \in K[y] \mid Q \text{ is monic, } \deg_y(Q) = d_\mu(\nu),\ \mu(Q) < \nu(Q)\right\}.$$
3. Put
$$\begin{aligned}\Psi_\mu(\nu) &:= \nu\left(\Phi_\mu(\nu)\right)\\ &= \left\{\nu(Q) \in K[y] \mid Q \text{ is monic, } \deg_y(Q) = d_\mu(\nu),\ \mu(Q) < \nu(Q)\right\}.\end{aligned}$$

Proposition 4.9. *The set $\Psi_\mu(\nu)$ is contained in $\Gamma_{\mu+}$ or in $\Gamma_{\mu+} \cup \{\alpha\}$, where α is a maximal element of $\Psi_\mu(\nu)$ if it exists. Moreover, $\Psi_\mu(\nu)$ is bounded below by $d\mu(y) \geq 0$.*

Proof. Let $d := d_\mu(\nu)$. First we will show that $\Psi_\mu(\nu)$ is bounded below by $d \cdot \mu(y)$. Let $\beta \in \Psi_\mu(\nu)$ and choose $Q \in \Phi_\mu(\nu)$ such that $\nu(Q) = \beta$. Suppose that $\beta < d\mu(y)$ and write $Q = y^d + g$, with $g \in K[y]$, $\deg_y(g) < d$. Since $\mu\left(y^d\right) = d\mu(y) > \beta = \nu(Q) > \mu(Q)$, we have $\mu(Q) = \mu(g)$. Since $\nu(y^d) \geq \mu(y^d) = d\mu(y) > \beta = \nu(Q)$, we have $\nu(Q) = \nu(g)$, but $\nu(g) = \mu(g)$ by definition of d, therefore $\nu(Q) = \mu(Q)$, which is a contradiction.

Now we will prove that any element $\beta \in \Psi_\mu(\nu)$ which is not a maximal element must be in Γ_μ.

Suppose that β and α are two elements of $\Psi_\mu(\nu)$ such that $\beta < \alpha$. Choose Q and Q' in $\Phi_\mu(\nu)$ such that $\nu(Q) = \beta$ and $\nu(Q') = \alpha$. Write $Q' = Q + z$ with $z \in K[y]$, $\deg_y(z) < d$. Since $\alpha > \beta$, we have $\nu(Q) = \nu(z)$. But $\nu(z) = \mu(z)$ by definition of $d_\mu(\nu)$. Hence $\beta \in \Gamma_\mu$. □

From Proposition 4.9 we see that $\Psi_\mu(\nu)$ is well ordered. We will denote by $\beta_\mu(\nu)$ the smallest element of $\Psi_\mu(\nu)$.

Choose $Q \in \Phi_\mu(\nu)$ such that $\nu(Q) = \beta_\mu(\nu)$. By Proposition 4.6, Q is a key polynomial for ν, the truncation ν_Q is a valuation with $\mu < \nu_Q \leq \nu$, and the augmented value group Γ_Q of ν_Q is $\Gamma_\mu + \beta_\mu(\nu)\mathbb{Z}$. Hence the set Γ_{Q+} of positive values of Γ_Q is a well ordered set.

If $\nu_Q < \nu$, we can repeat the same process as above with μ replaced by ν_Q.

Moreover, the valuation ν_Q does not depend on the choice of Q, as we will prove in the following proposition.

Proposition 4.10. *With the notation as above, if Q' is another polynomial in $\Phi_\mu(\nu)$ such that $\nu(Q') = \beta_\mu(\nu)$ then $\nu_Q = \nu_{Q'}$.*

Proof. Let f be a polynomial of minimal degree such that $\nu_Q(f) \neq \nu_{Q'}(f)$ and suppose that $\nu_Q(f) < \nu_{Q'}(f)$. Clearly $\deg_y(f) \geq d_\mu(\nu)$. Let $f = a_sQ^s + \cdots + a_0$ be the Q-expansion of f and let

$$g = a_{s-1}Q^{s-1} + \cdots + a_0.$$

By definition of ν_Q we have $\nu(a_iQ^i) \geq \nu_Q(f)$ for each $0 \leq i \leq s$.

Suppose first that $\nu(a_sQ^s) > \nu_Q(f)$. Then $\nu_Q(f) = \nu_Q(g)$.

Since $\deg_y(g) < \deg_y(f)$, we have $\nu_Q(g) = \nu_{Q'}(g)$. Therefore

$$\nu_{Q'}(a_sQ^s) = \nu(a_sQ^s) > \nu_Q(g) = \nu_{Q'}(g).$$

This implies that $\nu_{Q'}(f) = \nu_{Q'}(g)$, that leads to $\nu_{Q'}(f) = \nu_Q(f)$ which is a contradiction. We have proved that $\nu_Q(f) = \nu(a_sQ^s)$.

Write $Q' = Q + z$, with $\deg_y(z) < d_\mu(\nu)$. Then $f = a_s(Q' - z)^s + \cdots + a_0$, and the Q'-expansion of f involves $a_sQ'^s$. Therefore we have $\nu(a_sQ'^s) = \nu_{Q'}(a_sQ'^s) \geq \nu_{Q'}(f) > \nu_Q(f) = \nu(a_sQ'^s)$, which is a contradiction. □

Lemma 4.11. *Let $f \in K[y]$ be such that*

$$\nu_Q(f) = \mu(f) \tag{4.2}$$

and let $f = qQ + r$ be the Euclidean division of f by Q in $K[y]$. Then $\nu_Q(f) = \nu_Q(r) < \nu_Q(qQ)$.

Proof. By definition of ν_Q, we have

$$\nu_Q(qQ) \geq \nu_Q(f). \tag{4.3}$$

Suppose we have equality, aiming for contradiction:

$$\nu_Q(qQ) = \nu_Q(f). \tag{4.4}$$

Then

$$\nu_Q(r) \geq \nu_Q(f). \tag{4.5}$$

By definition of ν_Q, we have

$$\mu(r) = \nu_Q(r). \tag{4.6}$$

Combining (4.2), (4.5) and (4.6), we obtain $\mu(f) \leq \mu(r)$. Then

$$\mu(f) \leq \mu(qQ). \tag{4.7}$$

We have

$$\nu_Q(qQ) > \mu(qQ). \tag{4.8}$$

Combining (4.2), (4.4), (4.7) and (4.8), we get $\nu_Q(qQ) > \mu(qQ) \geq \mu(f) = \nu_Q(f) = \nu_Q(qQ)$, which is a contradiction. □

Proposition 4.12 (Theorem 9.4 [10], Theorem 1.11 [17]). *Let Q' be a monic polynomial of minimal degree among those satisfying*

$$\nu_Q(Q') < \nu(Q').$$

Then the Q-expansion of Q' is given by $Q' = Q^s + a_{s-1}Q^{s-1} \cdots + a_0$ with

$$\nu_Q(Q') = s\beta_\mu(\nu) = \nu(a_0).$$

Proof. First, let $Q' = qQ + a_0$ be the Euclidean division of Q' by Q. We have

$$\nu(Q') > \nu_Q(Q') = \inf\{\nu_Q(qQ), \nu(a_0)\}$$

by definition of ν_Q. Hence $\nu(qQ) = \nu(a_0) = \nu_Q(Q')$.

Since $\deg_y(a_s) < d_\mu(\nu)$ and Q is irreducible in $K[y]$ by Proposition 4.4, there exist g and h in $K[y]$ with $deg_y(g) < d_\mu(\nu)$ and $ga_s + hQ = 1$. Now $ga_s = -hQ + 1$, $\nu_Q(a_s) = \mu(a_s)$ and $\nu_Q(g) = \mu(g)$, therefore by Lemma 4.11 we have $\nu(ga_s) = \nu(1) < \nu(-hQ)$.

Now for each j, $0 \leq j \leq s-1$, let $ga_j = q_jQ + r_j$ be the Euclidean division of ga_j by Q in $K[y]$. Since $\nu_Q(a_j) = \mu(a_j)$ and $\nu_Q(g) = \mu(g)$, by Lemma 4.11 we have $\nu(ga_j) = \nu(r_j) < \nu(q_jQ)$.

Consider the polynomial $Q'' = Q^s + r_{s-1}Q^{s-1} + \cdots + r_0$. We have $Q'' - gQ' = (r_s - ga_s)Q^s + (r_{s-1} - ga_{s-1})Q^{s-1} + \cdots + r_0 - ga_0$, with $r_s = 1$. Therefore

$$\begin{aligned}\nu_Q(Q'' - gQ') &\geq \inf_{0\leq j\leq s}\{\nu(r_j - ga_j) + j\beta_\mu(\nu)\}\\ &> \inf_{0\leq j\leq s}\{\nu(r_j) + j\beta_\mu(\nu)\} = \nu_Q(Q'')\end{aligned}$$

and $\nu_Q(Q'' - gQ') > \nu_Q(Q'') = \nu_Q(gQ')$.

If $\nu(Q'') = \nu_Q(Q'')$ then $\nu(Q''-gQ') \geq \nu_Q(Q''-gQ') > \nu_Q(Q'') = \nu(Q'')$ and we have $\nu(Q'') = \nu(gQ') > \nu_Q(gQ') = \nu_Q(Q'')$, which is impossible. Hence $\nu(Q'') > \nu_Q(Q'')$.

Since Q' is chosen of minimal degree we must have $\deg_y(Q'') \geq \deg_y(Q')$, but this implies that $\deg_y a_s = 0$ and $a_s = 1$.

We still have to prove that $\nu_Q(Q') = \nu(Q^s)$. By definition of ν_Q we have $\nu(Q^s) \geq \nu_Q(Q')$.

Suppose that $\nu(Q^s) > \nu_Q(Q')$, then $\nu_Q(Q') = \nu_Q(f)$, where $f = Q' - Q^s$. But since $\deg_y(f) < \deg_y(Q')$, we have $\nu_Q(f) = \nu(f)$. We obtain

$$\nu(Q^s) = \nu_Q(Q^s) > \nu_Q(Q') = \nu_Q(f) = \nu(f).$$

This implies that $\nu(Q') = \nu(f)$, which leads to $\nu(Q') = \nu_Q(Q')$, a contradiction. □

Corollary 4.13. *If* $Q' \in K[y]$ *is monic with* $\deg_y(Q') = d_\mu(\nu)$ *and* $\nu(Q') > \nu_Q(Q')$ *then* $\nu_Q(Q') = \beta_\mu(\nu)$.

Proof. This is a special case of Proposition 4.12 when $\deg_y(Q') = d_\mu(\nu)$. □

Proposition 4.14. *Let* $\{Q_i\}_{i\in I}$ *be a set of key polynomials for* ν, *with* I *a well ordered set and* $\deg_y(Q_i) \leq \deg_y(Q_j)$ *for* $i < j$ *in* I. *Let* $\nu_i := \nu_{Q_i}$ *be the truncation associated with each* Q_i. *Then the set* $\{Q_i\}_{i\in I}$ *is a complete set of key polynomials if and only if for each polynomial* $f \in K[y]$ *there exists an element* $i \in I(\nu)$ *such that* $\nu(f) = \nu_i(f)$ *and* $\deg_y Q_i \leq \deg_y f$.

Proof. To say that $\{Q_i\}_{i\in I(\nu)}$ is a complete set of key polynomials is equivalent to saying that every $f \in K[x]$ can be written in the form

$$f = \sum_\gamma a_\gamma \prod_{j=1}^{s} Q_j^{\gamma_j}, \tag{4.9}$$

where s is a strictly positive integer, $\gamma = (\gamma_1, \ldots, \gamma_s)$ ranges over a finite subset of $\mathbb{N}^s$, $a_\gamma \in K$ and

$$\sum_{j=1}^{s} \gamma_j \nu(Q_j) + \nu(a) \geq \nu(f).$$

Proof of "if". Assume that for each $f \in K[x]$ there exists an integer i as in the Proposition. We will construct the expression (4.9) recursively in $\deg_y f$. Assume that an expression of the form (4.9) exists for every polynomial of degree strictly less than $\deg_y f$. Put $\beta = \nu(f)$ and let i be such that $\beta = \nu_i(f)$ and $\deg_y Q_i \leq \deg_y f$. Write

$$f = \sum_{j=0}^{s_i} c_j Q_i^j, \tag{4.10}$$

where each $c_j Q_i^j \in P_\beta$ and $\deg_y c_j < \deg_y Q_i \leq \deg_y f$. By the induction assumption, each of the c_j admits an expansion of the form (4.9). Substituting all of these expansions into (4.10), we obtain the desired expansion

(4.9) of f. This completes the proof of "if".

Proof of "only if". Conversely, take $f \in K[y]$. Let $\beta = \nu(f)$. Write f in the form (4.9). Then

$$\beta = \nu(f) \geq \nu_s(f) \geq \min\left\{\nu_s\left(c_\gamma \prod Q_j^{\gamma_j}\right)\right\} = \min\left\{\nu\left(c_\gamma \prod Q_j^{\gamma_j}\right)\right\} \geq \beta.$$

Thus all the inequalities in the above formula are equalities, so the natural number $i := s$ satisfies the conclusion of "only if". □

4.4. *Construction of a Complete Set of Key Polynomials*

First we put $Q_1 := y$ and $d_1(\nu) = 1$. By Proposition 4.5, Q_1 is a key polynomial for ν. Consider the valuation $\nu_1 := \nu_{Q_1}$. We have $\nu_1 \leq \nu$. If $\nu_1 = \nu$ then the algorithm stops here, we put $I(\nu) = \{1\}$ and $\{Q_i\}_{i\in I(\nu)} = \{Q_1\}$. We will prove in Proposition 4.16 below that $\{Q_i\}_{i\in I(\nu)}$ is complete for ν.

Now suppose that $\nu_1 < \nu$. Then we can apply the results of §4.3 to $\mu = \nu_1$. Put $d_2(\nu) = d_{\nu_1}(\nu)$, $\Phi_1(\nu) = \Phi_{\nu_1}(\nu)$, $\Psi_1(\nu) = \Psi_{\nu_1}(\nu)$ and $\beta_2(\nu) = \beta_{\nu_1}(\nu)$.

Choose $Q_2 \in \Phi_1(\nu)$ such that $\nu(Q_2) = \beta_2(\nu)$ and let $\nu_2 := \nu_{Q_2}$.

We have $\nu_1 < \nu_2 \leq \nu$. If $\nu_2 = \nu$, then the algorithm stops here, we put $I(\nu) = \{1, 2\}$ and $\{Q_i\}_{i\in I(\nu)} = \{Q_1, Q_2\}$. By Proposition 4.16 below, $\{Q_i\}_{i\in I(\nu)}$ is complete for ν.

Otherwise, if $\nu_2 < \nu$, we can apply the results of §4.3 to $\mu = \nu_1$ and repeat the same process with ν_1 replaced by ν_2.

Assume that for a certain natural number $n \geq 2$ a set $\{Q_i\}_{i\geq n}$ has been constructed. If $\nu_n = \nu$ then, by Proposition 4.14 and Proposition 4.16 below, $\{Q_i\}_{i\geq n}$ is complete for ν. The construction stops here.

Otherwise, we have $\nu_n < \nu$. Let us apply the results of §4.3 to $\mu = \nu_n$. Put

$$d_{n+1}(\nu) = d_{\nu_n}(\nu),$$

$\Phi_n(\nu) = \Phi_{\nu_n}(\nu)$, $\Psi_n(\nu) = \Psi_{\nu_n}(\nu)$ and $\beta_{n+1}(\nu) = \beta_{\nu_n}(\nu)$.

Choose $Q_{n+1} \in \Phi_n(\nu)$ such that $\nu(Q_{n+1}) = \beta_{n+1}(\nu)$ and let $\nu_{n+1} := \nu_{Q_{n+1}}$.

Repeating this process, there are two possibilities. The first is that we find valuations $\{\nu_i\}_{i\leq n}$ and key polynomials $\{Q_i\}_{i\leq n}$ such that $\nu_n = \nu$.

The second is that we construct an infinite set $\{Q_i\}_{i\in\mathbb{N}}$ of key polynomials and valuations $\{\nu_i\}_{i\in\mathbb{N}}$. We will study this case after Proposition 4.16.

Remark 4.15. Let $\{Q_i\}_{i\in I(\nu)}$ be the set constructed above, with $I(\nu) = \{1, 2, \dots\}$ (possibly infinite). Even though the polynomials $\{Q_i\}_{i\in I(\nu)}$ are not uniquely determined, their degrees $\{d_i(\nu)\}_{i\in I(\nu)}$, their values $\{\beta_i(\nu)\}_{i\in I(\nu)}$, the associated valuations $\{\nu_i\}_{i\in I(\nu)}$ are uniquely determined by ν, from the construction above and by Proposition 4.10. As well, the sets $\Phi_i(\nu)$ and $\Psi_i(\nu)$, for $i \in I(\nu)$, are uniquely determined by ν by construction.

Proposition 4.16. *Let $\{Q_i\}_{i\in I(\nu)}$ be the set constructed above, with $I(\nu)$ possibly infinite. Let $f \in K[y]$ and suppose there exists $i_0 \in I(\nu)$ such that* $\deg_y(f) < \deg_y(Q_{i_0})$. *There exists $i < i_0$ such that* $\deg_y(Q_i) \le \deg_y(f)$ *and $\nu_i(f) = \nu(f)$.*

Proof. Multiplying f by a non-zero element of K does not change the problem, therefore we may assume that f is monic.

Now let i be such that

$$(d_i(\nu), \beta_i(\nu)) = \max\{j \in I(\nu) \mid (d_j(\nu), \beta_j(\nu)) \le_{lex} (\deg_y(f), \nu(f))\}.$$

Now by construction, if $i + 1$ does not exist, then $\nu(f) = \nu_i(f)$. Otherwise, if $i + 1 \in I(\nu)$, then $(d_{i+1}(\nu), \beta_{i+1}(\nu)) > (\deg_y(f), \nu(f))$, and by definition of Q_{i+1} we have $\nu(f) = \nu_i(f)$. □

Suppose that $\{Q_i\}_{i\in I(\nu)}$ is constructed as above with $I(\nu) = \mathbb{N}$. Two cases are possible.

Case 1: The set $\Phi_i(\nu)$ is finite for each $i \in I(\nu)$.
Case 2: There exists $i \in I(\nu)$ such that $\Phi_i(\nu)$ is infinite.

Proposition 4.17. *If $\beta_{i+1}(\nu)$ is a maximal element of $\Psi_i(\nu)$ then $\Psi_i(\nu) = \{\beta_{i+1}(\nu)\}$ and*

$$\Psi_i(\nu) \cap \Psi_{i+1}(\nu) = \emptyset. \tag{4.11}$$

Moreover, if $\Psi_{i+1}(\nu) \neq \emptyset$ then $d_{i+2}(\nu) > d_{i+1}(\nu)$ and $\beta_{i+2}(\nu) > \beta_{i+1}(\nu)$. If $\beta_{i+1}(\nu)$ is not maximal in $\Psi_i(\nu)$, then $\Psi_{i+1}(\nu) = \Psi_i(\nu) \setminus \{\beta_{i+1}(\nu)\}$ and $d_{i+2}(\nu) = d_{i+1}(\nu)$.

Proof. First suppose that $\beta_{i+1}(\nu)$ is maximal in $\Psi_i(\nu)$. Since by definition $\beta_{i+1}(\nu)$ is the minimal element of $\Psi_i(\nu)$ and $\Psi_i(\nu)$ is totally ordered, we have $\Psi_i(\nu) = \{\beta_{i+1}\}$.

Now if $\Psi_{i+1}(\nu) = \emptyset$, the equality (4.11) holds trivially. Thus we will assume that

$$\Psi_{i+1}(\nu) \neq \emptyset.$$

We will first prove that $d_{i+2}(\nu) > d_{i+1}(\nu)$. By construction, we have $d_{i+2}(\nu) \geq d_{i+1}(\nu)$. Aiming for contradiction, suppose that we have equality. Take an element $Q \in \Phi_{i+1}(\nu)$. We have $\deg_y(Q) = d_{i+2}(\nu) = d_{i+1}(\nu)$ and $\nu(Q) > \nu_{i+1}(Q) \geq \nu_i(Q)$, therefore $Q \in \Phi_i(\nu)$, by definition of $\Phi_i(\nu)$. Hence $\nu(Q) \in \Psi_i(\nu) = \{\beta_{i+1}(\nu)\}$. On the other hand, Q satisfies the hypothesis on Q' in Corollary 4.13. Therefore

$$\nu_{i+1}(Q) = \beta_{i+1}(\nu).$$

This implies that $\nu(Q) = \nu_{i+1}(Q)$ which contradicts the fact that $Q \in \Phi_{i+1}(\nu)$.

We have proved that $d_{i+1}(\nu) < d_{i+2}(\nu)$. Now the fact that $\beta_{i+2}(\nu) > \beta_{i+1}(\nu)$ follows from Proposition 4.12, since $\beta_{i+2}(\nu) > \nu_{i+1}(Q_{i+2}) = s\beta_{i+1}$, where $s = \deg_{Q_i} Q_{i+1}$.

Since $\beta_{i+2}(\nu) > \beta_{i+1}(\nu)$, we have $\Psi_i(\nu) \cap \Psi_{i+1}(\nu) = \emptyset$.

Next, suppose that $\beta_{i+1}(\nu)$ is not a maximal element in $\Psi_i(\nu)$. Let $\beta \in \Psi_i(\nu) \setminus \{\beta_{i+1}(\nu)\}$. Choose $Q \in \Phi_i(\nu)$ such that $\nu(Q) = \beta$ and write $Q = Q_{i+1} + z$. Since $\beta > \beta_{i+1}(\nu)$, we have $\nu(z) = \beta_{i+1}(\nu)$ and $\nu_{i+1}(Q) = \beta_{i+1}(\nu) < \beta = \nu(Q)$. We have proved that $Q \in \Phi_{i+1}(\nu)$, therefore $d_{i+2}(\nu) = d_{i+1}(\nu)$ and $\beta \in \Psi_{i+1}(\nu)$. Thus $\Psi_i(\nu) \setminus \{\beta_{i+1}(\nu)\} \subset \Psi_{i+1}(\nu)$.

Now let $\beta \in \Psi_{i+1}(\nu)$. By Corollary 4.13, we have $\beta > \beta_{i+1}(\nu)$. Take an element

$$Q \in \Phi_{i+1}(\nu)$$

such that $\nu(Q) = \beta$. We have $\deg_y(Q) = d_{i+2}(\nu) = d_{i+1}(\nu)$, and $\nu(Q) > \nu_{i+1}(Q) \geq \nu_i(Q)$, hence $Q \in \Phi_i(\nu)$ and $\beta \in \Psi_i(\nu) \setminus \{\beta_{i+1}(\nu)\}$. □

Corollary 4.18. *If $\Psi_i(\nu)$ is infinite for some $i \in I(\nu)$, then $\Psi_{i+1}(\nu)$ is infinite and*

$$\Psi_{i+1}(\nu) \subset \Psi_i(\nu).$$

Corollary 4.19. *Let $i_0 = \inf\{i \in \mathbb{N} \mid \#\Psi_i(\nu) = \infty\}$, then for each $i \geq i_0$, the value group Γ_i of ν_i is equal to the value group Γ_{i_0} of ν_{i_0}.*

Proof. This is a direct consequence of Proposition 4.9 (2) and Corollary 4.18. □

First, suppose we are in **Case 1**:

Proposition 4.20. *The degrees $d_i(\nu)$ are unbounded in $\mathbb{N}$.*

Proof. Take an element $i \in I(\nu)$. We will prove that there exists $j \in I(\nu)$ such that

$$d_j(\nu) > d_{i+1}(\nu).$$

Since $\Psi_i(\nu)$ is finite, it admits a maximal element α. If $\beta_{i+1}(\nu) = \alpha$ then, by Proposition 4.17, we have $d_{i+2}(\nu) > d_{i+1}(\nu)$. Suppose that $\beta_{i+1}(\nu)$ is not maximal. By Proposition 4.17 we have $\Psi_{i+1}(\nu) = \Psi_i(\nu) \setminus \{\beta_{i+1}(\nu)\}$, therefore α is also the maximal element of $\Psi_{i+1}(\nu)$. Now repeat the same reasoning: if $\beta_{i+2}(\nu) = \alpha$, then $d_{i+3}(\nu) > d_{i+2}(\nu) = d_{i+1}(\nu)$, otherwise we have $\Psi_{i+2}(\nu) = \Psi_{i+1}(\nu) \setminus \{\beta_{i+2}(\nu)\}$ and α is the maximal element of $\Psi_{i+2}(\nu)$. The process must end since $\Psi_i(\nu)$ is finite. □

Theorem 4.21. *The set of key polynomials $\{Q_i\}_{i\in I(\nu)}$ is complete for ν.*

Proof. Let $f \in K[y]$. By Proposition 4.20, there exists $i \in I(\nu)$ such that $d_i(\nu) > \deg_y(f)$. Now the result follows from Proposition 4.14 and Proposition 4.16. □

Suppose we are in **Case 2**.

Theorem 4.22.

1. *If $\nu^{-1}(\infty) = \{0\}$ then the set $\{Q_i\}_{i\in I(\nu)}$ is complete for ν.*
2. *If $\nu^{-1}(\infty) \neq \{0\}$ then there exists a key polynomial Q_ω for ν, that generates the ideal $\nu^{-1}(\infty)$ and it is of minimal degree such that $\nu_i(Q_\omega) < \nu(Q_\omega)$ for all $i \in \mathbb{N}$. Moreover, the set $\{Q_i\}_{i\in I(\nu)} \cup \{Q_\omega\}$ is complete for ν.*

Proof. Fix a polynomial $f \in K[y]$ such that $\nu(f) < \infty$. We have the following inequalities:

$$\nu_1(f) \leq \cdots \leq \nu_i(f) \leq \cdots \leq \nu(f) < \infty.$$

By Corollary 4.19, all those values belong to the value group Γ_{i_0} of ν_{i_0}, where $i_0 \in I(\nu)$ is defined in Corollary 4.19. Now $\Gamma_{i_0} = \beta_0\mathbb{Z} + \beta_1\mathbb{Z} + \cdots + \beta_{i_0}\mathbb{Z}$ is discrete, hence there exists a certain integer j such that $\nu_i(f) = \nu(f)$ for any $i \geq j \in \mathbb{N}$.

Now 1) follows from Proposition 4.14.

To prove 2), suppose that $\nu^{-1}(\infty) \neq \{0\}$. The set $\nu^{-1}(\infty)$ is an ideal in $K[y]$, it can be generated by one element. Choose Q_ω to be a monic polynomial that generates $\nu^{-1}(\infty)$. The polynomial Q_ω has minimal degree among the polynomials in $\nu^{-1}(\infty)$.

We have $\epsilon_\nu(Q_\omega) = \infty$, and Q_ω of minimal degree with this property, hence Q_ω is a key polynomial for ν.

We have $\nu = \nu_\omega := \nu_{Q_\omega}$, and for any polynomial $f \in K[y]$, if $f \notin \nu^{-1}(\infty)$, by the discussion at the beginning of the proof, there exists $i \in I(\nu)$ such that $\nu_i(f) = \nu(f)$, otherwise, if $f \in \nu^{-1}(\infty)$, then $\nu(f) = \nu_\omega(f) = \infty$. Hence the set $\{Q_i\}_{i\in\mathbb{N}} \cup \{Q_\omega\}$ is complete for ν by Proposition 4.14. □

For the rest of the paper if Q_ω exists, we put $I(\nu) = \mathbb{N} \cup \{\omega\}$.

We denote:

1. $D(\nu) := \max_{i\in I(\nu)}\{d_i(\nu)\}$, if this maximum exists; otherwise, we put $D(\nu) = \infty$.
2. $N(\nu)$ the maximal element of $I(\nu)$, if this maximum exists, otherwise, we put $N(\nu) = \infty$.

Remark 4.23. From the construction above we see that:

1. $N(\nu) = \infty$ if and only if $I(\nu) = \mathbb{N}$.
2. If $D(\nu) = \infty$ then $I(\nu) = \mathbb{N}$.
3. If $D(\nu) < \infty$ and ($N(\nu) = \infty$ or $N(\nu) = \omega$), we are in the case where there exists $i \in I(\nu)$ such that $\#\Psi_i(\nu) = \infty$.

5. The order relation on $\mathcal{V}$

5.1. *Invariants of comparable valuations*

Let $\tilde{\mu}$ and $\tilde{\nu}$ be two elements of $\mathcal{V}$ with $\tilde{\mu} < \tilde{\nu}$. Choose local coordinates x and y such that $\tilde{\nu}(x) = \tilde{\mu}(x) = 1$.

Put $K = k(x)$ and let μ and ν be the valuations of $K(y)$, corresponding to $\tilde{\mu}$ and $\tilde{\nu}$, respectively.

Let $\{Q_i\}_{i\in I(\nu)}$ be a complete set of key polynomials associated to ν.

Lemma 5.1. *There exists $i \in I(\nu)$ such that $\mu(Q_i) < \nu(Q_i)$.*

Proof. Suppose that for all $i \in I(\nu)$ we have $\mu(Q_i) = \nu(Q_i)$.

Since $\nu > \mu$, there exists $f \in K[y]$ such that $\nu(f) > \mu(f)$. Choose $f \in K[y]$ of minimal degree among the polynomials having this property.

Since $\{Q_i\}_{i\in I(\nu)}$ is complete for ν, there exists $i \in I(\nu)$ such that $\nu(f) = \nu_i(f)$.

Let $f = qQ_i + r$ be the Euclidean division of f by Q_i.

We have $\nu(f) > \mu(f) \geq \inf\{\mu(qQ_i), \mu(r)\} = \inf\{\nu(qQ_i), \nu(r)\} \geq \nu_i(f)$, which is a contradiction. $\square$

Let $i_0 := \min\{i \in I(\nu) \mid \mu(Q_i) < \nu(Q_i)\}$.

Proposition 5.2. *If $i_0 = 1$, then $I(\mu) = \{1\}$ and $\mu = \mu_1 < \nu_1$.*

Proof. Since $i_0 = 1$, we have $\mu(y) < \nu(y)$. It is sufficient to prove that $\mu = \mu_1$.

Suppose there exists $f \in K[y]$, with $\mu(f) > \mu_1(f)$. Choose f of minimal degree satisfying $\mu(f) > \mu_1(f)$ and let $f = qy + r$ be the Euclidean division of f by y.

We have $\mu(f) > \mu_1(f) = \inf\{\mu(qy), \mu(r)\}$, therefore, $\mu(f) > \mu(qy) = \mu(r)$. Since $\nu(f) \geq \mu(f)$, we have $\nu(f) > \mu(r) = \nu(r)$. This implies that $\nu(f) > \nu(qy) = \nu(r)$. Finally, we get $\nu(qy) = \nu(r) = \mu(r) = \mu(qy)$. But $\nu(qy) = \nu(q) + \nu(y) > \mu(q) + \mu(y) = \mu(qy)$, and we have a contradiction. $\square$

Proposition 5.3. *If $i_0 > 1$ then for any $i \in I(\nu)$ with $i < i_0$, we have $i \in I(\mu)$, $\mu_i = \mu_{Q_i} = \nu_i$, $\beta_i(\mu) = \beta_i(\nu)$ and $d_i(\mu) = d_i(\nu)$.*

Proof. Since $i_0 > 1$, we have $\mu(y) = \nu(y)$, hence $\mu_1 = \nu_1$, $\beta_1(\mu) = \beta_1(\nu)$ and $d_1(\mu) = d_1(\nu)$.

Take an integer i, $1 < i < i_0$ (in particular $i \in \mathbb{N}$), and suppose inductively that for all j, $1 \leq j < i$, we have $\nu_j = \mu_j$, $\beta_j(\mu) = \beta_j(\nu)$ and $d_j(\mu) = d_j(\nu)$.

We will first prove that $d_i(\mu) = d_i(\nu)$.

If f is a monic polynomial with $\mu(f) > \mu_{i-1}(f) = \nu_{i-1}(f)$, then $\nu(f) \geq \mu(f) > \nu_{i-1}(f)$, hence $d_i(\nu) \leq d_i(\mu)$.

To prove the equality, we will prove that $\mu(Q_i) > \mu_{i-1}(Q_i)$. Indeed, by definition of i_0 we have $\mu(Q_i) = \nu(Q_i)$ since $i < i_0$. Hence $\mu(Q_i) = \nu(Q_i) > \nu_{i-1}(Q_i) = \mu_{i-1}(Q_i)$. Therefore $d_i(\nu) = d_i(\mu)$.

Now to prove that $\beta_i(\mu) = \beta_i(\nu)$, we still have to prove that if f is a monic polynomial with $\mu(f) > \mu_{i-1}(f)$ and $\deg_y(f) = d_i(\mu)$, then $\mu(f) \geq \mu(Q_i)$. In this case, we will have $\beta_i(\mu) = \mu(Q_i)$, and since by definition of i_0, $\mu(Q) = \nu(Q)$, we get the desired equality.

Let f be such a polynomial. Write $f = Q_i + g$ with $\deg_y(g) < d_i(\mu)$.

If $\mu(f) < \mu(Q_i)$ then $\mu(f) = \mu(g) = \mu_{i-1}(g) = \nu_{i-1}(g) = \nu(g)$ and

$$\nu(Q_i) \geq \mu(Q_i) > \mu(f) = \nu(g).$$

Hence $\nu(g) = \nu(f)$ and $\nu(Q_i) > \nu(f)$.

We have proved that $\beta_i(\nu) > \nu(f)$ and $\nu(f) \geq \mu(f) > \mu_{i-1}(f) = \nu_{i-1}(f)$, which contradicts the definition of $\beta_i(\nu)$.

Since Q_i is monic of degree $d_i(\mu)$ with $\mu(Q_i) = \beta_i(\mu)$, we have $\mu_i = \mu_{Q_i}$. Since $\beta_i(\mu) = \beta_i(\nu)$ then $\mu_i = \nu_i$. □

Proposition 5.4. *We have $i_0 < \omega$. In other words, $i_0 \in \mathbb{N}$.*

Proof. Suppose that $\omega \in I(\nu)$ and $i_0 = \omega$.

By Remark 4.23 (3) and Corollary 4.19, there exists $i_1 \in \mathbb{N}$ such that $\beta_i(\nu) \in \Gamma_{i_1}$ for all $i \geq i_1$, with $\Gamma_{i_1} = \mathbb{Z} + \beta_1\mathbb{Z} + \cdots + \beta_{i_1}\mathbb{Z} \subset \mathbb{R}$. Hence for all $i \in \mathbb{N}$, we have $\nu_i(Q_\omega) \in \Gamma_{i_1}$.

Let us show that

$$\nu_1(Q_\omega) < \nu_2(Q_\omega) < \cdots < \nu(Q_\omega) = \infty.$$

Indeed, assume that there exists $i \in I(\nu) \setminus \{\omega\}$, such that $\nu_i(Q_\omega) = \nu_{i+1}(Q_\omega)$, aiming for contradiction. Write $Q_\omega = qQ_{i+1} + r$ the Euclidean division of Q_ω by Q_{i+1}. We have $\nu_{i+1}(qQ_{i+1}) > \nu_i(qQ_{i+1}) \geq \inf\{\nu_i(Q_\omega), \nu_i(r)\} = \inf\{\nu_{i+1}(Q_\omega), \nu_{i+1}(r)\}$, hence

$$\nu_{i+1}(qQ_{i+1}) > \nu_{i+1}(Q_\omega), \nu_{i+1}(r).$$

This implies that $\nu(qQ_{i+1}) > \nu(r)$. Hence $\nu(Q_\omega) = \nu(r)$ and $\nu(r) = \infty$, then r must be equal to 0 and $q = 1$, since $\nu^{-1}(\infty) = (\mathbb{Q}_\omega)$. But $Q_{i+1} \neq Q_\omega$ and we have a contradiction.

By Proposition 5.3 we have $\nu_i = \mu_i$, hence

$$\mu_1(Q_\omega) < \mu_2(Q_\omega) < \cdots < \mu(Q_\omega).$$

We have a strictly increasing sequence in Γ_{i_1}, it most be unbounded in $\mathbb{R}$, hence $\mu(Q_\omega) = \infty$. This contradicts the fact that $\mu(Q_\omega) < \nu(Q_\omega)$. □

Proposition 5.5. *Either $\mu = \nu_{i_0-1}$ or $\mu = \mu_{i_0} < \nu_{i_0}$.*

Proof. As in the proof of Proposition 5.3, we have $d_{i_0}(\nu) \leq d_{i_0}(\mu)$.

Suppose first that $\mu(Q_{i_0}) = \mu_{i_0-1}(Q_{i_0})$. We will prove that in this case we have $\mu = \nu_{i_0-1}$.

Suppose, aiming for contradiction, that there exists $f \in K[y]$ such that $\mu(f) > \mu_{i_0-1}(f)$, and choose f of minimal degree among all the polynomials having this property. Since

$$d_{i_0}(\nu) \leq d_{i_0}(\mu),$$

we have $\deg_y(f) \geq d_{i_0}(\nu)$.

Let $f = qQ_{i_0} + r$ be the Euclidean division of f by Q_{i_0}. By the minimality of deg f, we have $\mu(q) = \mu_{i_0-1}(q)$ and $\mu(r) = \mu_{i_0-1}(r)$.

We have

$$\nu_{i_0-1}(r) = \nu(r) \geq \inf\{\nu(f), \nu(qQ_{i_0})\} > \inf\{\nu_{i_0-1}(f), \nu_{i_0-1}(qQ_{i_0})\}.$$

Hence $\nu_{i_0-1}(r) > \nu_{i_0-1}(f) = \nu_{i_0-1}(qQ_{i_0})$. But $\mu(qQ_{i_0}) = \nu_{i_0-1}(qQ_{i_0})$ and $\mu(f) > \nu_{i_0-1}(f)$, hence $\mu(f) > \mu(qQ_{i_0}) = \mu(r)$, therefore $\nu_{i_0-1}(qQ_{i_0}) = \mu(qQ_{i_0}) = \mu(r) = \nu_{i_0-1}(r)$ which is a contradiction.

Now suppose that $\mu(Q_{i_0}) > \mu_{i_0-1}(Q_{i_0})$. We have $d_{i_0}(\mu) = d_{i_0}(\nu)$. We will prove that $\mu(Q_{i_0}) = \beta_{i_0}(\mu)$. Suppose that there exists a monic polynomial Q such that $\deg_y(Q) = d_{i_0}(\mu)$, $\mu_{i_0-1}(Q) < \mu(Q)$ and $\mu(Q) < \mu(Q_{i_0})$.

Write $Q = Q_{i_0} + g$ with $\deg_y(g) < d_{i_0}(\mu)$. We have $\mu(Q_{i_0}) > \mu(Q) = \mu(g)$. Therefore $\nu(Q_{i_0}) > \mu(Q_{i_0}) = \mu(g) = \nu(g)$. Hence $\nu(Q_{i_0}) > \nu(g) = \nu(Q)$, in particular,

$$\beta_{i_0}(\nu) = \nu(Q_{i_0}) > \nu(Q)$$

which contradicts the definition of $\beta_{i_0}(\nu)$.

We have $\beta_{i_0}(\mu) = \mu(Q_{i_0})$, $d_{i_0}(\mu) = d_{i_0}(\nu)$ and $\mu_{i_0} = \mu_{Q_{i_0}}$. It remains to prove that $\mu = \mu_{i_0}$.

Take any polynomial f in $K[y]$. If $\deg_y(f) < d_{i_0}(\mu)$ then $\mu_{i_0}(f) = \mu(f)$. Suppose that $\deg_y(f) \geq d_{i_0}(\mu)$ and let $f = qQ_{i_0} + r$ be the Euclidean division of f by Q_{i_0}.

If $\mu(f) > \mu_{i_0}(f)$ then $\mu(f) > \mu(qQ_{i_0}) = \mu(r)$. But $\nu(f) \geq \mu(f)$ and $\nu(r) = \mu(r)$, therefore $\nu(f) > \nu(r) = \nu(qQ_{i_0})$. Then $\nu(qQ_{i_0}) = \mu(qQ_{i_0})$ which is impossible, hence $\mu(f) = \mu_{i_0}(f)$. □

Corollary 5.6. *The valuations ν with $N(\nu) = \infty$ or ($N(\nu) \neq \infty$ and $\beta_{N(\nu)}(\nu) = \infty$) are maximal elements of the set of valuations μ of $K(y)$ with $\mu(x) = 1$.*

From the preceding results we also deduce

Remark 5.7.

1. $N(\mu) \leq N(\nu)$ and $D(\mu) \leq D(\nu)$.
2. Either μ is the y-adic valuation with $\mu(y) < \nu(y)$, or there exists $i \in I(\nu)$ such that for each $j \leq i$, $\mu_j = \nu_j$, $I(\mu) = \{1, \ldots, i+1\}$, $\{Q_j\}_{j \in I(\mu)}$ is a complete set of key polynomials for μ and $\mu = [\nu_i, Q_{i+1}, \mu(Q_{i+1})]$.
3. $N(\mu) = N < \infty$ and μ and ν have the same sets of first N key polynomials. More precisely, any set $\{Q_i\}_{i \in \{1,\ldots,N\}}$ of first N key polynomials for μ is also a set of first N key polynomials for ν and vice versa.

5.2. *Structure Theorems*

Theorem 5.8. *Let $\tilde{\mu}$ and $\tilde{\nu}$ be two valuations in $\mathcal{V}$. Then there exists an infimum of $\tilde{\mu}$ and $\tilde{\nu}$ (that is, the greatest element that is less than or equal to $\tilde{\mu}$ and $\tilde{\nu}$) in the poset $\mathcal{V}$.*

Proof. Fix local coordinates x and y such that $\tilde{\mu}(x) = \tilde{\nu}(x) = 1$. Let μ and ν be the corresponding valuations on $k(x, y)$ under the correspondence in Theorem (3.12).

To prove the Theorem, we will prove that the infimum of μ and ν exists.

First we will define a valuation $\mu \wedge \nu$ and then prove that it is the infimum of μ and ν.

Let $\{\nu_i\}_{i \in I(\nu)}$ and $\{\mu_i\}_{i \in I(\mu)}$ be the truncations associated to ν and μ respectively.

Suppose first that for each $i \in I(\nu) \cap I(\mu)$ we have $\nu_i = \mu_i$. If $I(\mu) \subseteq I(\nu)$ then $\mu \leq \nu$ and $\mu \wedge \nu = \mu$, otherwise, if $I(\mu) \subset I(\nu)$ then $\nu < \mu$ and $\mu \wedge \nu = \nu$.

Now suppose that there exists $i \in I(\nu) \cap I(\mu)$ such that $\nu_i \neq \mu_i$. Let

$$i_0 = \inf\{i \in I(\nu) \cap I(\mu) \mid \nu_i \neq \mu_i\}.$$

Suppose first that ν_{i_0} and μ_{i_0} are comparable. Without loss of generality, we may assume that $\mu_{i_0} < \nu_{i_0}$. In this case put $\mu \wedge \nu = \mu_{i_0}$. Clearly $\mu_{i_0} \leq \nu$ and $\mu_{i_0} \leq \nu$.

Let ν' be a valuation of $k(x, y)$ such that $\nu' \leq \mu$ and $\nu' \leq \nu$. Since $\nu' \leq \nu$, we have $\nu'(x) = 1$. We know from Remark 5.7 (1) that $N(\nu') < N(\mu)$ and $N(\nu') < N(\nu)$. Let $\{\nu'_i\}_{i \leq N(\nu')}$ be the truncations associated to ν'. From Remark 5.7 (2) we know that for each $i < N(\nu')$ we have $\nu'_i = \mu_i$ and $\nu'_i = \nu_i$, therefore $N(\nu') \leq i_0$. We have $\nu' = \nu'_{N(\nu')} \leq \mu_{N(\nu')} \leq \mu_{i_0} = \mu \wedge \nu$.

Next, suppose that ν_{i_0} and μ_{i_0} are not comparable. In particular, we have $i_0 > 1$ (since μ_1 and ν_1 are always comparable). Put $\mu \wedge \nu = \mu_{i_0 - 1} = \nu_{i_0 - 1}$. Choose ν' as in the paragraph above and let $\{\nu'_i\}_{i \leq N(\nu')}$ be the truncations associated to ν'. By Remark 5.7 (3) the valuations ν' and ν have the same set of $N(\nu')$ key polynomials, and the valuations ν' and μ have the same set of $N(\nu')$ key polynomials. Therefore if, $N(\nu') = i_0$, we would have $\nu_{i_0} = [\nu_{i_0-1}, Q_{i_0}, \beta_{i_0}(\nu)]$ and $\mu_{i_0} = [\mu_{i_0-1}, Q_{i_0}, \beta_{i_0}(\mu)]$. The latter two valuations are comparable, hence $N(\nu') \leq i_0 - 1$. We have $\nu' = \nu'_{N(\nu')} \leq \mu_{N(\nu')} \leq \mu_{i_0-1} = \mu \wedge \nu$. □

Theorem 5.9. *Let $\tilde{S}$ be a totally ordered convex subset of $\mathcal{V}$. Then $\tilde{S}$ has a majorant in $\mathcal{V}$.*

Remark 5.10. A short proof of a more general version of this result — one for rings of arbitrary dimension — is given in Lemma 3.9 (i) of [13] using elementary properties.

Proof. Since we are searching for a majorant, we may assume that $\tilde{S}$ contains $\tilde{\nu}_{\mathfrak{m}}$. Since $\tilde{S}$ is totally ordered, we can fix local coordinates x and y such that $\tilde{\nu}(y) \geq \tilde{\nu}(x) = 1$ for all $\tilde{\nu} \in \tilde{S}$.

By Theorem 3.12, there exists a totally ordered convex subset S of the set of valuations over $k(x, y)$, satisfying $1 = \nu(x) \leq \nu(y)$ for all $\nu \in S$, and such that S contains $\nu_{\mathfrak{m}}$. Also by Theorem 3.12 the set $\tilde{S}$ has a majorant in $\mathcal{V}$ if and only if the set S has a majorant in the set of valuations over $k(x, y)$, satisfying $1 = \nu(x) \leq \nu(y)$.

By Corollary 5.6, if S contains an element ν with $N(\nu) = \infty$ or it contains an element ν with $\beta_{N(\nu)}(\nu) = \infty$ then S has a maximal element. Suppose that S does not contain a maximal element.

By Remark 5.7 (1), $N(\nu)$ and $D(\nu)$ define increasing functions on S.

We claim that there exists an initial segment $I \subset \mathbb{N}$ and a set of monic polynomials $\{Q_i\}_{i \in I(S)}$ such that for every valuation $\nu \in S$ the set $\{Q_i\}_{i \in I(\nu)}$ is complete for ν (the fact that $I \subset \mathbb{N}$ follows from the fact that S does not contain a maximal element).
Indeed, take $\nu \in S$, $N \in \mathbb{N}$ and let $\{Q_i\}_{i \leq N}$ be a complete set of key polynomials for ν. Let $\nu' \in S$. If $\nu' < \nu$, then by Remark 5.7 (3) the set $\{Q_i\}_{i \leq N(\nu')}$ is a complete set of key polynomials for ν'. Otherwise, if $\nu' > \nu$, then, again by Remark 5.7 (3), we can add to $\{Q_i\}_{i \leq N}$ the key polynomials $\{Q_i\}_{N < i \leq N(\nu)}$ to obtain a complete set of key polynomials for ν'.

Suppose first that $N(\nu)$ is bounded from above. In this case there exists $N \in \mathbb{N}$ with $I(S) = \{1, \ldots, N\}$, and a valuation $\nu \in S$ with $N(\nu) = N$.

The set $\{\beta_N(\nu) \mid \nu \in S,\ N(\nu) = N\}$ is bounded in $\bar{\mathbb{R}}$. Let $\bar{\beta}$ be a majorant for this set in $\bar{\mathbb{R}}$. If $N = 1$, let μ be the y-adic valuation with $\mu(y) = \bar{\beta}$. Otherwise, if $N > 1$, let $\mu = [\nu_{i-1}, Q_N, \bar{\beta}]$. Then μ is a majorant for S.

Now suppose that $N(\nu)$ is unbounded in $\mathbb{N}$, that is, $I(S) = \mathbb{N}$.

We have $D(\nu) < \infty$ for all $\nu \in S$ since S does not contain a maximal element. Consider the set $D(S) = \{D(\nu) \mid \nu \in S\}$. Again, we have two cases, either $D(S)$ has a maximal element D, or it is unbounded in $\mathbb{N}$.

Suppose first that $D(S)$ is unbounded in $\mathbb{N}$. For each $f \in K[y]$, put

$$\mu(f) := \max\{\nu(f) \mid \nu \in S\}.$$

Note first that this maximum is well-defined. Indeed, let $f \in K[y]$. Let

$\nu \in S$ with

$$D(\nu) > \deg_y(f).$$

For every $\nu' \in S$ with $\nu \leq \nu'$ we have $\nu(f) = \nu'(f)$.
It is not difficult to verify that μ is a valuation on $k(x, y)$ and that μ is a majorant for S.

Now suppose that $D(S)$ has a maximal element D. There exists a cofinal sequence $\{\nu_i\}_{i \in I(S)}$ of valuations in S with $\nu_i = [\nu_{i-1},\ Q_i,\ \nu_i(Q_i)]$ for each $i > 1$. Therefore, if we write $\beta_i = \nu_i(Q_i)$, the value group Γ_i of ν_i is $\beta_0\mathbb{Z} + \cdots + \beta_i\mathbb{Z}$, with $\beta_i \in \mathbb{Q}$, by Proposition 4.6.

We claim that for every $f \in K[y]$, if there exists $i \in I(S)$ with $\nu_i(f) = \nu_{i+1}(f)$ then $\nu_j(f) = \nu_i(f)$ for all $j \in I(S)$, $j \geq i$. Indeed, let $i \in I(S)$ be such that $\nu_i(f) = \nu_{i+1}(f)$. By construction, we have

$$\nu_i(Q_{i+1}) < \nu_{i+1}(Q_{i+1}) = \nu_j(Q_{i+1}) \quad \text{for all } j > i.$$

Now let $f = qQ_{i+1} + r$ be the Euclidean division of f by Q_{i+1}. Since $\nu_{i+1}(f) = \nu_i(f)$, we have $\nu_{i+1}(qQ_{i+1}) > \nu_{i+1}(f) = \nu_i(r)$. Now for all $j > i$ we have

$$\nu_j(qQ_{i+1}) \geq \nu_{i+1}(Q_{i+1}) > \nu_i(r) = \nu_j(r).$$

Therefore $\nu_j(f)$ must be equal to $\nu_j(r) = \nu_i(r) = \nu_i(f)$.

If for all $f \in K[y]$ there exists $i \in I(S)$ with $\nu_i(f) = \nu_{i+1}(f)$, we put

$$\mu(f) := \max_{i \in I(S)} \{\nu_i(f)\}. \tag{5.1}$$

Otherwise, if there exists $f \in K[y]$ with

$$\nu_i(f) < \nu_{i+1}(f) \quad \text{for all } i \in I(S), \tag{5.2}$$

take f monic of minimal degree, satisfying (5.2). We have $\deg_y(f) > D$ by definition of the polynomials Q_i and the valuations ν_i. Put $\mu(f) = \infty$. For a polynomials $g \in K[y]$, let $g = qf + r$ be the Euclidean division of g by f, and put $\mu(g) = \max_{i \in I(S)}\{\nu_i(r)\}$. Then μ is a valuation of $k(x, y)$ which is a majorant for S. □

6. Nonmetric Tree Structure on $\mathcal{V}$

We will now define rooted non-metric trees.

Definition 6.1. A rooted non-metric tree is a poset $(\mathcal{T}, \leq)$ such that:

(T1) Every set of the form $I_\tau = \{\sigma \in \mathcal{T} \mid \sigma \leq \tau\}$ is isomorphic (as an ordered set) to a real interval.

(T2) Every totally ordered convex subset of $\mathcal{T}$ is isomorphic to a real interval.

(T3) Every non-empty subset S of $\mathcal{T}$ has an infimum in $\mathcal{T}$.

Let us consider the following special case of the condition (T3):

(T3′) There exists a (unique) smallest element $\tau_0 \in \mathcal{T}$.

Lemma 6.2 (Lemma 3.4 [13]). *Under the conditions* (T1) *and* (T3′), *the following conditions are equivalent:*

(T3) *Every non-empty subset* $S \subset \mathcal{T}$ *has an infimum.*

(T3″) *Given two elements* $\tau, \sigma \in \mathcal{T}$, *the set* $\{\tau, \sigma\}$ *has an infimum* $\tau \wedge \sigma$.

Definition 6.3. A rooted nonmetric tree $\mathcal{T}$ is complete if every increasing sequence $\{\tau_i\}_{i\geq 1}$ in $\mathcal{T}$ has a majorant, that is, an element τ_∞, with $\tau_i \leq \tau_\infty$ for every i.

Theorem 6.4. *The valuation space* $\mathcal{V}$ *is a complete nonmetric tree rooted at* $\tilde{\nu}_{\mathfrak{m}}$.

Proof.

(T3′) It is clear that $(\mathcal{V}, \leq)$ is a partially ordered set with unique minimal element $\tilde{\nu}_{\mathfrak{m}}$.

(T1) Fix $\tilde{\nu}$ in $\mathcal{V}$, with $\tilde{\nu} > \nu_{\mathfrak{m}}$. We will show that the set $S = \{\tilde{\mu} \in \mathcal{V} \mid \nu_{\mathfrak{m}} \leq \tilde{\mu} \leq \tilde{\nu}\}$ is a totally ordered set isomorphic to an interval in $\bar{\mathbb{R}}_+$.
Choose local coordinates x and y such that $1 = \tilde{\nu}(x) \leq \tilde{\nu}(y)$.
Let ν be the valuation of $k(x, y)$ corresponding to $\tilde{\nu}$ and let $\{Q_i\}_{i\in I(\nu)}$ be a complete sequence of key polynomials for ν. The sequence $\frac{\beta_i(\nu)}{d_i(\nu)}$ is strictly increasing. If $I(\nu)$ has a maximal element ℓ, put $I = \left[1, \frac{\beta_\ell(\nu)}{d_\ell(\nu)}\right] \subset \bar{\mathbb{R}}$. Otherwise, put $I = [1, \infty) \subset \bar{\mathbb{R}}$. We will prove that S is isomorphic to I as an ordered set.
To each $t \in I$ we will associate a valuation $\tilde{\nu}_t$ in S.
Let $t \in I$. If $t = 1$, put $\tilde{\nu}_t = \tilde{\nu}_{\mathfrak{m}}$. If $I(\nu)$ has a maximal element ℓ and $t = \frac{\beta_\ell(\nu)}{d_\ell(\nu)}$, put $\tilde{\nu}_t = \tilde{\nu}$.
Now suppose that $1 < t < \frac{\beta_\ell(\nu)}{d_\ell(\nu)}$. There exists a unique element $u \in I(\nu)$ such that $\frac{\beta_{u-1}(\nu)}{d_{u-1}(\nu)} < t \leq \frac{\beta_u(\nu)}{d_u(\nu)}$. Let $\nu_t := [\mu_{u-1}, Q_u, td_u]$ and $\tilde{\nu}_t$ the corresponding valuation in $\mathcal{V}$. That the resulting map is a bijection follows from Remark 5.7.

(T2) By Theorem 5.9, every totally ordered convex subset $\tilde{S}$ of $\mathcal{T}$ has a majorant in $\mathcal{T}$. With (T3′) and (T1) this proves (T2). This also proves that $\mathcal{T}$ is complete.

(T3) is an immediate consequence of Theorem 5.8 and Lemma 6.2.

□

Remark 6.5. Let $(R, \mathfrak{m}, k)$ and $(R', \mathfrak{m}', k')$ be two regular two-dimensional local rings such that the residue fields k and k' have the same cardinality. Let $\iota : k \cong k'$ be a bijection between the two fields (as sets, that is, ι need not be a homomorphism of fields). Using the results of this paper it can be shown that ι induces a homeomorphism of the respective valuative trees, associated to R and R'. Thus, up to homeomorphism, a valuative tree associated to a regular local ring $(R, ,k)$ depends only on the cardinality of the residue field k.

References

[1] S. Abhyankar, *On the valuations centered in a local domain*, Amer. J. Math. 78 (1956), 321–348.

[2] N. Bourbaki, *Algèbre Commutative*, Chapitres 5 à 7, Masson, 1985.

[3] J. Decaup, M. Spivakovsky, W. Mahboub, *Abstract key polynomials and comparison theorems with the key polynomials of Mac Lane – Vaquie.* arXiv: 1611.06392v1 [math.AC].

[4] C. Favre and M. Jonsson, *The Valuative Tree*, Lec. Notes in Math. 1853. Springer-Verlag, 2004.

[5] A. Granja , *The valuative tree of a two-dimensional regular local ring*, Math. Res. Lett. 14 (2004), 19–34.

[6] F. J. Herrera Govantes, M. A. Olalla Acosta, M. Spivakovsky, *Valuations in algebraic field extensions*, Journal of Algebra 312 (2007), no. 2, 1033–1074.

[7] F. J. Herrera Govantes, W. Mahboub, M. A. Olalla Acosta, M. Spivakovsky, *Key polynomials for simple extensions of valued fields.* arXiv:1406.0657v3 [math.AG].

[8] O. Kashcheyeva, *Constructing examples of semigroups of valuations*, Journal of Pure and Applied Algebra 220 (12) (2016), 3826–3860.

[9] S. MacLane, *A construction for prime ideals as absolute values of an algebraic field*, Duke Math. J. 2 (1936), 492–510.

[10] S. MacLane, *A construction for absolute values in polynomial rings*, Trans. Amer. Math. Soc. 40 (1936), 363–395.

[11] S. MacLane and O.F.G Schilling, *Zero-dimensional branches of rank one on algebraic varieties*, Ann. of Math. 40 (1939), 507–520.

[12] W. Mahboub, *Key Polynomials*, Journal of Pure and Applied Algebra 217 (6) (2013), 989–1006.

[13] J. Novacoski, *Valuations centered at a two-dimensional regular local ring: Infima and Topologies*, Valuation theory in interaction, 389–403, EMS Ser. Congr. Rep., Eur. Math. Soc., Zürich, 2014.

[14] J. Novacoski and M. Spivakovsky, *Key polynoimals and pseudo-convergent sequences*, Journal of Algebra 495 (2018), 199–219.

[15] M. Spivakovsky, *Valuations in Function Fields of Surfaces*, Amer. J. Math. 112 (1) (1990), 107–156.
[16] M. Vaquié, *Valuations and local uniformization*, Advanced Studies in Pure Mathematics 43, 2006, Singularity Theory and Its Applications, 477–527.
[17] M. Vaquié, *Extension d'une valuation*, Trans. Amer. Math. Soc. 359 (2007), no. 7, 3439–3481.
[18] M. Vaquié, *Famille admise associée à une valuation de K[x]*, Singularités Franco-Japonaises, 391–428, Sémin. Congr., 10, Soc. Math. France, Paris, 2005.
[19] M. Vaquié, *Famille admissible de valuations et défaut d'une extension*, Journal of Algebra 311 (2007), no. 2, 859–876.
[20] O. Zariski, and P. Samuel, Commutative Algebra, Vol. II, New York/Heidelberg/Berlin, 1960.

The higher topological complexity of configuration spaces of odd-dimensional spheres

Nguyen Viet Dung

Institute of Mathematics,
Vietnam Academy of Science and Technology
E-mail: vietdung@math.ac.vn

Nguyen Van Ninh

Department of Mathematics,
Thai Nguyen University of Education
E-mail: nguyenninhsp@gmail.com

For each $n = 2, 3, \ldots$, the higher topological complexity $TC_n(X)$ of a topological space X is a numerical homotopy invariant of the topological space X, defined by Y. Rudyak in [12]. It is a generalization of the topological complexity $TC(X)$ defined by M. Farber in [5] in relation to the problem of motion planning algorithm. In this paper, we compute the higher topological complexity TC_n of configuration spaces of spheres of odd dimension. Our main tool is a method developed in [9] giving the upper bound for the higher topological complexity TC_n. The same method will also be applied to compute TC_n of the 2-dimensional torus $\mathbb{T}$.

Keywords: Topological complexity, Schwarz genus, configuration space

2010 MSC: Primary 52C35, 55R80

1. Introduction

For $n \in \mathbb{N}$ let J_n denote the wedge sum of n closed unit intervals $[0,1]_i$, $i = 1, ..., n$, where each interval $[0,1]_i$ has 0 as its base point. For a path-connected space X let X^{J_n} denote the space of all continuous maps from J_n to X with compact-open topology.

Definition 1 (see [12]). The higher topological complexity $TC_n(X)$ of

Acknowledgement: The first author was supported in part by the grant *Some selected topics in Analysis and Geometry #ICRTM01_2020.01.*

the space X is the Schwarz genus of the fibration

$$\begin{array}{rcl} e_n^X : X^{J_n} & \longrightarrow & X^n \\ \gamma & \longmapsto & (\gamma(1_1), \ldots, \gamma(1_n)) \end{array}.$$

In other words, it is the smallest number k such that there is an open covering $\{X_i, i = 1, \ldots, k\}$ of X^n and on each X_i there exists a continuous section $s_i : X_i \longrightarrow X^{J_n}$ of e_n^X (i.e., $e_n^X \circ s_i = id_{X_i}$).

In the case $n = 2$, $TC_2(X)$ coincides with the topological complexity $TC(X)$ defined by M. Farber in [5]. The higher topological complexity TC_n is an invariant of homotopy type of the space X. It has been computed for spheres by Y. Rudyak in [12], for any k-dimensional torus and for the product of spheres by I. Babase et al. in [1].

Interested in computing TC_n for the complement of a hyperplane arrangement we succeeded in computing TC_n for the complement of fiber type arrangements in [9] and of generic arrangements in [10]. The higher topological complexity for the complement of reflection arrangements has also been computed by S. Yuzvinsky in [14]. In this paper, using the method developed in [9] we compute the higher topological complexity of configuration spaces of odd-dimensional spheres. We can also compute the higher topological complexity of the 2-dimensional torus $\mathbb{T}$ (see Theorem 2) using our method and of orientable surfaces of higher genus (see Theorem 5), using the same method as in [2]. This generalization of results of [2], has been obtained in [4]. However our arguments in the proof of the case of the 2-dimensional torus $\mathbb{T}$ is quite new.

Let X be a topological space. Recall that the configuration space of k distinct ordered points in X is

$$F(X, k) = \{(x_1, \ldots, x_k) \in X^k \mid x_i \neq x_j \text{ with } 1 \leq i \neq j \leq k\}.$$

Our main results are:

Theorem 1. *Let $\mathbb{S}^m$ be the unit sphere in $\mathbb{R}^{n+1}$, $F(\mathbb{S}^m, k)$ the configuration space of k distinct ordered points in $\mathbb{S}^m$. Assume that $m > 1$ is an odd integer. Then, for $k \geq 2$ we have*

$$TC_n(F(\mathbb{S}^m, k+1)) = nk.$$

Theorem 2. *The higher topological complexity of the configuration space of k distinct ordered points on the two-dimensional torus $\mathbb{T}$ is*

$$TC_n(F(\mathbb{T}, k)) = n(k+1) - 1.$$

Below are some important properties of the higher topological complexity TC_n which will be used in our computations in the rest of the paper.

1. Let X be a finite path-connected CW-complex. Suppose that $X^n = X_1 \cup \ldots \cup X_k$, where each X_i is an ENR and accepts a continuous section of e_n^X. Then, by similar argument as those in [12] we have $TC_n(X) \leq k$.

   Next properties are consequences of those of the Schwarz genus for fibration e_n^X (see [13]).
2. If X is a path-connected CW-complex, then
$$TC_n(X) \leq n \dim X + 1. \tag{1}$$
3. For path-connected spaces X and Y we have
$$TC_n(X \times Y) \leq TC_n(X) + TC_n(Y) - 1. \tag{2}$$
4. Given a topological space X, let $C = \{C_1, \ldots, C_p\}$ and $D = \{D_1, \ldots, D_q\}$ be open coverings of X^n such that on each $C_i \cap D_j$ there exists a continuous section of e_n^X. Then (see [13], Proposition 20)
$$TC_n(X) \leq p + q - 1. \tag{3}$$
5. Let $cat(X)$ denote the Lusternik - Schnirelman category of a space X. It is defined as the smallest integer k such that X may be covered by k open subsets $X = V_1 \cup \ldots \cup V_k$ with each inclusion $V_i \hookrightarrow X$ null-homotopic. Then we have
$$TC_n(X) \leq cat(X^n) \leq n(cat(X) - 1) + 1. \tag{4}$$
6. Since the map e_n^X is a fibrational substitute of the diagonal map $d_n : X \to X^n$ (i.e., there exists a homotopy equivalence $h : X \to X^{J_n}$ such that $d_n = e_n^X \circ h$), one also call $TC_n(X)$ the Schwarz genus of the map d_n (see [12]).

   Suppose that m is an integer such that there exist cohomology classes $u_i \in H^*(X^n, \mathbb{F})$, where $\mathbb{F}$ is a field, for $i = 1, \ldots, m$, satisfying $d_n^* u_i = 0, \forall i$ and $u_1.u_2 \ldots u_m \neq 0 \in H^*(X^n, \mathbb{F})$. The largest number m with this property is called *d_n-cup-length* of X, denoted by $cl(X, n)$ (see [1]). Then, following Theorem 4 in [13] we have the lower bound for the nth-topological complexity
$$TC_n(X) \geq cl(X, n) + 1 \tag{5}$$
   for path-connected topological space X.

7. For path-connected (finite) polyhedrons X and Y we have

$$cl(X \times Y, n) \geq cl(X, n) + cl(Y, n). \tag{6}$$

Remark 1. In this paper, we consider the case $\mathbb{F} = \mathbb{Q}$ and for the sake of brevity we will write $H^*(X)$ instead of $H^*(X, \mathbb{Q})$. Suppose that X is a CW-complex and u is a cohomology class in $H^*(X)$. Let us denote

$$\overline{(u)}_t = 1 \otimes 1 \otimes \ldots \otimes 1 \otimes \overset{\overset{t}{\vee}}{u} \otimes 1 \otimes \ldots \otimes 1 - u \otimes 1 \otimes \ldots \otimes 1 \otimes 1, \ \ t = 2, \ldots, n.$$

These are cohomology classes in $H^*(X^n) \cong \underbrace{H^*(X) \otimes \ldots \otimes H^*(X)}_{n \ \ times}$ satisfying $d_n^* \overline{(u)}_t = 0$.

The paper is organized as follow. As examples, the higher topological complexity of the bouquet of arbitrary spheres in Theorem 3 and of orientable surfaces has been computed in Theorem 4 in Section 2 using properties of TC_n listed above. In Section 3, using arguments similar to those in [9], we first construct a family of homotopy equivalences between a certain bouquet of spheres and the fiber of the projection of configuration spaces on its first $k - 1$ coordinates

$$\pi = \pi_k : F(X, k) \longrightarrow F(X, k - 1),$$

where X is an odd dimensional sphere or the 2-dimensional torus. Combining this with the inequality (3) we get an upper bound for the higher topological complexity of these configuration spaces. Together with lower bound for the TC_n determined by knowledge of the cohomology of configuration spaces we obtain our main results, Theorems 1 and 2.

2. Examples

In this section, using the properties of the higher topological complexity given in the previous section we will compute explicitly the numerical value of TC_n of the bouquet of spheres and of orientable surfaces.

Theorem 3. *Let X be the bouquet of arbitrary spheres $X = \mathbb{S}^{k_1} \vee \ldots \vee \mathbb{S}^{k_m}$, $k_i \geq 1, m > 1$. Then $TC_n(X) = n + 1$.*

Proof. It is known that the cohomology algebra $H^*(X)$ is generated by m elements $u_i \in H^{k_i}(X), i = 1, \ldots, m$ satisfying $u_i u_j = 0, \forall i, j$. Since $m > 1$, we can choose two distinct generators, say, u_i, u_j, $i \neq j$. For $t = 2, \ldots, n$, put

$$\overline{(u_i)}_t = 1 \otimes 1 \otimes \ldots \otimes 1 \otimes \overset{\overset{t}{\vee}}{u_i} \otimes 1 \otimes \ldots \otimes 1 - u_i \otimes 1 \otimes \ldots \otimes 1 \otimes 1.$$

Then we have

$$\overline{(u_j)}_n \prod_{t=2}^{n} \overline{(u_i)}_t = \pm u_j \otimes u_i \otimes \ldots \otimes u_i \pm u_i \otimes \ldots u_i \otimes u_j \neq 0.$$

Moreover, by means of the above remark, $d_n^* \overline{(u_i)}_t = d_n^* \overline{(u_j)}_n = 0$ for $t = 2, \ldots, n$. Hence, the inequality (5) implies $TC_n(X) \geq n + 1$.

So, it suffices to show $TC_n(X) \leq n + 1$. Let P denote the base point of the bouquet $X = \mathbb{S}^{k_1} \vee \ldots \vee \mathbb{S}^{k_m}$, $P_i \in \mathbb{S}^{k_i}$ the antipodes of P in $\mathbb{S}^{k_i}$ respectively, for $i = 1, \ldots, m$. Let us fix a path γ_i from P to P_i and γ_i^{-1} denote the inverse path.

Let us choose a decomposition $\mathbb{S}^{k_i} = U_i \cup V_i$, where U_i, V_i are $ENRs$ and $U_i \cap V_i = \emptyset$, $P \in U_i$, $P_i \in V_i$. Obviously $X \times X$ is the disjoint union of $U_i \times U_j, U_i \times V_j, V_i \times U_j, V_i \times V_j$, $1 \leq i, j \leq m$. For $(A, B) \in X \times X$, we define a path $[A, B]$ from A to B as follows

+ If $(A, B) \in U_i \times U_j$ then $[A, B] = \ell_{[A,P]} * \ell_{[P,B]}$, a path going from A to B through P.
+ If $(A, B) \in U_i \times V_j$ then $[A, B] = \ell_{[A,P]} * \gamma_j * \ell_{[P_j,B]}$, a path going from A to B through P and P_j.
+ If $(A, B) \in V_i \times U_j$ then $[A, B] = \ell_{[A,P_i]} * \gamma_i^{-1} * \ell_{[P,B]}$, a path going from A to B through P_i and P
+ If $(A, B) \in V_i \times V_j$ then $[A, B] = \ell_{[A,P_i]} * \gamma_i^{-1} * \gamma_j * \ell_{[P_j,B]}$, a path going from A to B through P_i, P and P_j.

Here, $\ell_{[A,P]}$, $\ell_{[A,P_i]}$, $\ell_{[P,B]}$, $\ell_{[P_j,B]}$ denote the shortest paths from A to P, P_i and from P and P_j to B respectively.

Let $U = \cup U_i$, $V = \cup V_i$. Then, for $K \subset \{1, 2, \ldots, n\}$ put

$$D_K = \{(A_1, \ldots, A_n) | A_i \in V \text{ if only if } i \in K\}, \text{ and}$$
$$D_k = \bigcup_{|K|=k} D_K.$$

Consider the map

$$\begin{array}{rccc} s: & X^n & \longrightarrow & X^{J_n}. \\ & (A_1, A_2, \ldots, A_n) & \longmapsto & ([A_1, A_1], [A_1, A_2], \ldots, [A_1, A_n]) \end{array}$$

By definition, the restriction $s|_{D_K} : D_K \longrightarrow X^{J_n}$ is a continuous section of e_n^X. Moreover, the sets D_k, $k = 0, \dots, n$, are ENRs and form a covering of X^n consisting of $n+1$ subsets. Therefore we have $TC_n(X) \leq n+1$.

The proof is complete. □

Remark 2. The proof of the upper bound $TC_n(X) \leq n+1$ can be simplified by using the inequality $TC_n(X) \leq n(cat(X)-1)+1$ (see the inequality (4)) and noticing that *cat* of any suspension is 2. We thank the referee for this remark. However, we need to use our construction of paths $[A, B]$ here in the proof of the Theorem 1.

Theorem 4. *Let Σ_g be an orientable surface of genus g. Then*

$$TC_n(\Sigma_g) = \begin{cases} n+1 & \text{if } g = 0, \\ 2n-1 & \text{if } g = 1, \\ 2n+1 & \text{if } g > 1. \end{cases}$$

Proof. The theorem has been obtained by M. Farber in [6] for the case of topological complexity TC. For the case of higher topological complexity, it is known from the work of Y. Rudyak (see [12]) and I. Babase et al. (see [1]) that $TC_n(\mathbb{S}^2) = n+1$ and $TC_n(\mathbb{T}) = 2n-1$. These are the cases $g = 0$ and $g = 1$ respectively.

So, we need only to consider the case $g > 1$. Let us choose cohomology classes $u_1, v_1, u_2, v_2 \in H^1(\Sigma_g)$ satisfying ${u_i}^2 = 0$, ${v_i}^2 = 0$, $u_i v_j = u_i u_j = v_i v_j = 0$ with $i \neq j$ and $0 \neq u_1 v_1 = u_2 v_2 \in H^2(\Sigma_g)$.

For $i = 1, 2$ and $t = 2, \dots, n$, put

$$\overline{(u_i)_t} = 1 \otimes 1 \otimes \dots \otimes 1 \otimes \overset{\overset{t}{\vee}}{u_i} \otimes 1 \otimes \dots \otimes 1 - u_i \otimes 1 \otimes \dots \otimes 1 \otimes 1,$$

$$\overline{(v_i)_t} = 1 \otimes 1 \otimes \dots \otimes 1 \otimes \overset{\overset{t}{\vee}}{v_i} \otimes 1 \otimes \dots \otimes 1 - v_i \otimes 1 \otimes \dots \otimes 1 \otimes 1.$$

Then:

$$\left(\overline{(u_2)_n} \prod_{t=2}^{n} \overline{(u_1)_t}\right) \left(\overline{(v_2)_n} \prod_{t=2}^{n} \overline{(v_1)_t}\right)$$
$$= \pm u_2 v_2 \otimes u_1 v_1 \otimes \dots \otimes u_1 v_1 \pm u_1 v_1 \otimes \dots \otimes u_1 v_1 \otimes u_2 v_2 \neq 0.$$

On the other hand $d_n^* \overline{(u_i)_t} = d_n^* \overline{(v_i)_t} = 0$ for any $i = 1, 2$ and $t = 2, \dots, n$. So, the inequality (5) implies that $TC_n(\Sigma_g) \geq 2n+1$.

Since Σ_g is a CW-complex of dimension 2, the inequality (1) yields $TC_n(\Sigma_g) \leq 2n+1$.

Therefore $TC_n(\Sigma_g) = 2n+1$. □

3. The higher topological complexity of configuration spaces

In this section we will compute the higher topological complexity for the configuration spaces of odd dimensional spheres, of the 2-dimensional torus as well as of orientable surfaces of genus $g \geq 2$.

3.1. *Configuration spaces on spheres*

We first consider the configuration spaces on spheres. For $k = 1$ we have $F(\mathbb{S}^m, 1) = \mathbb{S}^m$. For $k = 2$, it is easy to check that the maps

$$\begin{aligned} f : F(\mathbb{S}^m, 2) &\longrightarrow \mathbb{S}^m \\ (x, y) &\longmapsto x \end{aligned}$$

and

$$\begin{aligned} g : \mathbb{S}^m &\longrightarrow F(\mathbb{S}^m, 2) \\ x &\longmapsto (x, -x) \end{aligned}$$

satisfy the conditions $f \circ g = id_{\mathbb{S}^m}$ and $g \circ f \simeq id_{F(\mathbb{S}^m,2)}$. So, $F(\mathbb{S}^m, 2)$ has the homotopy type of $\mathbb{S}^m$. Hence

$$TC_n(F(\mathbb{S}^m, k)) = TC_n(\mathbb{S}^m) = \begin{cases} n & \text{if } m \text{ odd} \\ n+1 & \text{if } m \text{ even} \end{cases}$$

for $k = 1, 2$ (see [12]).

Next, we consider the case $k > 2$. Let us denote $\pi = \pi_k : F(\mathbb{S}^m, k+1) \longrightarrow F(\mathbb{S}^m, k)$ the projection on the first k coordinates. It easy to see that the map

$$\begin{aligned} p_x : \ \pi^{-1}(x_1, \ldots, x_k) &\longrightarrow \mathbb{S}^m \setminus \{x_1, \ldots, x_k\} \\ (x_1, \ldots, x_k, x_{k+1}) &\longmapsto \qquad x_{k+1} \end{aligned}$$

is a homeomorphism for all $x = (x_1, \ldots, x_k) \in F(\mathbb{S}^m, k)$. In order to prove Theorem 1, we need following lemmas.

Lemma 1. *Suppose that m is an odd integer. For each k, there exists a section $\sigma = \sigma_k : F(\mathbb{S}^m, k) \longrightarrow F(\mathbb{S}^m, k+1)$ of π_k.*

Proof. Since m is odd, we can fix v to be a nowhere zero continuous tangent vector field on $\mathbb{S}^m$. For $x \in \mathbb{S}^m$, denote $\gamma_x : [0, 1] \longrightarrow \mathbb{S}^m$ the spherical arc joining x to $-x$, which has $v(x)$ as its tangent vector at x. Precisely, γ_x is defined by

$$\gamma_x(t) = \cos \pi t.x + \sin \pi t.v(x), \quad t \in [0, 1].$$

For $r > 0$, $z \in \mathbb{S}^m$, put

$$\mathbf{S}^m_{(z,r)} = \{y \in \mathbb{S}^m : \|x - y\| = r\}.$$

For $(x_1, \dots, x_k) \in F(\mathbb{S}^m, k)$, let us denote

$$r(x_1, \dots, x_k) = \frac{1}{2} \min\{\|x_1 - x_2\|, \dots, \|x_1 - x_k\|\} > 0.$$

Then the map $\sigma = \sigma_k : F(\mathbb{S}^m, k) \longrightarrow F(\mathbb{S}^m, k+1)$ defined by

$$\sigma(x_1, \dots, x_k) = (x_1, \dots, x_k, \varphi(x_1, \dots, x_k)),$$

with $\varphi(x_1, \dots, x_k) = \mathbf{S}^m_{(x_1, r(x_1, \dots, x_k))} \cap \gamma_{x_1}$, is a section of $\pi = \pi_k$. □

Lemma 2. *Let X_{k-1} be the bouquet of $k-1$ spheres $\mathbb{S}^{m-1}$ with the base point P. Then, there exists a family of homotopy equivalences $h_x : \pi^{-1}(x_1, \dots, x_k) \longrightarrow X_{k-1}$ depending continuously on $x = (x_1, \dots, x_k) \in F(\mathbb{S}^m, k)$, such that $h_x(\sigma(x)) = P$.*

Proof. Suppose that $K_1 = \{x_1, x_2, \dots, x_k\}$ and $K_2 = \{y_1, y_2, \dots, y_k\}$ are two subsets of the interior $\mathring{\mathbb{B}}^m$ of the unit ball $\mathbb{B}^m$ and $f_0 : K_1 \to K_2$ is a bijection. Then there exists a homeomorphism $f : \mathbb{B}^m \longrightarrow \mathbb{B}^m$ such that

$$\begin{aligned} f|_{K1} &= f_0 \\ f|_{\partial \mathbb{B}^m} &= id. \end{aligned}$$

We can assume without loss of generality that $f_0(x_i) = y_i$ for $i = 1, 2, \dots, k$. We consider the first pair (x_1, y_1). Let U be an open neighborhood of x_1 and y_1 in $\mathbb{B}^m$, satisfying $U \cap K_1 = \{x_1\}$, $U \cap K_2 = \{y_1\}$, and $U \cong \mathring{\mathbb{B}}^m$. We then can construct a homeomorphism $g_1 : U \longrightarrow U$ such that $g_1(x_1) = y_1$ and $g_1|\partial U = id$. This homeomorphism g_1 can be easily extended to a homeomorphism $g_1 : \mathbb{B}^m \longrightarrow \mathbb{B}^m$ by putting $g_1 = id$ on $\mathbb{B}^m \setminus U$. Using the same construction for all other pairs $(x_i, y_i), i = 2, 3, \dots, k$, we get homeomorphisms $g_i : \mathbb{B}^m \longrightarrow \mathbb{B}^m$ for all $i = 1, 2, \dots, k$. Then the composition of all these homeomorphisms g_i is the required homeomorphism.

For each $x = (x_1, \dots, x_k) \in F(\mathbb{S}^m, k)$ we have seen that $\pi^{-1}(x_1, \dots, x_k)$ is homeomorphic to $\mathbb{S}^m \setminus \{x_1, \dots, x_k\}$ via the map $g(x_1, \dots, x_k, x_{k+1}) = x_{k+1}$. Obviously, there is a homotopy equivalence f_x, depending continuously on x from $\mathbb{S}^m \setminus x_k$ to the ball $\mathbb{B}^m$ sending $\sigma(x)$ to the center of the ball. This ball $\mathbb{B}^m$ contains $k-1$ points $P_j = f_x(x_j)$ for $j = 1, \dots k-1$. Then, by the above argument, there is a homeomorphism ψ from $\mathbb{B}^m$ to $\mathbb{B}^m$, sending the centre of $\mathbb{B}^m$ to that of $\mathbb{B}^m$ and P_j to Q_j, respectively, for $1 \leq j \leq k-1$. Here

$$Q_1 = \frac{1}{2}Q,$$
$$Q_j = \frac{1}{2}\left(\cos\left(\frac{2\pi(j-1)}{k-1}\right)Q + \sin\left(\frac{2\pi(j-1)}{k-1}\right)v(Q)\right),$$

$j = 2, \dots, k-1$, where $Q = (1, 0, \dots, 0) \in \mathbb{S}^m$ and v is the fixed vector field as in the proof of Lemma 1.

There is a canonical homotopy from $\mathbb{B}^m \backslash \{Q_1, \dots, Q_{k-1}\}$ to the bouquet X_{k-1} of $k-1$ spheres $\mathbb{S}^{m-1}$ with the base point P, sending the center of $\mathbb{B}^m$ to the base point P. Then the composition $h_x = h \circ \psi \circ f_x \circ g$ is the required homotopy equivalence. □

Proof of Theorem 1. Since m is odd, according to [7, Theorem 8.2, p. 114], we have

$$H^*(F(\mathbb{S}^m, k+1)) \cong H^*(F(\mathbb{R}^m, k)) \otimes H^*(\mathbb{S}^m).$$

So, $cl(F(\mathbb{S}^m, k+1), n) \geq cl(F(\mathbb{R}^m, k), n) + cl(\mathbb{S}^m, n)$. On the other hand, $cl(\mathbb{S}^m, n) = n-1$ (see [1]) and $cl(F(\mathbb{R}^m, k), n) = n(k-1)$ (see [3]). Hence, $cl(F(\mathbb{S}^m, k+1), n) \geq nk-1$. By means of the inequality (5), we get

$$TC_n(F(\mathbb{S}^m, k+1)) \geq nk.$$

Next, we prove that

$$TC_n(F(\mathbb{S}^m, k+1)) \leq nk. \tag{7}$$

According to Lemma 2, there exists a family of homotopy equivalences $h_x : \pi^{-1}(x_1, \dots, x_k) \longrightarrow X_{k-1}$ depending continuously on x satisfying $h_x(\sigma(x)) = P$.

Suppose that $TC_n(F(\mathbb{S}^m, k)) = p$, and $U = \{U_1, \dots, U_p\}$ is an open covering of $(F(\mathbb{S}^m, k))^n$ such that there exists a section of $e_n^{F(\mathbb{S}^m,k)}$ on each U_i. Put $C_i = \{(A_1, \dots, A_n) \in (F(\mathbb{S}^m, k+1))^n \mid (\pi(A_1), \dots, \pi(A_n)) \in U_i\}$. Then $C = \{C_1, \dots, C_p\}$ is an open covering of $(F(\mathbb{S}^m, k+1))^n$.

Suppose that $TC_n(X_{k-1}) = q$, and $V = \{V_1, \dots, V_q\}$ is an open covering of $(X_{k-1})^n$ such that there exists a section of $e_n^{X_{k-1}}$ on each V_j. Let $D_j = \{(A_1, \dots, A_n) \in (F(\mathbb{S}^m, k+1))^n \mid (h_{\pi(A_1)}(A_1), \dots, h_{\pi(A_n)}(A_n)) \in V_j\}$. Then $D = \{D_1, \dots, D_q\}$ is an open covering of $(F(\mathbb{S}^m, k+1))^n$.

We will now construct a section of $e_n^{F(\mathbb{S}^m,k+1)}$ on each $C_i \cap D_j$, $i = 1, \dots, p$, $j = 1, \dots, q$. Suppose that $(A_1, \dots, A_n) \in C_i \cap D_j \subset F(\mathbb{S}^m, k+1)^n$.

By definition $(A_1, \dots, A_n) \in C_i$ means $(\pi(A_1), \dots, \pi(A_n)) \in U_i$. Since there exists a section s_1 of $e_n^{F(\mathbb{S}^m,k)}$ on U_i, there is a path going from $\pi(A_1)$ to $\pi(A_i)$, for $i = 1, \dots, n$, determined by this section s_1.

Similarly, $(A_1, \dots, A_n) \in D_j$ means $(h_{\pi(A_1)}(A_1), \dots, h_{\pi(A_n)}(A_n)) \in V_j$ and there exists a section s_2 of $e_n^{X_{k-1}}$ on V_j. Let us choose the section s_2 to be the one defined in the proof of Theorem 3. Then there is a path going from $h_{\pi(A_i)}(A_i)$ to P for any $i = 1, \dots, n$, determined by this section.

Now we denote by $[A_1, A_i]$ the path $\gamma_1 * \gamma_2 * \gamma_3$, where

- γ_2 is the image under the section σ of the path connecting $\pi(A_1)$ and $\pi(A_i)$ in $F(\mathbb{S}^m, k)$, determined by s_1 as above.
- γ_1 is the image of the path connecting $h_{\pi(A_1)}(A_1)$ to P in X_k, determined by s_2 as above, under the homotopy inverse of $h_{\pi(A_1)}$.
- γ_3 is the image of the path connecting P to $h_{\pi(A_i)}(A_i)$ in X_k, determined by s_2 as above, under the homotopy inverse of $h_{\pi(A_i)}$.

Now for $(A_1, \dots, A_n) \in C_i \cap D_j$ we can define the map s by $s(A_1, \dots, A_n) = ([A_1, A_1], [A_1, A_2], \dots, [A_1, A_n])$. Obviously this is a continuous section of $e_n^{F(\mathbb{S}^m,k+1)}$.

Applying the inequality (3) we have

$$TC_n(F(\mathbb{S}^m, k+1)) \le p + q - 1 = TC_n(F(\mathbb{S}^m, k)) + TC_n(X_{k-1}) - 1.$$

Moreover, it follows from Theorem 3 that $TC_n(X_{k-1}) \le n + 1$. So,

$$TC_n(F(\mathbb{S}^m, k+1)) \le TC_n(F(\mathbb{S}^m, k)) + n.$$

By induction on k, we get

$$TC_n(F_n(\mathbb{S}^m, k+1)) \le TC_n(F(\mathbb{S}^m, 2)) + n(k-1) = n + n(k-1) = nk.$$

So, the inequality (7) is proved. Thus, $TC_n(F(\mathbb{S}^m, k+1)) = nk$. □

3.2. *Configuration spaces of orientable surfaces*

Suppose that $A = \oplus_{i=0}^{l} A_i, B = \oplus_{j=0}^{k} B_j$ are anti-commutative graded algebras of finite type over a field $\mathbb{F}$. That is A_i, B_j are of finite dimension (over $\mathbb{F}$) for each i and j. The tensor product $A \otimes B$ is again a graded algebra with the multiplication given by $(a \otimes b)(a' \otimes b') = (1)^{|b|\,|a'|} aa' \otimes bb'$. Inductively, $A^{\otimes n} = A^{\otimes(n-1)} \otimes A$ is also a graded algebra. The multiplication then defines a homomorphism of graded algebras

$$\mu : A^{\otimes n} \to A.$$

We define the d_n-zero divisor cup length of A, denoted by $cl(A,n)$, to be the maximal number k such that there exist $u_1,\dots,u_k \in \ker\mu$ satisfying $u_1\dots u_k \neq 0$.

It is easy to show the following property of the d_n-zero divisor cup length.

Lemma 3. *Let A be a graded algebra with unit over a field $\mathbb{F}$. If B is a subalgebra of A, then $cl(A,n) \geq cl(B,n)$.*

Put $A = H^*(X)$, we have $cl(X,n) = cl(A,n)$. By means of the inequality (5) we have

$$TC_n(X) \geq cl(H^*(X),n) + 1.$$

As is well known, the knowledge of the cohomology of $F(X,k)$ will give us the lower bound for its higher topological complexity. Suppose that X is an oriented manifold of dimension m. Let $\Delta \in H^m(X\times X)$ be the diagonal cohomology class. Let $p_{i,j} : X^k \to X\times X$ denote the natural projections, $p_{i,j}(x_1,\dots,x_k) = (x_i,x_j)$, where $1 \leq i,j \leq k$. Then B. Totaro (see [11]) shows that if X is a complex projective variety, the Leray spectral sequence of the inclusion $F(X,k) \hookrightarrow X^k$, which converges to $H^*(F(X,k))$, will degenerate after the first nontrivial differential and that the cohomology ring $H^*(F(X,k))$ is determined by $H^*(X)$. Precisely, we have

Lemma 4 (see [11]). *Let X be a smooth projective variety on $\mathbb{C}$ of real dimension m, let $H = H^*(X^k)$, and let I be the ideal in H generated by the elements*

$$\Delta_{i,j} = p_{i,j}^*\Delta \in H^m(X^k)$$

for all $1 \leq i,j \leq k$. Then the quotient $A = H/I$ is a subalgebra of the rational cohomology ring of the configuration space $F(X,k)$.

Therefore, Lemma 3 and Lemma 4 imply that $cl(H/I,n)$ is a lower bound for $TC_n(F(X,k))$.

In order to prove Theorem 2 we first denote $\pi = \pi_k : F(\mathbb{T},k) \longrightarrow F(\mathbb{T},k-1)$ the projection on the first $k-1$ coordinates. The map

$$\begin{aligned} p_x : \pi^{-1}(x_1,\dots,x_{k-1}) &\longrightarrow \mathbb{T}\setminus\{x_1,\dots,x_{k-1}\} \\ (x_1,\dots,x_{k-1},x_k) &\longmapsto \qquad x_k \end{aligned}$$

is a homeomorphism for all $x = (x_1,\dots,x_{k-1}) \in F(\mathbb{T},k-1)$.

Lemma 5. *For each $k > 1$, there exists a section $\sigma = \sigma_k : F(\mathbb{T},k-1) \longrightarrow F(\mathbb{T},k)$ of π_k.*

Proof. Consider $\mathbb{S}^1$ as the set of complex numbers with modulus 1 and $\mathbb{T} = \mathbb{S}^1 \times \mathbb{S}^1$. For a positive integer $r > 0$ and a point $x = \left(e^{i\varphi_1}, e^{i\varphi_2}\right) \in \mathbb{T}$, put $\mathbf{T}_{(x,r)} = \{y \in \mathbb{T} \mid \|x - y\| = r\}$. Let γ_x denote the path defined by $\gamma_x(t) = \{\left(e^{i\varphi_1}, e^{i(\varphi_2 + t\pi)}\right) \mid t \in [0,1]\}$. Then we define a map $\sigma = \sigma_k : F(\mathbb{T}, k-1) \longrightarrow F(\mathbb{T}, k)$ as follow. For $x = (x_1, ..., x_{k-1}) \in F(\mathbb{T}, k-1)$

1. If $k > 2$, let $r = \dfrac{1}{2}\min\{1, \|x_1 - x_2\|, \ldots, \|x_1 - x_{k-1}\|\}$, $x_k = \mathbf{T}_{(x_1,r)} \cap \gamma_{x_1}$. Then we put $\sigma(x) = \sigma(x_1, \ldots, x_{k-1}) = (x_1, \ldots, x_{k-1}, x_k)$.
2. If $k = 2$, then we put $\sigma(x)\sigma(x_1) = (x_1, \gamma_{x_1}(1))$.

Obviously, the map σ is a section of π for any $k > 1$. □

Lemma 6. *Let X_k be the bouquet of k circles $\mathbb{S}^1$ with the base point P. Then, there exists a family of homotopy equivalences $h_x : \pi^{-1}(x_1, \ldots, x_{k-1}) \longrightarrow X_k$ depending continuously on x, such that $h_x(\sigma(x)) = P$.*

Proof. Using arguments similar to those in the proof of Lemma 2 we can prove that if $K_1 = \{x_1, x_2, \ldots, x_k\}$ and $K_2 = \{y_1, y_2, \ldots, y_k\}$ are two subsets of $\mathbb{T}$ and $f_0 : K_1 \to K_2$ is a bijection, then there exists a homeomorphism $f : \mathbb{T} \longrightarrow \mathbb{T}$ such that

$$f|_{K1} = f_0.$$

For each $x = (x_1, \ldots, x_{k-1}) \in F(\mathbb{T}, k-1)$ we have seen that $\pi^{-1}(x_1, \ldots, x_{k-1})$ is homeomorphic to $\mathbb{T} \setminus \{x_1, \ldots, x_{k-1}\}$ via the map $g(x_1, \ldots, x_{k-1}, x_k) = x_k$. Obviously, there is a homeomorphism f_x, depending continuously on x from $\mathbb{T}$ to the $\mathbb{T}$ having $f_x(\sigma(x)) = (e^0, e^0)$ and therefore, $f_x(x_j) \neq (e^0, e^0)$ for $j = 1, \ldots k-1$. Then, by the above argument, there is a homeomorphism ψ from $\mathbb{T}$ to $\mathbb{T}$, sending (e^0, e^0) to (e^0, e^0) and $f_x(x_j)$ to Q_j, respectively. Here $Q_j = \left(e^{i\frac{2\pi j}{k}}, e^{i\frac{2\pi j}{k}}\right) \in \mathbb{T}$ for $1 \leq j \leq k-1$.

There is a canonical homotopy from $\mathbb{T} \backslash \{Q_1, \ldots, Q_{k-1}\}$ to the bouquet of k circles X_k, sending the point (e^0, e^0) to the based point P. Then the composition $h_x = h \circ \psi \circ f_x \circ g$ is the required homotopy equivalence. □

Proof of Theorem 2. We follow the same scheme of the proof of Theorem 1. By means of Lemma 6, there exists a family of homotopy equivalences $h_x : \pi^{-1}(x_1, \ldots, x_{k-1}) \longrightarrow X_k$ depending continuously on $x = (x_1, \ldots, x_{k-1})$, such that $h_x(\sigma(x)) = P$.

Suppose that $TC_n(F(\mathbb{T}, k-1)) = p$. Let $U = \{U_1, \ldots, U_p\}$ be an open covering of $(F(\mathbb{T}, k-1))^n$ with a section of $e_n^{F(\mathbb{T},k-1)}$ on each U_i.

Put $C_i = \{(A_1, \ldots, A_n) \in (F(\mathbb{T}, k))^n \mid (\pi(A_1), \ldots, \pi(A_n)) \in U_i\}$. Then $C = \{C_1, \ldots, C_p\}$ is an open covering of $(F(\mathbb{T}, k))^n$.

Suppose that $TC_n(X_k) = q$. Let $V = \{V_1, \ldots, V_q\}$ is an open covering of $(X_k)^n$ with a section of $e_n^{X_k}$ on each V_j. Let $D_j = \{(A_1, \ldots, A_n) \in (F(\mathbb{T}, k))^n \mid (h_{\pi(A_1)}(A_1), \ldots, h_{\pi(A_n)}(A_n)) \in V_j\}$. Then $D = \{D_1, \ldots, D_q\}$ is an open covering of $(F(\mathbb{T}, k))^n$.

We will now construct a section of $e_n^{F(\mathbb{T},k)}$ on each $C_i \cap D_j$, $i = 1, \ldots, p$, $j = 1, \ldots, q$.

Suppose that $(A_1, \ldots, A_n) \in C_i \cap D_j \subset F(\mathbb{T}, k)^n$.

Since $(A_1, \ldots, A_n) \in C_i$ we have $(\pi(A_1), \ldots, \pi(A_n)) \in U_i$. Because there is a section s_1 of $e_n^{F(\mathbb{T},k)}$ on U_i we can choose a path going from $\pi(A_1)$ to $\pi(A_i)$, for $i = 1, \ldots, n$, determined by this section s_1.

Since $(A_1, \ldots, A_n) \in D_j$ we have $(h_{\pi(A_1)}(A_1), \ldots, h_{\pi(A_n)}(A_n)) \in V_j$ and there exists a section s_2 of $e_n^{X_{k-1}}$. Let us choose the section s_2 to be the one defined in the proof of Theorem 3. Then there is a path going from $h_{\pi(A_i)}(A_i)$ to P for any $i = 1, \ldots, n$, determined by this section.

Now we define the path denoted $[A_1, A_i]$ by putting $[A_1, A_i] = \gamma_1 * \gamma_2 * \gamma_3$, where

- γ_2 is the image under the section σ of the path connecting $\pi(A_1)$ and $\pi(A_i)$ in $F(\mathbb{T}, k-1)$, determined by s_1 as above.
- γ_1 is the image of the path connecting $h_{\pi(A_1)}(A_1)$ to P in X_k, determined by the section s_2 as above, under the homotopy inverse of the homotopy equivalence $h_{\pi(A_1)}$.
- γ_3 is the image of the path connecting P to $h_{\pi(A_i)}(A_i)$ in X_k, determined by the section s_2 as above, under the homotopy inverse of $h_{\pi(A_i)}$.

Now for $(A_1, \ldots, A_n) \in C_i \cap D_j$ we define a map s by putting $s(A_1, \ldots, A_n) = ([A_1, A_1], [A_1, A_2], \ldots, [A_1, A_n])$. Obviously we obtain a continuous section of $e_n^{F(\mathbb{T},k)}$. Applying the inequality (3) we have

$$TC_n(F(\mathbb{T}, k)) \leq p + q - 1 = TC_n(F(\mathbb{T}, k-1)) + TC_n(X_k) - 1.$$

Moreover, it follows from Theorem 3, $TC_n(X_k) \leq n + 1$ for $k \geq 2$. So,

$$TC_n(F(\mathbb{T}, k)) \leq TC_n(F(\mathbb{T}, k-1)) + n \text{ for all } k \geq 2.$$

By induction on k, we get

$$TC_n(F_n(\mathbb{T}, k)) \leq TC_n(F(\mathbb{T}, 1)) + n(k-1) = 2n - 1 + n(k-1) = n(k+1) - 1.$$

So in order to prove Theorem 2, it suffices to show that $cl(F(\mathbb{T}, k), n) \geq n(k+1) - 2$.

By means of Lemma 4, $H^*(F(\mathbb{T},k))$ has a subalgebra A as a quotient of the cohomology $H^*(\mathbb{T}^k)$. We denote this algebra by A_T. Suppose that $a, b \in H^1(\mathbb{T})$ are generators of $H^*(\mathbb{T})$. Then $H^*(\mathbb{T}^k)$ has generators a_i, b_i, $1 \le i \le k$, where $a_i = 1 \times \ldots \times \overset{\overset{i}{\vee}}{a} \times \ldots \times 1$ and $b_i = 1 \times \ldots \times \overset{\overset{i}{\vee}}{b} \times \ldots \times 1$. From these generators, in [2], D. Cohen and M. Farber have constructed a new basis $\{x_j, y_j |\ 1 \le j \le k\}$ for $H^*(\mathbb{T}^k)$ by putting $x_j = a_1$ if $j = 1$ and $x_j = a_j - a_1$ if $2 \le j \le k$; $y_j = b_1$ if $j = 1$ and $y_j = b_j - b_1$ if $2 \le j \le k$. They showed that A_T has generators, which are also denoted, with an abuse of notation, by $x_1, \ldots, x_k, y_1, \ldots, y_k$ with relations

$$x_j y_j = 0 \ \ (2 \le j \le k), \ \ x_j y_i + x_i y_j = 0 \ \ (2 \le i < j \le k).$$

Therefore, A_T has a basis

$$\{x_1^{\epsilon_x} y_1^{\epsilon_y} x_J y_K | \epsilon_x, \epsilon_y \in \{0,1\}, J, K \subset \{2,3,\ldots,k\}, \text{ with } \max J < \min K\}.$$

Here $x_J = x_{j_1} \ldots x_{j_p}$ if $J = \{j_1, \ldots, j_p\}$ and $y_K = y_{k_1} \ldots y_{k_q}$ if $K = \{k_1, \ldots, k_q\}$.

Now we choose the following elements

$$\overline{(x_i)}_t = 1 \otimes 1 \otimes \ldots \otimes 1 \otimes \overset{\overset{t}{\vee}}{x_i} \otimes \ldots \otimes 1 - x_i \otimes 1 \otimes \ldots \otimes 1 \otimes 1 \in [A_T]^{\otimes n}, \quad t = 2, \ldots, n,$$

$$\overline{(y_i)}_t = 1 \otimes 1 \otimes \ldots \otimes 1 \otimes \overset{\overset{t}{\vee}}{y_i} \otimes \ldots \otimes 1 - y_i \otimes 1 \otimes \ldots \otimes 1 \otimes 1 \in [A_T]^{\otimes n}, \quad t = 2, \ldots, n.$$

Put

$$\varphi = \prod_{t=2}^{n} \left(\overline{(x_1)}_t \overline{(y_1)}_t \right) \left(\prod_{j=2}^{k} \left(\overline{(y_j)}_n \prod_{t=2}^{n} \overline{(x_j)}_t \right) \right).$$

Using the above basis of A_T we see that in the expansion of φ, the summand $\pm y_2 \ldots y_k \otimes \underbrace{x_1 y_1 x_2 \ldots x_k \otimes \ldots \otimes x_1 y_1 x_2 \ldots x_k}_{(n-1)}$ can not be exterminated by other summands. So we have φ is nonzero. On the other hand, $d_n^*(\overline{(x_i)}_t) = d_n^*(\overline{(y_i)}_t) = 0$. So, $cl(A_T) \ge n(k+1) - 2$. It follows that $cl(F(\mathbb{T},k), n) \ge cl(A_T) \ge n(k+1) - 2$, and we have $TC_n(F(\mathbb{T},k)) \ge n(k+1) - 1$.

Thus, $TC_n(F(\mathbb{T},k)) = n(k+1) - 1$. □

Next we consider the case of orientable surfaces of genus $g \geq 2$. Modifying arguments in [2] we obtain the following.

Theorem 5. *Suppose that Σ is an orientable surface of genus $g \geq 2$. Then the higher topological complexity of the configuration space of k distinct ordered points in Σ is given by*

$$TC_n(F(\Sigma,k)) = n(k+1)+1.$$

Proof. For $k=1$, we have $F(\Sigma,1)=\Sigma$ and $TC_n(F(\Sigma,1)) = 2n+1$ as computed in Theorem 4. So we need only to consider the case $k \geq 2$.

It is known that $F(\Sigma,k)$ has the homotopy type of a cell complex of dimension $k+1$ (see [2]). Therefore, the inequality (1) implies that $TC_n(F(\Sigma,k)) \leq n(k+1)+1$.

So, in order to prove Theorem 5 we need to show $cl(F(\Sigma,k),n) \geq n(k+1)$. By means of Lemma 4, $H^*(F(\Sigma,k))$ has a subalgebra A as a quotient of the cohomology $H^*(\Sigma^k)$. We denote this algebra by A_Σ. We want to show that $cl(A_\Sigma,n) \geq n(k+1)$.

Let us choose generators $a(q), b(q)$, $1 \leq q \leq g$ of $H^1(\Sigma)$ with $a(p)a(q) = a(p)b(q) = b(p)b(q) = 0$ for $p \neq q$ and $a(q)b(q) = \omega$, where ω is the generator of $H^2(\Sigma)$. Then $H^*(\Sigma^k)$ has generators $u_i = 1 \times \ldots \times 1 \times \overset{\overset{i}{\vee}}{u} \times 1 \times \ldots \times 1$, $1 \leq i \leq k$, with $u = a(q)$ or $b(q)$.

Recall from Lemma 4 that A_Σ is the quotient of $H^*(\Sigma^k)$ by the ideal generated by $\Delta_{i,j} = p_{i,j}^*\Delta \in H^*(\Sigma^k)$, denoted by I_Σ, where

$$\Delta = \omega \times 1 + 1 \times \omega - \sum_{q=1}^{g}(b(p) \times a(p) - a(p) \times b(p)),$$

$$\Delta_{i,j} = \omega_i + \omega_j - \sum_{q=1}^{g}(b(q)_j \times a(q)_i - a(q)_i \times b(q)_j).$$

Let J_Σ denote the ideal $\langle a(q)_i a(q)_j, a(q)_i b(q)_j, b(q)_i b(q)_j \mid 1 \leq i,j \leq k, i \neq j, 2 \leq q \leq g \rangle$ of $H^*(\Sigma^k)$. An easy computation shows that

$$B_\Sigma = H^*(\Sigma^k)/(I_\Sigma + J_\Sigma)$$

is a quotient of A_Σ.

Put $x_j = a(1)_j - a(1)_1$ and $y_j = b(1)_j - b(1)_1$, $j \geq 2$. For sets $J, K \subset \{2,3,\ldots,k\}$ with $\max J < \min K$ we have $x_J y_K = u_1 \times \ldots \times u_k$ with $u_i \in \{1\} \cup \{x_j \mid j \in J\} \cup \{y_j \mid j \in K\}$. It implies that the set

$$\{x_J y_K \mid J, K \subset \{2,3,\ldots,k\}, \max J < \min K\}$$

is linearly independent in B_Σ.

Now for any $u \in B_\Sigma$ put

$$\overline{(u)_t} = 1 \otimes 1 \otimes \ldots \otimes 1 \otimes \overset{\overset{t}{\vee}}{u} \otimes 1 \otimes \ldots \otimes 1 - u \otimes 1 \otimes \ldots \otimes 1 \otimes 1,\ t = 2, \ldots, n.$$

Put

$$\Phi = \left(\overline{(a(1)_1)_n} \prod_{t=2}^{n} \overline{(a(2)_2)_t}\right)\left(\overline{(b(1)_1)_n} \prod_{t=2}^{n} \overline{(b(2)_2)_t}\right) \prod_{j=2}^{k} \left(\overline{(x_j)_n} \prod_{t=2}^{n} \overline{(y_j)_t}\right).$$

Using the above linearly independent system in B_Σ we imply that the expansion of the product Φ contains the summand $\pm 2\omega_1 y_2 y_3 \ldots y_k \otimes \omega_1 x_2 x_3 \ldots x_k \otimes \ldots \otimes \omega_1 x_2 x_3 \ldots x_k$ which can not be exterminated by any other summand. That means the above product Φ is nonzero.

Consider the canonical projection $\pi : A_\Sigma \to B_\Sigma$. For $u \in B_\Sigma$, choose an element $u^* \in A_\Sigma$ such that $\pi(u^*) = u$. For $t = 2, \ldots, n$, put

$$\overline{(u^*)_t} = 1 \otimes 1 \otimes \ldots \otimes 1 \otimes \overset{\overset{t}{\vee}}{u^*} \otimes \ldots \otimes 1 - u^* \otimes 1 \otimes \ldots \otimes 1 \otimes 1.$$

Then, we have $\pi^{\otimes n}(\overline{u^*_t}) = \bar{u}_t$, where $\pi^{\otimes n} : A_\Sigma^{\otimes n} \to B_\Sigma^{\otimes n}$ is the induced mapping. Let

$$\Phi^* = \left(\overline{(a(1)_1^*)_n} \prod_{t=2}^{n} \overline{(a(2)_2^*)_t}(1)\right)\left(\overline{(b(1)_1^*)_n} \prod_{t=2}^{n} \overline{(b(2)_2^*)_t}\right) \prod_{j=2}^{k} \left(\overline{(x_j^*)_n} (\prod_{t=2}^{n} \overline{(y_j^*)_t}\right).$$

Since $\pi^{\otimes n}(\Phi^*) = \Phi \neq 0 \in B_\Sigma^{\otimes n}$ we have $\Phi^* \neq 0 \in A_\Sigma^{\otimes n}$. On other hand, the elements $\overline{(u^*)_t}, t = 2, \ldots, n$ are d_n-zero divisors of A_Σ. Therefore, $cl(F(\Sigma, k), n) = cl(H^*(F(\Sigma, k)), n) \geq cl(A_\Sigma, n) \geq n(k+1)$.

Thus, $TC_n(\Sigma, k) = n(k+1) + 1$. □

Remark 3. The results of Theorem 2 and of Theorem 5 have been presented in [8] together with some other results. They also have been obtained independently in [4]. However, the proof of Theorem 2 here is quite different, based on a method developed by the authors in [9]. It should be noted that in the proof of Theorem 5, in order to prove $cl(F(\Sigma, k), n) = cl(H^*(F(\Sigma, k)), n) \geq cl(A_\Sigma, n) \geq n(k+1)$, we have chosen $n(k+1)$ d_n-divisors of A_Σ having non-zero product instead of using the surjectivity of the canonical projection π which is a loose argument of the paper [2] as indicated in [4].

References

[1] I. Basabe, J. González, Y. Rudyak, and D. Tamaki, *Higher topological complexity and its symmetrization*, Algebr. Geom. Topology 14 (2014), 223–244.

[2] D. Cohen and M. Farber, *Topological complexity of collision - free motion planning on surfaces*, Compositio Mathematica 147 (2011), 649–660.

[3] J. González and M. Grant *Sequential motion planning of non-colliding particles in Euclidean spaces*, Proc. Amer. Math. Soc. 143 (2015), 4503–4512.

[4] J. González and B. Gutiérrez, *Topological complexity of collision-free multitasking motion planning on orientable surfaces*, Contemp. Math. 702, AMS (2018), p. 151–163.

[5] M. Farber, *Topological complexity of motion planning*, Discrete Comput. Geom. 29 (2003), 211–221.

[6] M. Farber, *Invitation to Topological Robotics*, Zurich Lectures in Advanced Mathematics, European Mathematical Society (EMS), Zürich, 2008.

[7] E. Fadell and S. Husseini, *Geometry and Topology of Configuration Spaces*, Springer-Verlag, 2001.

[8] Nguyen Viet Dung and Nguyen Van Ninh, *Higher topological complexity and hyperplane arrangements*, Invited talk at the Asian Mathematical Conference, Busan 30/6 - 4/7, 2013.

[9] Nguyen Viet Dung and Nguyen Van Ninh, *The Higher Topological Complexity of Complement of Fiber Type Arrangements*, Acta Math. Vietnam. 42 (2017), 249–256. doi:10.1007/s40306-016-0199-3.

[10] Nguyen Viet Dung and Nguyen Van Ninh, *Higher topological complexity complement of a generic hyperplane arrangement*, Submitted.

[11] B. Totaro, *Configuration spaces of algebraic varieties*, Topology 35 (1996), 1057–1067.

[12] Y. Rudyak, *On higher analogs of topological complexity*, Topology and its Application 157 (2010) 916–920.

[13] A. Schwarz, *The genus of fiber space*, Amer. Math. Sci, Transl. 55 (1966) 49–140.

[14] S. Yuzvinsky, *Higher Topological Complexity of Artin Type Groups*, Configuration Spaces, Volume 14 of the Springer INdAM Series (2016), 119–128.

Tropical analogies of the delta invariant and degree-genus formula

Takuhiro Takahashi

Mathematical Institute, Tohoku University,
Aoba, Sendai, Miyagi, 980-8578, Japan
E-mail: ttaku32@gmail.com

In this paper, the tropical delta number $\delta^{\mathrm{trop}}(T, V)$ of a vertex V of a tropical plane curve T is defined by the maximal number of locally irreducible components of T at V minus the number of interior lattice points in the polygon $\Xi(V)$ dual to V. It will be proved that if $\Xi(V)$ is prime then the delta invariants of singularities of reduced polynomial maps with Newton polygon $\Xi(V)$ is bounded above by $\delta^{\mathrm{trop}}(T, V)$. We then show that a tropical version of the degree-genus formula holds for tropical plane curves that are irreducible and unweighted. We will verify these results using several examples, including our previous works concerning tropical tacnodal curves.

Keywords: plane tropical curve; tropical singular point; degree-genus formula

1. Introduction

Tropical geometry is a rapidly developing area of combinatorial algebraic geometry. In the epoch-making article by G. Mikhalkin [3], tropical versions of several classical notions such as the genus and smoothness of points are defined. A *tropical plane curve* T is a 1-dimensional polyhedral complex $(T^{[0]}, T^{[1]})$ realized in $\mathbb{R}^2$ having a weight function $w : T^{[1]} \to \mathbb{Z}_{\geq 1}$, where $T^{[i]}$ is the set of i-dimensional polyhedra of T. The *genus* $g^{\mathrm{trop}}(T)$ of the tropical curve T is defined by the minimal number of the rank of $H_1(\Gamma) := H_1(\Gamma, \mathbb{Z})$, where Γ is an open graph which parametrizes the tropical curve (see [3, Definition 2.9] or Definition 2.5 of this paper). Using this definition, he proved the equivalence of the enumeration of nodal complex curves of genus g on toric surfaces and that of simple tropical curves with $g^{\mathrm{trop}}(T) = g$.

In this paper, we introduce a tropical version of the delta invariant of a plane curve singularity, called the *tropical delta number*, and show a tropical version of the degree-genus formula for tropical plane curves that are irreducible and unweighted. The tropical delta number is defined as follows. An irreducible decomposition of a tropical plane curve T at a vertex V will be introduced in Section 3. This decomposition is not unique unlike the local irreducible decomposition of classical plane curve singularities. Therefore, we define the number of irreducible components of T at V by the maximal number of locally irreducible components among all local irreducible decompositions of T at V, and denote it by $r(V)$. Then, we define the *tropical delta number* of V by

$$\delta^{\text{trop}}(T, V) := r(V) - 1 + \sharp(\text{Int}(\Xi(V)) \cap \mathbb{Z}^2) \geq 0,$$

where $\Xi(V)$ is the dual polygon of V by Duality Theorem. Note that, if V is trivalent then the tropical delta number is $(m(V) - 1)/2$, where $m(V)$ is the Mikhalkin's multiplicity.

We prove the following inequality between the tropical delta number and the delta invariant of a plane curve singularity.

Theorem I. *For a prime polygon $P \subset \mathbb{R}^2$, the inequality*

$$\delta^{\text{trop}}(T_P, V_P) \geq \max_{f \in \mathcal{F}(P)} \{\delta(C_f, p) \ ; \ Sing(C_f) = \{p\}\}$$

holds, where T_P is the dual tropical curve of P with exactly one vertex V_P, $\mathcal{F}(P)$ is the set of reduced polynomials $f \in \mathbb{C}[z, w]$ whose Newton polygon is P and C_f is the toric compactification of $\{(z, w) \in (\mathbb{C}^*)^2 \ ; \ f(z, w) = 0\}$ in the toric surface $X(P)$ associated with P.

Notice that, in case of A_2 and A_3, P is a triangle if and only if the above inequality is strict. More precisely, the equality holds

- for the polygons Δ_1 and Δ_2 in [5] in the case of A_2, and
- for the polygons $\Delta_{\text{I}}, \Delta_{\text{II}}, \Delta_{\text{VI}}, \Delta_{\text{VII}}$ and Δ_{VIII} in [6] in the case of A_3.

In the second half of this paper, we study global properties of singular unweighted tropical curves. For a polygon $\Delta \subset \mathbb{R}^2$ and an irreducible curve C in the toric surface $X(\Delta)$ associated with Δ whose toric degree is Δ, the equality

$$g(C) = \sharp(\text{Int}\Delta \cap \mathbb{Z}^2) - \sum_{p \in C} \delta(C, p)$$

holds, where $g(C)$ is the geometric genus of C and $\delta(C, p)$ is the delta number of C at p. This equality is known as the *degree-genus formula* (see [3, Subsection 5.2] and [7, Corollary 7.1.3]). By using the tropical delta number, we may obtain a tropical version of this formula.

Theorem II (Tropical degree-genus formula). *For an unweighted irreducible tropical curve T whose Newton polygon is Δ, the following equality holds:*

$$g^{\mathrm{trop}}(T) = \sharp(\mathrm{Int}\Delta \cap \mathbb{Z}^2) - \sum_{V \in T^{[0]}} \delta^{\mathrm{trop}}(T, V).$$

It is known that the classical formula does NOT hold for reducible curves. After the proof of Theorem II, we verify the formula for an irreducible tropical curve (Example 5.2) and show that there is a counterexample of reducible curves (Example 5.3). These examples support the assumption of the irreducibility in Theorem II.

We organize this paper as follows. In Section 2, we introduce some basic notions and known results on tropical curves. In Section 3, we define a tropical analogy of plane curve singularities and study its local property such as a local irreducible decomposition. In Section 4, we introduce the notion of maximal delta number of a polygon and study a relation between the tropical delta number and the maximal delta number. At the end of this section, we prove Theorem I. In Section 5, we prove Theorem II.

The author would like to thank the organizers of FJV2017 Kagoshima for giving him the opportunity to give a poster presentation in the symposium. He is also grateful to the referee for careful reading of the manuscript and for giving useful comments. This work is partially supported by the Grant-in-Aid for Scientific Research (S), JSPS KAKENHI Grant Number JP17H06128.

Terminology

A set in $\mathbb{R}^d$ is a *(lattice) polyhedron* if it is the intersection of a finite number of closed rational half-spaces of $\mathbb{R}^d$ whose vertices are contained in the lattice $\mathbb{Z}^d \subset \mathbb{R}^d$. The *dimension* of a polyhedron is defined to be the dimension of the affine span of the polyhedron. A polyhedron is a *polytope* if it is compact. We call a 2-dimensional polytope a *polygon*. A subset of a polygon is a *sub-polygon* if it is a polygon as a subset of $\mathbb{R}^2$.

2. Basics of tropical plane curves

Let $\bar{\Gamma} = (\bar{\Gamma}^{[0]}, \bar{\Gamma}^{[1]})$ be a geometric realization of a weighted finite graph without divalent vertices and $w_{\bar{\Gamma}} : \bar{\Gamma}^{[1]} \to \mathbb{Z}_{\geq 1}$ be the weight function. We denote the set of univalent vertices as $\bar{\Gamma}^{[0]}_{\infty} \subset \bar{\Gamma}^{[0]}$. Set $\Gamma := \bar{\Gamma} \setminus \bar{\Gamma}^{[0]}_{\infty}$. This open graph has the structure of a weighted graph $(\Gamma^{[0]}, \Gamma^{[1]})$ with the weight function $w_\Gamma : \Gamma^{[1]} \to \mathbb{Z}_{\geq 1}$.

Definition 2.1 (Parametrized tropical curve [3, Deifinition 2.2], [4, Definition 1.1]). A proper map $h : \Gamma \to \mathbb{R}^n$ is a *parametrized tropical curve* if the following conditions hold:

1. For any edge $E \in \Gamma^{[1]}$, the restriction $h|_E$ is an embedding and its image is a segment of rational slope.
2. Let $V \in \Gamma^{[0]}$ be a vertex of Γ, $E_1, \ldots, E_m \in \Gamma^{[1]}$ be edges adjacent to V and $u_j \in \mathbb{Z}^n$ be primitive integral vector emanating from $h(V)$ in the direction of $h(E_j)$. Then $\sum_{j=1}^{m} w_\Gamma(E_j)u_j = 0$ holds.

Two parametrized tropical curves $h : \Gamma \to \mathbb{R}^n$ and $h' : \Gamma' \to \mathbb{R}^n$ are *isomorphic* if there exists a homeomorphism $\Phi : \Gamma \to \Gamma'$ respecting the weights such that $h = h' \circ \Phi$.

A 1-dimensional *polyhedral complex* is a finite graph consisting of vertices, edges and half-lines. The image of a parametrized tropical curve has a structure of 1-dimensional polyhedral complex as follows:

Proposition 2.2 (Structure Theorem [3, Theorem 3.15]). *The image of a parametrized tropical curve is a 1-dimensional polyhedral complex $T \subset \mathbb{R}^n$ with a weight function $w : T^{[1]} \to \mathbb{Z}_{>0}$ such that the pair (T, w) satisfies the following conditions:*

- *each edge $E_i \in T^{[1]}$ is a segment or a half-ray in $\mathbb{R}^n$ which has a rational slope $v_i \in \mathbb{Z}^n$,*
- *for any vertex $V \in T^{[0]}$, and the edges $E_1, \ldots E_m \in T^{[1]}$ of T adjacent to V, the balancing condition $\sum_{j=1}^{m} w(E_j)v_j = 0$ holds,*

where $T^{[i]}$ is the set of i-dimensional polyhedra of T.

Such a 1-dimensional polyhedral complex of a tropical curve is exclusively called a *1-dimensional balanced rational polyhedral complex* [2] or a *balanced rectilinear graph* [1]. In this note, we call the image of a parametrized tropical curve a *tropical curve.* A tropical curve T is *reducible* if there exist two proper tropical sub-curves T_1 and T_2 of T such that $T = T_1 \cup T_2$ with multiplicity. A tropical curve T is *irreducible* otherwise. For example, a double tropical line is reducible.

Hereafter, we restrict the discussion to the case of tropical curves in $\mathbb{R}^2$, called *plane tropical curves.* We may omit the word "plane" if there is no confusion.

A plane tropical curve can be given by a non-linear locus of a tropical (Laurent) polynomial.

Theorem 2.3 (Mikhalkin [2, Proposition 2.4]). *Let T be a plane tropical curve. There exists a Laurent polynomial function over the Max-Plus semi-field*

$$\tau(X,Y) := \max_{(i,j)\in A}(a_{ij} + iX + jY) : \mathbb{R}^2 \to \mathbb{R}$$

whose non-linear locus coincides with T, where $A \subset \mathbb{Z}^2$ is a finite set.

Here, the *non-linear locus* of a polynomial $\tau(X,Y)$ over the Max-Plus algebra is the set

$$\{(X,Y) \in \mathbb{R}^2\,;\ \tau \text{ is non-smooth at } (X,Y)\}.$$

Notice that, a polynomial τ is not unique in general. However, the Newton polygon N_τ of τ is determined uniquely, where the Newton polygon N_τ of τ is defined by

$$N_\tau := \mathrm{Conv}(A) = \mathrm{Conv}\left(\{(i,j) \in \mathbb{Z}^2\,;\ a_{ij} \neq -\infty\}\right).$$

For a tropical polynomial τ, the Newton polytope of τ is called the *degree* of T, denoted by $\deg(T)$.

Next, we discuss a combinatorial property of tropical curves. For a tropical polynomial

$$\tau(X,Y) := \max_{(i,j)\in A}(a_{ij} + iX + jY)$$

which defines a tropical curve T, the 3-dimensional polyhedron

$$\hat{\Delta}_T := \mathrm{Conv}\left(\{(i,j,t) \in \mathbb{R}^2 \times \mathbb{R}\,;\ (i,j) \in A, a_{ij} \geq t\}\right)$$

induces a lattice subdivision of $\Delta := N_T$ as follows. Let $\mathrm{proj} : \mathbb{R}^2 \times \mathbb{R} \to \mathbb{R}^2$ be the first projection, and $\hat{\Delta}_1, \ldots, \hat{\Delta}_N$ be the compact facets of $\hat{\Delta}_T$. Then, for each $i = 1, \ldots, N$, the image $\Delta_i := \mathrm{proj}(\hat{\Delta}_i)$ is a sub-polygon of Δ. We denote the set of k-dimensional faces of Δ_i's by $S_T^{[k]}$. We can easily check that the union $S_T := S_T^{[2]} \cup S_T^{[1]} \cup S_T^{[0]}$ makes a polyhedral subdivision of Δ. Such a subdivision satisfies the following theorem.

Theorem 2.4 (Duality Theorem [3, Proposition 3.11]). *There exists a bijection*

$$\Xi : T^{[1]} \cup T^{[0]} \to S_T^{[1]} \cup S_T^{[2]}$$

such that

(1) *the edges of T are in 1-to-1 correspondence with the edges of S_T so that for each edge $E \in T^{[1]}$, the edge $\Xi(E) \in S_T^{[1]}$ is orthogonal to E and has the lattice length $w(E)$,*

(2) *the vertices of T are in 1-to-1 correspondence with the polytopes $\Delta_1, \dots, \Delta_N$ of S_T so that the valency of a vertex $V_i \in T^{[0]}$ is equal to the number of sides of the corresponding polytope $\Xi(V_i) = \Delta_i \in S_T^{[2]}$,*

(3) *the components of $\mathbb{R}^2 \setminus T$ are in 1-to-1 correspondence with the vertices of the subdivision $S_T^{[0]}$.*

Because of this theorem, the subdivision S_T of Δ is called the *dual subdivision* of T.

Next, we introduce the geometric genus of a tropical curve according to the paper of Mikhalkin. For a plane graph Γ, we define the *genus* $g(\Gamma)$ of Γ as the first Betti number rank $H_1(\Gamma)$ of Γ.

Definition 2.5 (Geometric genus [3, Definition 2.9]). For a plane tropical curve T, the (geometric) *genus* $g^{\text{trop}}(T)$ of T is defined by

$$g^{\text{trop}}(T) := \min_{\Gamma} \left\{ g(\Gamma) \,;\, \Gamma \to \mathbb{R}^2 \text{ is a parametrization of } T \right\}.$$

3. Tropical delta number

Let V be a trivalent vertex of a tropical curve, and $v_1, v_2, v_3 \in \mathbb{Z}^2$ and $w_1, w_2, w_3 \in \mathbb{Z}_{\geq 1}$ be the slopes and weights of edges adjacent to V, respectively. Then the *Mikhalkin's multiplicity* of V is defined by

$$m(V) := w_1 w_2 |v_1 \times v_2| = w_2 w_3 |v_2 \times v_3| = w_3 w_1 |v_3 \times v_1|,$$

where $|u \times v|$ means the area of the parallelogram spanned by u and v.

Definition 3.1 (Smooth/Singular vertex). A vertex $V \in T$ is a *tropically smooth vertex* (or *smooth vertex* for simplicity) of T if the following two conditions are satisfied:

- V is trivalent,
- the Mikhalkin's multiplicity of V is 1.

A vertex $V \in T$ is a *tropically singular vertex* (or *singular vertex* for simply) of T if it is not smooth.

For a singular vertex, there is a reasonable way to define locally irreducible components at the singular vertex. Let $V \in T$ be a vertex and $E_1, \dots, E_n$ be edges adjoint to V. The n-valent vertex V is a *locally reducible singularity* if there are positive integers $w_1, \dots, w_n$ such that

- $0 \leq w_i \leq w(E_i)$ for $i = 1, \dots, n$,
- $(w_1, \dots, w_n) \neq (w(E_1), \dots, w(E_n))$ and $(w_1, \dots, w_n) \neq (0, \dots, 0)$, and

- $E_1 \cup E_2 \cup \ldots \cup E_n$ satisfies the balancing condition $w_1 E_n + w_2 E_n + \cdots + w_n E_n = 0$.

Otherwise, it is called *locally irreducible.* We may decompose a singular vertex V into locally irreducible components. The set $\mathcal{D}_V$ of locally irreducible components at V is called an *irreducible decomposition at* V. An example of an irreducible decomposition is given in Figure 1. Remark that, after the decomposition, we need to remove the divalent vertex V_2 and unite the connecting two edges so that the component becomes a line since a tropical curve has no divalent vertices.

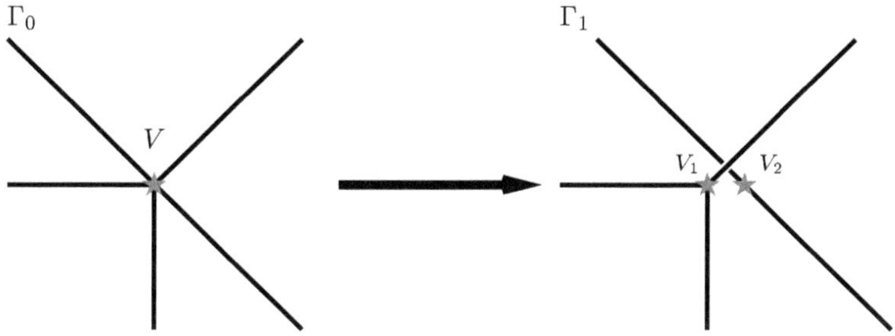

Fig. 1. An example of an irreducible decomposition at the 5-valent vertex V

Remark 3.2. We remark that a choice of a local irreducible decomposition is NOT unique. For example, we consider the triple point $V = (0, 0) \in \mathbb{R}^2$ which consists of the following three lines:

$$\ell_1 := \{x = 0\}, \quad \ell_2 := \{y = 0\}, \quad \ell_3 := \{x = y\} \subset \mathbb{R}^2.$$

Then $\mathcal{D}_1 := \{\ell_1, \ell_2, \ell_3\}$ is a local irreducible decomposition at V. On the other hand, $\mathcal{D}_2 := \{L_1, L_2\}$ is also a local irreducible decomposition at V, where

$$L_1 := (\ell_1 \cap \{y \leq 0\}) \cup (\ell_2 \cap \{x \leq 0\}) \cup (\ell_3 \cap \{x \geq 0\}),$$
$$L_2 := (\ell_1 \cap \{y \geq 0\}) \cup (\ell_2 \cap \{x \geq 0\}) \cup (\ell_3 \cap \{x \leq 0\}).$$

For a local irreducible decomposition $\mathcal{D}$ at V, we call a locally irreducible component of valency ≥ 3 as a *multivalent component* of $\mathcal{D}$, and denote the number of multivalent components of $\mathcal{D}$ as $M_{\mathcal{D}}$. We also call a locally irreducible component which is a line a *line component* of $\mathcal{D}$ and denote the number of line components of $\mathcal{D}$ as $N_{\mathcal{D}}$. We set

$$r_{\mathcal{D}}(V) := N_{\mathcal{D}} + M_{\mathcal{D}} = \sharp\mathcal{D}.$$

We denote the maximal number of $r_{\mathcal{D}}(V)$ among local irreducible decompositions at V as $r(V)$, that is,

$$r(V) := \max_{\mathcal{D}} r_{\mathcal{D}}(V).$$

Hereafter, we only deal with unweighted tropical curves. Here, a tropical curve is *unweighted* if the image of the weight function is $\{1\}$.

Definition 3.3 (Tropical delta number). For a vertex $V \in T$, we define the *tropical delta number* $\delta^{\mathrm{trop}}(T, V)$ as

$$\delta^{\mathrm{trop}}(T, V) := r(V) - 1 + \sharp(\mathrm{Int}(\Xi(V)) \cap \mathbb{Z}^2) \geq 0.$$

For an unweighted tropical curve T, the *total tropical delta number* $\delta^{\mathrm{trop}}_{\mathrm{total}}(T)$ of T is

$$\delta^{\mathrm{trop}}_{\mathrm{total}}(T) := \sum_{V \in T^{[0]}} \delta^{\mathrm{trop}}(T, V).$$

We remark that, for a trivalent vertex $V \in T^{[0]}$, by the Pick's formula, we have

$$m(V) = 2\delta^{\mathrm{trop}}(T, V) + 1.$$

Proposition 3.4.

(1) *The tropical delta number of a vertex of an unweighted tropical curve is a local invariant over any isomorphism $\Phi \in \mathbb{R}^2 \rtimes \mathrm{GL}(\mathbb{Z}^2)$ of tropical curves. Moreover, the total delta number is a global invariant over any isomorphism of tropical curves.*

(2) *Let $V \in T$ be a vertex of an unweighted tropical curve T.*
   (i) *V is smooth if and only if $\delta^{\mathrm{trop}}(T, V) = 0$.*
   (ii) *$\delta^{\mathrm{trop}}_{\mathrm{total}}(T) = 0$ if and only if T is a smooth curve in Mikhalkin's meaning [3, Definition 2.18].*

Proof. (1) This claim is proved by Definition 3.3 directly.
(2) Clearly, it is true that $\delta^{\mathrm{trop}}(T, V) = 0$ if and only if

$$r(V) = 1 \quad \text{and} \quad \sharp(\mathrm{Int}(\Xi(V)) \cap \mathbb{Z}^2) = 0.$$

Therefore we obtain (i). The claim (ii) follows from (i) immediately. □

Note that, Proposition 3.4.(2) is an analogy of the corresponding known facts in the theory of plane curve singularity.

4. Classical singularities vs. tropical singularities

In this section, we study relation between tropical delta numbers and the classical delta invariant. To study the relation, we first introduce the maximal delta number of a polygon, which corresponds to the delta invariant of a regular singularity of a classical curve degenerated by a tropicalization method (see [5] for the definition of a regular singularity). For a polygon $P \subset \mathbb{R}^2$, we set

$$\mathcal{F}(P) := \{f \in \mathbb{C}[z, w]\,;\, f \text{ is reduced and the Newton polygon of } f \text{ is } P\}.$$

We denote the curve in $X(P)$ defined by a polynomial $f \in \mathcal{F}(P)$ as C_f.

Definition 4.1 (Maximal delta number). For a polygon $P \subset \mathbb{R}^2$, we define the *maximal delta number of* P, denoted by $\delta(P)$, as

$$\delta(P) := \max_{f \in \mathcal{F}(P)} \{\delta(C_f, p)\,;\, \mathrm{Sing}(C_f) = \{p\}\},$$

where $\delta(C_f, p)$ is the delta invariant of $p \in C_f$.

Definition 4.2 (Local model of singular vertex). Suppose that for a polygon $P \subset \mathbb{R}^2$, we can obtain a tropical curve T_P which is dual to P with only one vertex V_P uniquely up to the action of $\mathbb{R}^2 \rtimes \mathrm{GL}(\mathbb{Z}^2)$. We call this tropical curve T_P the *local model of singular vertex* defined by P.

Example 4.3 (Classical vs. tropical singularities). We compare classical plane curve singularities and tropical ones in the following examples.

(1) (Tropical normal crossing) Set $P_1 := \mathrm{Conv}\{(0,0), (1,0), (0,1), (1,1)\} \subset \mathbb{R}^2$. Then, the local model defined by P_1 consists of two lines ℓ_1 and ℓ_2 intersecting transversally (see Figure 2). We call this singularity (T_{P_1}, V_{P_1}) the *tropical node*. It satisfies $\delta^{\mathrm{trop}}(T_{P_1}, V_{P_1}) = \delta(P_1) = 1$.

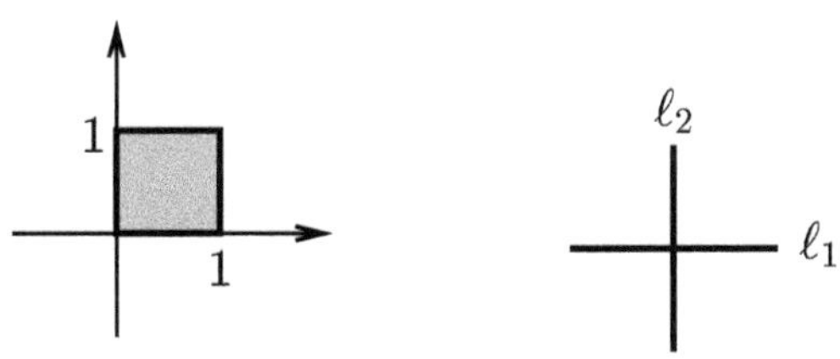

Fig. 2. The polygon P_1 and the tropical normal crossing T_{P_1}

(2) (Tropical triple point) Set $P_2 := \mathrm{Conv}\{(0,0),(1,0),(0,1),(2,1),(1,2),(2,2)\} \subset \mathbb{R}^2$. Then, the local model defined by P_2 consists of three lines ℓ_1, ℓ_2 and ℓ_3 intersecting at one point (see Figure 3). We call this singularity (T_{P_2}, V_{P_2}) the *tropical triple point*. It satisfies $\delta^{\mathrm{trop}}(T_{P_2}, V_{P_2}) = \delta(P_2) = 3$.

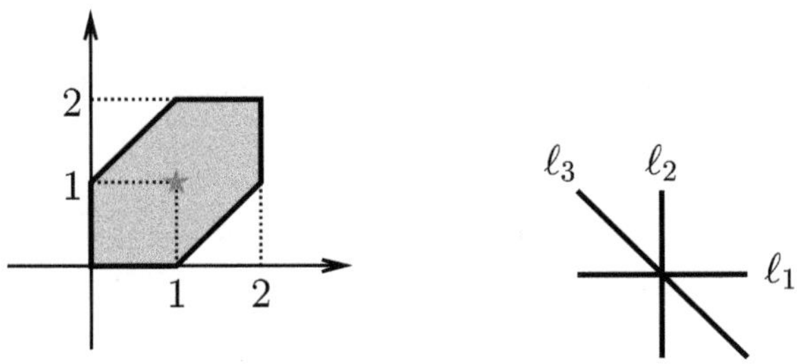

Fig. 3. The polygon P_2 and the tropical triple point T_{P_2}

(3) (Tropical ordinal cusp [5]) Set $P_3 := \mathrm{Conv}\{(1,0),(2,0),(0,1),(1,2)\} \subset \mathbb{R}^2$. Then, the local model defined by P_3 is a tree with a 4-valent vertex (see Figure 4). We remark that this polygon appears in [5, Section 4.1], where it is denoted as Δ_1. It satisfies $\delta^{\mathrm{trop}}(T_{P_3}, V_{P_3}) = \delta(P_3) = 1$, where the latter equality follows from [5, Lemma 4.2 (i) and (iv)].

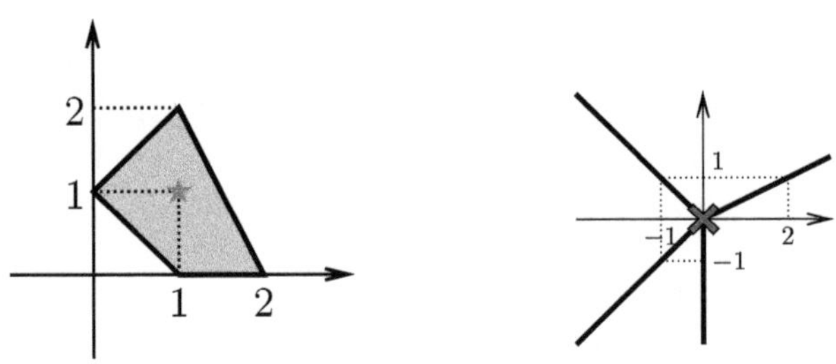

Fig. 4. The polygon P_3 and the local model T_{P_3}

(4) (Tropical tacnode [6]) Set $P_4 := \mathrm{Conv}\{(0,0),(1,0),(0,1),(2,1),(1,2)\} \subset \mathbb{R}^2$. Then, the local model defined by P_4 is a graph obtained as the union of a line ℓ and a tropical line C, that is, a smooth trivalent graph (see Figure 5). We remark that this polygon appears in [6], where it

is denoted as Δ_{VII}. It satisfies $\delta^{\mathrm{trop}}(T_{P_4}, V_{P_4}) = \delta(P_4) = 2$, where the latter equality follows from [6, Lemma 3.5].

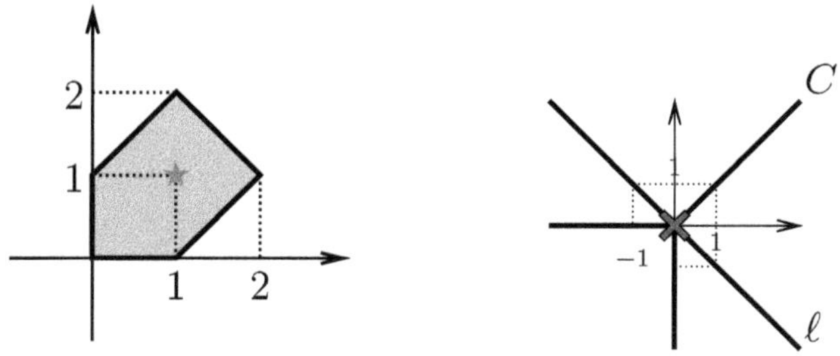

Fig. 5. The polygon P_4 and the local model T_{P_4}

Example 4.4 (Ordinal node, cusp and tacnode). The local models of the following examples are non-smooth trivalent graphs.

(a) (Tropical ordinal node [3]) Set $P_a := \mathrm{Conv}\{(0,0),(1,2),(2,1)\} \subset \mathbb{R}^2$. It satisfies $\delta^{\mathrm{trop}}(T_{P_a}, V_{P_a}) = \delta(P_a) = 1$.
(b) (Tropical ordinal cusp [5]) Set $P_b := \mathrm{Conv}\{(0,0),(3,2),(2,3)\} \subset \mathbb{R}^2$. It satisfies $\delta^{\mathrm{trop}}(T_{P_b}, V_{P_b}) = 2$ though $\delta(P_b) = 1$ by [5, Lemma 4.2 (ii)].
(c) (Tropical ordinal tacnode [6]) Set $P_c := \mathrm{Conv}\{(0,0),(4,3),(3,4)\} \subset \mathbb{R}^2$. It satisfies $\delta^{\mathrm{trop}}(T_{P_c}, V_{P_c}) = 3$ though $\delta(P_b) = 2$ by [6, Lemma 3.4].

A polygon $P \subset \mathbb{R}^2$ is *prime* if there does NOT exist 1- or 2-dimensional polytopes $P_1, \ldots, P_n \subset \mathbb{R}^2$ for some $n \in \mathbb{Z}_{\geq 2}$ such that $P = P_1 + \cdots + P_n$, where the notation $+$ means the Minkowski sum of polygons. Note that if V_P is an irreducible singular vertex then P is a prime polygon. But the converse is not true in general.

From the above Examples 4.3 and 4.4, it can be inferred that the inequality

$$\delta^{\mathrm{trop}}(T_P, V_P) \geq \delta(P) \tag{4.1}$$

holds. This is true if the polygon is prime but we do not know if it holds or not in general.

Theorem I. *Inequality* (4.1) *holds for a prime polygon* $P \subset \mathbb{R}^2$.

Proof. By the assumption, for any polynomial $f \in \mathcal{F}(P)$, the curve $C_f \subset X(P)$ defined by f is irreducible. For a polynomial $f \in \mathcal{F}(P)$ such that the curve C_f has exactly one singular point $p \in C_f$, we have

$$\begin{aligned} g(C_f) &= \sharp(\mathrm{Int}(P) \cap \mathbb{Z}^2) - \delta(C_f, p) \\ &= \delta^{\mathrm{trop}}(T_P, V_P) - (r(V) - 1) - \delta(C_f, p) \geq 0, \end{aligned}$$

by the classical toric degree-genus formula. As a result, we obtain

$$\delta^{\mathrm{trop}}(T_P, V_P) \geq \delta^{\mathrm{trop}}(T_P, V_P) - (r(V) - 1) \geq \delta(P).$$

□

5. Tropical degree-genus formula

We recall the assertion in the second theorem.

Theorem II (Tropical degree-genus formula). *For an unweighted irreducible tropical curve T whose degree is Δ, the following equality holds:*

$$g^{\mathrm{trop}}(T) = \sharp(\mathrm{Int}\Delta \cap \mathbb{Z}^2) - \delta^{\mathrm{trop}}_{\mathrm{total}}(T). \tag{5.1}$$

Before proving the theorem, we characterize the genus of a graph combinatorially:

Lemma 5.1. *The genus of an open graph Γ can be obtained as*

$$\begin{aligned} g(\Gamma) =& \sharp(\textit{edges of}\,\Gamma) - \sharp(\textit{vertices of}\,\Gamma) \\ &+ \sharp(\textit{connected components of}\,\Gamma) - \sharp(\textit{open edges of}\,\Gamma). \end{aligned}$$

Proof. Let $\bar{\Gamma}$ be a finite graph such that $\Gamma = \bar{\Gamma} \setminus \bar{\Gamma}_\infty$, where $\bar{\Gamma}_\infty$ is the set of univalent vertices of $\bar{\Gamma}$. If there is a line component, we add a divalent vertex in the middle. Then we have

$$\begin{aligned} g(\Gamma) &= \mathrm{rank} H_1(\bar{\Gamma}) \\ &= \big(\mathrm{rank} H_1(\bar{\Gamma}) - \mathrm{rank} H_0(\bar{\Gamma}) + 1\big) + \mathrm{rank} H_0(\bar{\Gamma}) - 1 \\ &= \sharp(\text{edges of } \bar{\Gamma}) - \sharp(\text{vertices of } \bar{\Gamma}) + \sharp(\text{connected components of } \bar{\Gamma}). \end{aligned}$$

Therefore, for the open graph Γ, one can obtain that

$$\begin{aligned} g(\Gamma) =& \sharp(\text{edges of } \Gamma) - \sharp(\text{vertices of } \Gamma) \\ &+ \sharp(\text{connected components of } \Gamma) - \sharp(\text{open edges of } \Gamma). \end{aligned}$$

□

Proof of Theorem II. Firstly, by Lemma 5.1, for a connected open graph Γ, the genus of Γ can be obtained as

$$g(\Gamma) = \sharp(\text{edges of } \Gamma) - \sharp(\text{vertices of } \Gamma) + 1 - \sharp(\text{open edges of } \Gamma).$$

We regard T as a plane graph and denote it as Γ_0. Let $V \in \Gamma_0$ be a vertex. As an irreducible decomposition at V, we take a decomposition $\mathcal{D}_V$ such that $r(V) = r_{\mathcal{D}_V}(V)$. We denote the graph with irreducible decomposition around V as Γ_1. Then the genera $g(\Gamma_0)$ and $g(\Gamma_1)$ satisfy the following relation:

$$\begin{aligned} g(\Gamma_1) &= \sharp(\text{edges of } \Gamma_0) - \{\sharp(\text{vertices of } \Gamma_0) + M_{\mathcal{D}_V} + N_{\mathcal{D}_V} - 1\} \\ &\quad + 1 - \sharp(\text{open edges of } \Gamma_0) \\ &= g(\Gamma_0) - (r(V) - 1), \end{aligned}$$

since the numbers $\sharp(\text{open edges of } \Gamma_0)$ and $\sharp(\text{open edges of } \Gamma_1)$ are the same. We denote the graph obtained by applying this operation to all vertices by Γ. Notice that the graph Γ may have divalent vertices. We regard the divalent vertex with adjacent two edges as one edge, and denote the new graph as $\widetilde{\Gamma}$. Note that $g(\Gamma) = g(\widetilde{\Gamma})$. Thus, we obtain a parametrization $h : \widetilde{\Gamma} \to \mathbb{R}^2$ of T which satisfies $g^{\text{trop}}(T) = g(\widetilde{\Gamma})$. We remark that, by Duality Theorem, we have

$$g(\Gamma_0) = \sharp(\text{Int}(\Delta) \cap \mathbb{Z}^2) - \sum_{P \in S_T^{[2]}} \sharp(\text{Int}(P) \cap \mathbb{Z}^2). \tag{5.2}$$

By using equality (5.2), we get

$$\begin{aligned} g(\widetilde{\Gamma}) &= g(\Gamma_0) - \sum_{V \in T^{[0]}} (r(V) - 1) \\ &= \sharp(\text{Int}(\Delta) \cap \mathbb{Z}^2) - \sum_{P \in S_T} \sharp(\text{Int}(P) \cap \mathbb{Z}^2) - \sum_{V \in T^{[0]}} (r(V) - 1) \\ &= \sharp(\text{Int}(\Delta) \cap \mathbb{Z}^2) - \sum_{V \in T^{[0]}} \{r(V) - 1 + \sharp(\text{Int}(\Xi(V)) \cap \mathbb{Z}^2)\} \\ &= \sharp(\text{Int}(\Delta) \cap \mathbb{Z}^2) - \sum_{V \in T^{[0]}} \delta^{\text{trop}}(T, V). \end{aligned}$$

Thus the formula in the assertion holds. □

Example 5.2 (Irreducible tropical curve). Let T_1 be a tropical quartic projective curve on the left in Figure 6. Note that, the tropical curve is irreducible and has a reducible singular vertex (the vertex is highlighted with $\star$ in the figure). The graph Γ_1 on the middle in Figure 6 is a parametrization of T_1 whose genus attains the genus of T_1. Therefore, by Lemma 5.1, we obtain

$$\begin{aligned} g^{\text{trop}}(T_1) &= \sharp(\text{edges of } \Gamma_1) - \sharp(\text{vertices of } \Gamma_1) \\ &\quad + \sharp(\text{connected components of } \Gamma_1) - \sharp(\text{open edges of } \Gamma_1) \\ &= 24 - 12 + 1 - 12 = 1. \end{aligned}$$

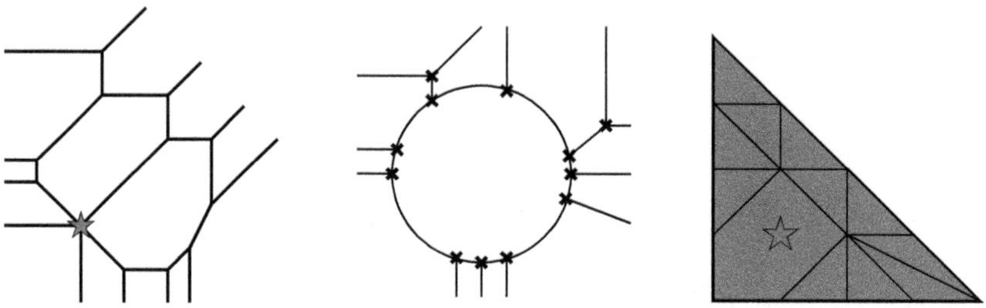

Fig. 6. An irreducible singular tropical quartic projective curve T_1, the minimal parametrization Γ_1 (the notation $\times$ means a vertex of Γ_1), and the dual subdivision of T (the notation $\star$ means an interior lattice point of a sub-polygon).

On the other hand, we can calculate that

$$\sharp(\text{Int}\Delta_4 \cap \mathbb{Z}^2) - \delta^{\text{trop}}_{\text{total}}(T) = 3 - 2 = 1,$$

where $\Delta_4 := \text{Conv}\{(0,0),(4,0),(0,4)\}$, and thus equality (5.1) in Theorem II is verified.

Example 5.3 (Reduced tropical curve). The irreducibility in the condition of the Theorem II is necessary. Actually, the following example does not satisfy equality (5.1). Let T_2 be a tropical quartic projective curve on the left in Figure 7. Note that, the dual subdivision of T_2 is as Figure 8. Note also that, the tropical curve has two components (their intersection is the vertex which is highlighted with $\star$ in the figure). The graph Γ_2 on the right in Figure 7 is a parametrization of T_2 whose genus attains the genus of T_2. By the same observation as in the previous example, we obtain $g^{\text{trop}}(T_2) = 21-10+2-11 = 2$ though $\sharp(\text{Int}\Delta_4 \cap \mathbb{Z}^2) - \delta^{\text{trop}}_{\text{total}}(T) = 3-3 = 0$. Thus equality (5.1) does not hold in this case.

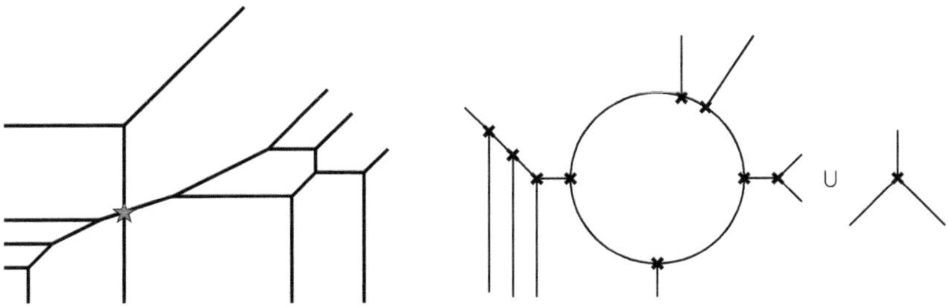

Fig. 7. A reduced singular tropical quartic projective curve T_2 and the minimal parametrization Γ_2 (the notation $\times$ means a vertex of Γ_2).

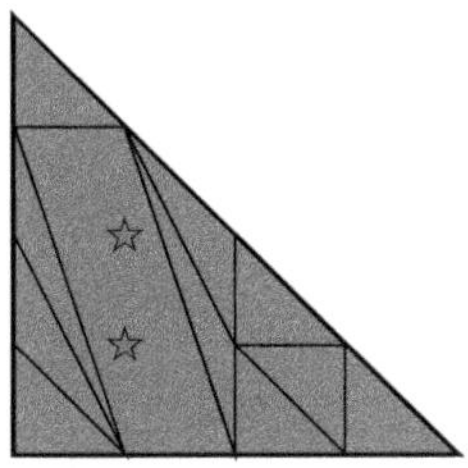

Fig. 8. The dual subdivision of T_2 (the notation $\star$ means an interior lattice point of a sub-polygon).

References

[1] E. Brugallé, I. Itenberg, G. Mikhalkin, K. Shaw, *Brief introduction to tropical geometry.* Proceedings of the Gökova Geometry-Topology Conference 2014, 1–75.

[2] G. Mikhalkin, *Decomposition into pair-of-pants for complex algebraic hypersurfaces.* Topology **43** (2004), 1035–1065.

[3] G. Mikhalkin, *Enumerative Tropical Algebraic Geometry in* $\mathbb{R}^2$. J. Amer. Math. Soc. **18** (2005), 313–377.

[4] T. Nishinou, B. Siebert, *Toric degenerations of toric varieties and tropical curves.* Duke. Math. J. **135** (2006), no. 1, 1–51.

[5] E. Shustin, *A tropical approach to enumerative geometry.* Algebra i Analiz **17** Issue 2 (2005), 170–214.

[6] T. Takahashi, *Tropicalization of 1-tacnodal curves on toric surfaces,* arXiv:1709.00788, to appear in Osaka J. Math.

[7] C. T. C. Wall, *Singular Points of Plane Curves*, London Math. Soc. Student Texts 63, 2004.

Maximal ideal cycles and multiplicities for cyclic coverings over normal surface singularities

Tadashi Tomaru

Gunma University, Aramaki campus, Room No GA-306,
Aramaki 4-2, Maebashi, Gunma 371-8510, Japan
E-mail: ttomaru@gunma-u.ac.jp

Let (X, o) be a normal complex surface singularity and h a reduced element of the maximal ideal $\mathfrak{m}_{X,o}$ of $\mathcal{O}_{X,o}$. Let $\pi : (\tilde{X}, E) \to (X, o)$ be the h-minimal good resolution (see Definition 3.1). Let (X_n, o) be the n-fold cyclic covering over (X, o) defined by $z^n = h$. Let $\pi_n : (\tilde{X}_n, E(n)) \to (X_n, o)$ be the h-minimal good resolution. In this paper, we study the maximal ideal cycle on $E(n)$ and embedded points of the pull-back of $\mathfrak{m}_{X,o}$ onto $\tilde{X}_n$. From them, we study the value of $\mathrm{mult}(X_n, o)$ (:=the multiplicity of (X_n, o)). Under a condition about h and a defining function of the maximal ideal cycle on E, we prove that $\mathrm{mult}(X_n, o)$ takes a constant value for any n with $n \geqq N_1(h)$, where $N_1(h)$ is a positive integer determined by (X, o) and h (see Theorem 3.6 (iv)).

Keywords: normal surface singularity, maximal ideal cycle, multiplicity

1. Introduction

In this paper, we consider normal complex surface singularities. Let $\pi\colon (\tilde{X}, E) \to (X, o)$ be a resolution of a normal complex surface singularity, where E denotes the exceptional set and $E = \bigcup_{i=1}^{r} E_i$ is the irreducible decomposition. Let Z_E be the fundamental cycle on E, (i.e., $Z_E := \min\{D = \sum_{i=1}^{r} a_i E_i \mid a_i > 0 \text{ and } DE_i \leqq 0 \text{ for any } i\}$). Then, the value Z_E^2 is independent of the choice of π; therefore we put $Z^2 := Z_E^2$.

Definition 1.1 (S.S.T. Yau [Y]). The maximal ideal cycle is defined by $M_E := \min\{(f \circ \pi)_E \mid f \in \mathfrak{m}_{X,o}, f \neq 0\}$, where $(f \circ \pi)_E := \sum_{i=1}^{r} \mathrm{ord}(f \circ \pi)_{E_i} E_i$ and $\mathrm{ord}(f \circ \pi)_{E_i}$ means the vanishing order of $f \circ \pi$ on E_i.

In [A], M. Artin defined the class of rational singularities and proved the following fundamental result.

Theorem 1.2 ([A]). *If (X, o) is a rational singularity, then $M_E = Z_E$ for any resolution and* $\mathrm{mult}(X, o) = -Z^2$.

Therefore, in the case of rational singularities, the value of $\mathrm{mult}(X, o)$ is determined by the topological structure of (X, o). However, in general case, the value of $\mathrm{mult}(X, o)$ and M_E^2 depend on analytic structure of (X, o). Also the later depends on the choice of a resolution π. Since $(f \circ \pi)_{\tilde{X}} \sim 0$ (linearly equivalent) and $(f \circ \pi)_E E_i \leqq 0$ for any i, we have $M_E \geqq Z_E$ and so $-M_E^2 \leqq -Z_E^2$. The following is a fundamental result for M_E and Z_E when $\mathrm{mult}(X, o)$ of a normal surface singularity (X, o) is discussed (for example, see [N, O]). In the following, we put $\mathfrak{m}\mathcal{O}_{\tilde{X}} := \mathfrak{m}_{X,o}\mathcal{O}_{\tilde{X}}$.

Theorem 1.3 (P. Wagreich [W]).

(i) $\mathrm{mult}(X, o) \geqq -M_E^2 \geqq -Z^2$.

(ii) $\mathfrak{m}\mathcal{O}_{\tilde{X}}$ *is locally principal* $\Leftrightarrow$ $\mathfrak{m}\mathcal{O}_{\tilde{X}} = \mathcal{O}_{\tilde{X}}(-M_E)$ $\Leftrightarrow$ $\mathrm{mult}(X, o) = -M_E^2$.

On the comparison of M_E and Z_E, the following result is well-known.

Theorem 1.4 (H. Laufer [La]). *Assume that $\pi\colon (\tilde{X}, E) \to (X, o)$ is the minimal resolution of a minimally elliptic singularity. Then*

(i) $M_E = Z_E$.

(ii) *If $Z^2 \leqq -2$, then $\mathfrak{m}\mathcal{O}_{\tilde{X}} = \mathcal{O}_{\tilde{X}}(-Z_E)$.*

(iii) $\mathrm{mult}(X, o) = \max(2, -Z^2)$.

If $\mathfrak{m}\mathcal{O}_{\tilde{X}}$ is not locally principal at $P \in \tilde{X}$, then P is called an *embedded point* of $\mathfrak{m}\mathcal{O}_{\tilde{X}}$. We remark that the embedded points of $\mathfrak{m}\mathcal{O}_{\tilde{X}}$ coincide the base points of the linear system $|\mathcal{O}_{\tilde{X}}(-M_E)|$. After suitable blowing-ups at embedded points, $\mathfrak{m}\mathcal{O}_{\tilde{X}}$ become to be locally principal (cf. [C, p. 43]).

Definition 1.5. Let $P \in \tilde{X}$ be an embedded point of $\mathfrak{m}\mathcal{O}_{\tilde{X}}$. Let $\delta_P(\tilde{X})$ (or $\delta(P)$) be the minimal number of blowing-ups needed to resolve the embedded point. Also, let $\delta(\tilde{X})$ be the total sum of $\delta_P(\tilde{X})$ for all embedded points.

In [KN], Konno-Nagashima studied the maximal ideal cycles for hypersurface singularities of Brieskorn type and proved a necessary and sufficient condition to be $M_E = Z_E$ on the minimal resolution $(\tilde{X}, E)$. Also, they determined $\delta_P(\tilde{X})$ for any embedded point P of $\mathfrak{m}\mathcal{O}_{\tilde{X}}$ ([KN, Proposition 3.9]). Further, Meng-Okuma [MO] generalized those results to complete intersection singularities of Brieskorn type. Those complete intersection singularities are very special objects among normal surface singularities. For those singularities, without the argument on maximal ideal cycles, the value of the multiplicity is immediately from the defining equations (see [MO, Lemma 6.3]). In our work, the computation of the maximal ideal cycle is essential to compute the multiplicity. However our work is inspired from the results in [KN] and [MO].

Let (X, o) be a normal surface singularity and h a reduced element of $\mathfrak{m}_{X,o}$. Let N be the embedded dimension of (X, o) and I_X the defining ideal of (X, o) in $\mathbb{C}^N$. Let (X_n, o) be the n-fold cyclic covering over (X, o) defined by $z^n = h$. Namely, the defining ideal of (X_n, o) is given by the ideal generated by I_X and $z^n - h$ in $\mathbb{C}\{z, x_1, \cdots, x_N\}$. Since h is reduced, (X_n, o) is a normal surface singularity by Tomari-Watanabe's result ([TW, Theorem 3.2]). Obviously, we have $\mathrm{mult}(X_n, o) \leqq n \cdot \mathrm{mult}(X, o)$. However, this is a too rough estimate. In this paper, under some conditions, we study the maximal ideal cycle and embedded points of $\mathfrak{m}\mathcal{O}_{\tilde{X}''_n}$, where $\mathfrak{m}\mathcal{O}_{\tilde{X}''_n}$ is the pull back of $\mathfrak{m}_{X_n,o}$ by a good resolution $\phi := \phi_1 \circ \phi_2 \circ \phi_3$ of (X_n, o) (see (3.1)). Using it, we prove our main result (Theorem 3.6 (iv)).

2. Preliminaries

For integers n, q with $1 \leqq q < n$ and $\gcd(n, q) = 1$, a cyclic quotient singularity of type $C_{n,q}$ is defined as the quotient space $\mathbb{C}^2/\langle e_n \rangle$ by the natural action $e_n \cdot (x, y) := (e_n x, e_n^q y)$ on $\mathbb{C}^2$, where $e_n := \exp(2\pi\sqrt{-1}/n)$. In the following, we prepare some facts. Let $\bar{W}$ be the analytic space $\{(x, y, z) \in U \times \mathbb{C} \mid z^n = x^a y^b\}$, where U is a small neighborhood of the origin of $\mathbb{C}^2$ and a, b, n are positive integers with $n \geqq 2$. Let us define three integers a_1, b_1 and n_1 as follows:

$$a_1 = \frac{a}{\gcd(a, \mathrm{lcm}(n, b))}, \; b_1 = \frac{b}{\gcd(b, \mathrm{lcm}(n, a))} \text{ and } n_1 = \frac{n}{\gcd(n, \mathrm{lcm}(a, b))}.$$

Lemma 2.1 ([T1], Lemma 2.5). *The normalization of $\bar{W}$ is a disjoint union of* $\gcd(a, b, n)$ *cyclic quotient singularities of type $C_{n_1,\mu}$, where μ is a non-negative integer defined by $a_1\mu + b_1 \equiv 0$ (mod n_1) and $0 \leqq \mu < n_1$.*

Let us prepare two lemmas for divisors on the minimal resolution of cyclic quotient singularities. Consider the continued fractional expansion $\dfrac{n}{q} = b_1 - \cfrac{1}{\ddots - \cfrac{1}{b_r}}$ (b_i is a positive integer with $b_i \geqq 2$), and consider the following configuration:

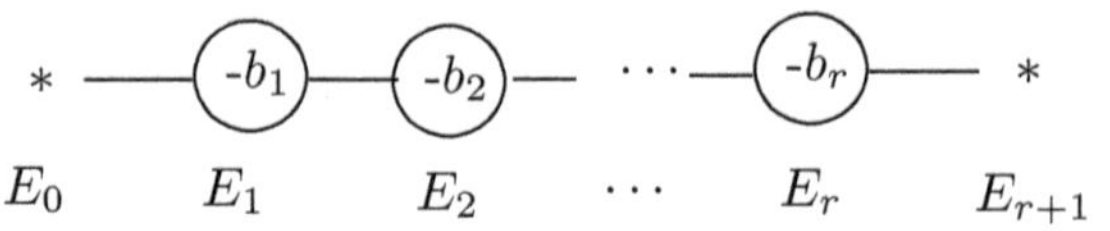

(2.1)

where E_0 (resp. E_{r+1}) is a smooth complex curve which intersects with $E := \bigcup_{i=1}^{r} E_i$ at only one point in E_1 (resp. E_r) transversally; also $\bigcup_{i=1}^{r} E_i$ is the exceptional set of the minimal resolution of $C_{n,q}$.

Lemma 2.2 ([T3], (3.3)). *Under the condition above, let* $D = \sum_{i=0}^{r+1} a_i E_i$ *be an effective divisor with* $a_i > 0$ *for any* i. *If* $DE_i = 0$ *for* $i = 1, \cdots, r$, *then* $a_1 = \dfrac{a_0 q + a_{r+1}}{n}$ *and* $a_r = \dfrac{a_{r+1} q' + a_0}{n}$, *where* q' *is an integer defined by* $qq' \equiv 1 \pmod{n}$ *and* $1 \leqq q' < n$.

Lemma 2.3 ([TT], Lemma 3.3). *Under the condition above, consider two effective divisors* $D_1 = \sum_{i=0}^{r+1} a_i E_i$ *and* $D_2 = \sum_{i=0}^{r+1} c_i E_i$ *such that* $D_k E_i = 0$ *for* $i = 1, \cdots, r$ *and* $k = 1, 2$. *Assume that* $a_0 \geqq c_0$. *Then we have the following:*

(i) *If* $a_{r+1} \geqq c_{r+1}$, *then* $D_1 \geqq D_2$.

(ii) *If* $a_{r+1} = 0$ *and* $c_{r+1} = 1$, *then* $D_1|_{\bigcup_{i=0}^{r} E_i} \geqq D_2|_{\bigcup_{i=0}^{r} E_i}$, *where* $D_k|_{\bigcup_{i=0}^{r} E_i}$ *means the restriction of* D_k *onto* $\bigcup_{i=0}^{r} E_i$ $(k = 1, 2)$.

3. Maximal ideal cycles and multiplicities of cyclic coverings over normal surface singularities

Let (X, o) be a normal surface singularity and h an element of $\mathfrak{m}_{X,o}$. Let $\sigma\colon (\tilde{X}, E) \to (X, o)$ be a good resolution such that $\mathrm{red}((h \circ \sigma)_{\tilde{X}})$ is a simple normal crossing divisor on $\tilde{X}$, where $\mathrm{red}(D)$ means the reduced form $\sum_{i=1}^{r} F_i$ for a divisor $D = \sum_{i=1}^{r} a_i F_i$ $(a_i \neq 0)$ on $\tilde{X}$.

Definition 3.1. If an irreducible component E_i of E intersects at least other three irreducible components of $\mathrm{supp}(h \circ \sigma)_{\tilde{X}}$, then we call E_i *an h-node component of E*. If E does not contain any (-1)-curve (i.e., a non-singular rational curve whose self-intersection number is -1) except for h-node components, then π is called *the h-minimal good resolution.* For a given h, such resolution always exists uniquely.

In the following, we assume that h is a reduced element of $\mathfrak{m}_{X,o}$. Let $\sigma\colon (\tilde{X}, E) \to (X, o)$ be the h-minimal good resolution. For the n-fold cyclic covering (X_n, o) over (X, o) defined by $z^n = h$, we construct the z-minimal good resolution of (X_n, o) through the following commutative diagram:

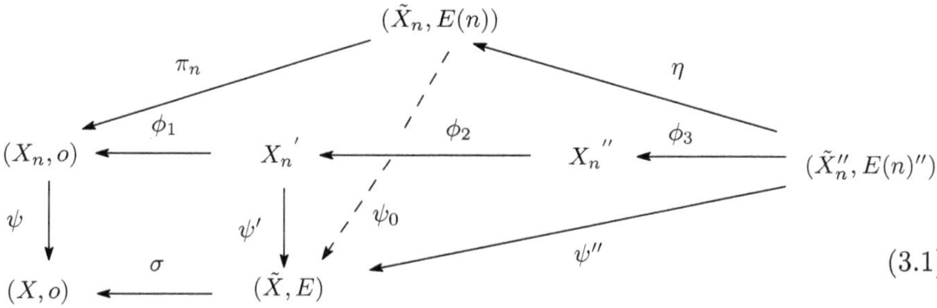

(3.1)

where $X'_n := X_n \times_X \tilde{X}$ and ψ is the natural projection; also ϕ_2 is the normalization. All singularities of X''_n are cyclic quotient singularities. Let ϕ_3 be the minimal resolution of all those singularities of X''_n. Then, $\phi := \phi_1 \circ \phi_2 \circ \phi_3 : (\tilde{X}''_n, E(n)'') \to (X_n, o)$ is a good resolution such that $\mathrm{red}((z \circ \phi)_{\tilde{X}''_n})$ is a simple normal crossing divisor. After a suitable succession η of contractions of (-1)-curves which are not z-nodes, we can obtain the z-minimal good resolution $\pi_n : (\tilde{X}_n, E(n)) \to (X_n, o)$ from ϕ. Furthermore, since $\psi'' := \psi' \circ \phi_2 \circ \phi_3$ is a holomorphic map, we obtain a generically finite meromorphic map ψ_0 such that $\psi_0|_{\tilde{X}_n \setminus E(n)}$ is a holomorphic map.

Let $E = \bigcup_{i=1}^{r} E_i$ be the irreducible decomposition of E. Let f be any element of $\mathfrak{m}_{X,o}$. Put $(f \circ \sigma)_{\tilde{X}} = \sum_{i=1}^{r} \mathrm{Coeff}_{E_i}(f \circ \sigma)_{\tilde{X}} E_i + \sum_{j=1}^{s} \mathrm{Coeff}_{G_j}(f \circ \sigma)_{\tilde{X}} G_j$, where $\bigcup_{j=1}^{s} G_j$ is the non-exceptional part of the curve defined by $f \circ \sigma = 0$ on $\tilde{X}$. Let $E_i{}'' := (\psi'')^{-1}_{*}(E_i)$ and $G_j{}'' := (\psi'')^{-1}_{*}(G_j)$ (i.e., the strict transforms) of E_i and G_j by ψ'' respectively. Tough they are not always irreducible curves, we have the following.

Lemma 3.2 ([T2], Lemma 4.3). *Let F be an irreducible component of* $\mathrm{supp}(f \circ \sigma)_{\tilde{X}}$ (i.e., *F is E_i or G_i*). *Let $F'' := (\psi'')^{-1}_{*}(F)$* (i.e., *strict transform of F by ψ''*). *Let $\alpha := \mathrm{Coeff}_F(h \circ \sigma)_{\tilde{X}}$ and $\beta := \mathrm{Coeff}_F(f \circ \sigma)_{\tilde{X}}$. Then, we have the following:*

(i) $\mathrm{Coeff}_{F''}(z \circ \phi)_{\tilde{X}''_n} = \dfrac{\alpha}{\gcd(n, \alpha)}$.

(ii) $\mathrm{Coeff}_{F''}(f \circ \sigma \circ \psi'')_{\tilde{X}''_n} = \dfrac{n\beta}{\gcd(n, \alpha)}$.

Now let us return to the situation of (3.1). Let $g \in \mathfrak{m}_{X,o}$ be an element satisfying $(g \circ \sigma)_E = M_E$ (maximal ideal cycle). Assume that $(h \circ \sigma)_{\tilde{X}} = \sum_{i=1}^{r} \operatorname{Coeff}_{E_i}(h \circ \sigma)_{\tilde{X}} E_i + C$, where C is the non-exceptional part.

Definition 3.3. Let $\bar{r} := \sharp\{i \mid E_i \cap C \neq \emptyset\}$; and let $C_{i,1}, \cdots, C_{i,s_i}$ be irreducible components of C intersecting E_i ($1 \leqq i \leqq \bar{r}$); thus $\bigcup_{i=1}^{\bar{r}} \bigcup_{j=\ell}^{s_i} C_{i,\ell}$ is the irreducible decomposition of C. We define several positive integers as follows:

(i) $a_i := \operatorname{Coeff}_{E_i}(h \circ \sigma)_E$ and $d_i := \operatorname{Coeff}_{E_i} M_E \quad (i = 1, \cdots, r)$.

(ii) $N_0(h) := \max\{a_i \mid i = 1, \cdots, \bar{r}\}$ and $N_1(h) := \max\{\lceil \frac{a_i}{d_i} \rceil \mid i = 1, \cdots, \bar{r}\}$, where $\lceil \alpha \rceil$ is the round up of $\alpha \in \mathbb{R}$.

In the following, we describe and prove our results.

Theorem 3.4. *If $n \geqq N_1(h)$, then $M_{E(n)} = (z \circ \pi_n)_{E(n)}$.*

Proof. Since η is a succesion of contractions of (-1)-curves and $\phi = \pi_n \circ \eta$, it is sufficient to prove the following:

$$M_{E(n)''} = (z \circ \phi)_{E(n)''}. \tag{3.2}$$

Let E_i, E_j be irreducible components of E with $E_i \cap E_j \neq \emptyset$. Let U be a small open neighborhood of the intersection point $E_i \cap E_j$. Let (u, v) be the coordinate on U such that $E_i \cap U = \{u = 0\}$ and $E_j \cap U = \{v = 0\}$; also let $a_k := \operatorname{Coeff}_{E_k}(h \circ \sigma)_E$ for $k = i, j$. Let $\bar{W} = \{(u, v, z) \in U \times \mathbb{C} \mid z^n = u^{a_i} v^{a_j}\}$; also let $\tilde{U}$ be a connected component of $\phi_2^{-1}(\bar{W})$. By Lemma 2.1, $\tilde{U}$ is a non-singular surface or a normal surface with only one cyclic quotient singularity. If we put $W := \phi_3^{-1}(\tilde{U})$, then W is a connected component of $(\psi'')^{-1}(U)$. Then W is a non-singular surface or the minimal resolution space of a cyclic quotient singularity. If $\tilde{U}$ contains a cyclic quotient singularity, the configuration of the exceptional set $\bigcup_{k=1}^{t} E''_{i,j,k}$ in W is given as follows:

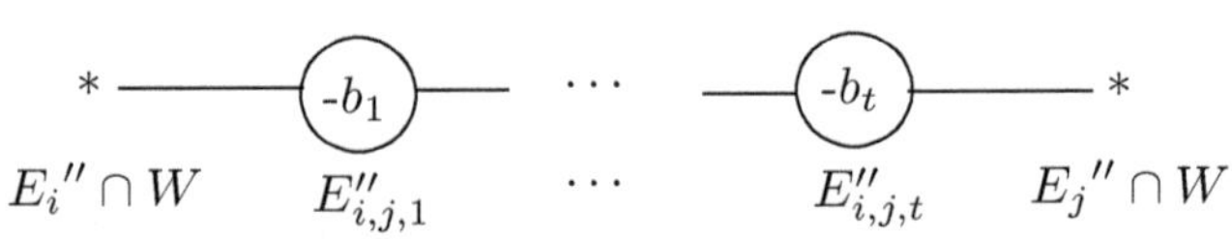

where E''_i and E''_j are the strict transforms of E_i and E_j by ψ'' respectively. On the other hand, from the assumption $n \geqq N_1(h)$, we have $n \geqq \lceil \frac{a_i}{d_i} \rceil \geqq$

$\dfrac{a_i}{d_i}$ for $i = 1, \cdots, \bar{r}$. Put $\bar{a}_i := \dfrac{a_i}{\gcd(a_i, n)}$ and $\bar{n} := \dfrac{n}{\gcd(a_i, n)}$. From Lemma 3.2, we have

$$\mathrm{Coeff}_{E_i''}(z \circ \phi)_{E(n)''} = \bar{a}_i \leqq d_i \bar{n} = \mathrm{Coeff}_{E_i''}(g \circ \sigma \circ \psi'')_{E(n)''} \tag{3.3}$$

for any i.

From Lemma 2.3 (i), we have

$$\mathrm{Coeff}_{E''_{i,j,k}}(z \circ \phi)_{E(n)''} \leqq \mathrm{Coeff}_{E''_{i,j,k}}(g \circ \sigma \circ \psi'')_{E(n)''} \tag{3.4}$$

for $k = 1, \cdots, t$.

From now on, let us consider (3.2) in the case of $E_i \cap C_{i,\ell} \neq \emptyset$ $(1 \leqq \ell \leqq s_i)$. As in above, let U' be a small open neighborhood of the point $E_i \cap C_{i,\ell}$. If we put $W' := (\psi'')^{-1}(U')$, then W' is a connected open set in $\tilde{X}_n''$ because h is a reduced element. Then, W' is a non-singular surface or the minimal resolution space of a cyclic quotient singularity. If W' is the former, then (3.2) holds on W' from (3.4). Hence, let us consider the case that W' is the later. The configuration of the exceptional set $\bigcup_{k=1}^{t_1} F''_{i,\ell,k}$ in W' is represented as follows:

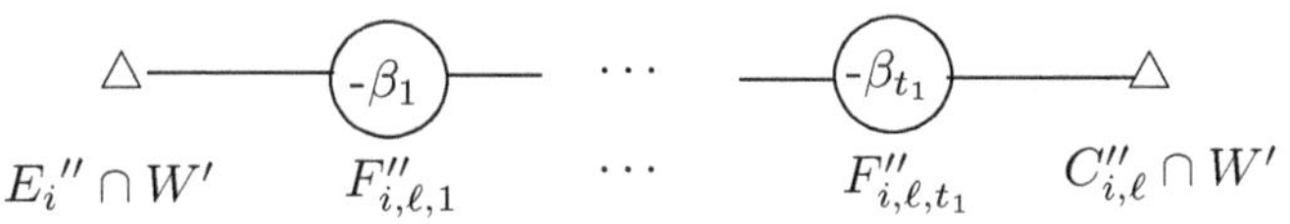

(3.5)

where $C''_{i,\ell} := (\psi'')^{-1}_*(C_{i,\ell})$ for $\ell = 1, \cdots, s_i$. By the same way as (3.4), we have $\mathrm{Coeff}_{E_i''}(z \circ \phi)_{E(n)''} \leqq \mathrm{Coeff}_{E_i''}(g \circ \sigma \circ \psi'')_{E(n)''}$. Then, from Lemma 2.3 (ii), we can easily check the following:

$$\mathrm{Coeff}_{F''_{i,\ell,k}}(z \circ \phi)_{E(n)''} \leqq \mathrm{Coeff}_{F''_{i,\ell,k}}(g \circ \sigma \circ \psi'')_{E(n)''} \tag{3.6}$$

for $k = 1, \cdots, t_1$. Since $M_{E(n)''} = \min\{(z \circ \phi)_{E(n)''}, (g \circ \sigma \circ \psi'')_{E(n)''})\}$ from $M_E = (g \circ \sigma)$, we complete the proof of (3.2). □

If $n \geqq N_0(h)$, then (X_n, o) is a Kodaira singularity (see [T2, Theorem 4.14]). Therefore the maximal ideal cycle coincides with the fundamental cycle on the minimal good resolution by the definition of Kodaira singularities ([K]).

Definition 3.5. We use notations in the diagram (3.1) and Definition 3.3. Put $C''_{i,\ell} := (\psi'')^{-1}_*(C_{i,\ell})$ and $E_i'' := (\psi'')^{-1}_*(E_i)$. We define the following:

$$Q_{i,\ell} := E(n)'' \cap C''_{i,\ell} \text{ for any } (i, \ell) \text{ with } 1 \leqq i \leqq \bar{r} \text{ and } 1 \leqq \ell \leqq s_i.$$

In the following, under the condition of $n \geqq N_1(h)$, we consider whether points $\{Q_{i,\ell}\}_{i=1,\cdots,\bar{r}\,;\,\ell=1,\cdots,s_i}$ are embedded points of $\mathfrak{m}_{X_n,o}\mathcal{O}_{\tilde{X}''_n}$ or not. Since $M_{E(n)''} = (z\circ\phi)_{E(n)''}$ and $\mathrm{red}(z\circ\phi)_{\tilde{X}''_n}$ is a simple normal crossing divisor, we can easily compute $\delta(Q_{i,\ell})$ when $Q_{i,\ell}$ is an embedded point. From this, we can obtain the value of $\mathrm{mult}(X_n, o)$.

Theorem 3.6. *Let (X_n, o) be the n-fold cyclic covering over a normal surface singularity (X, o) defined by $z^n = h$. Also, let $\phi := \phi_1 \circ \phi_2 \circ \phi_3 \colon (\tilde{X}''_n, E(n)'') \to (X_n, o)$ be a good resolution given in (3.1). Let $a_i, d_i, N_1(h)$ and $\bar{r}$ be integers in Definition 3.3. Suppose that there exists $g \in \mathfrak{m}_{X,o}$ which satisfies $M_E = (g\circ\sigma)_E$ and $\mathrm{supp}(g\circ\sigma)_{non-exc} \cap \mathrm{supp}(h\circ\sigma)_{non-exc} = \emptyset$. If $n \geq N_1(h)$, then we have the following.*

(i) $-M^2_{E(n)''} = \sum_{i=1}^{\bar{r}} \lceil \frac{a_i}{n} \rceil s_i$.

(ii) *Assume that there exists i satisfying $2 \leqq d_i$ $(1 \leqq i \leqq \bar{r})$. Then,*

$$Q_{i,\ell} \text{ is an embedded point for any } \ell \text{ with } 1 \leqq \ell \leqq s_i \iff \frac{a_i}{d_i - 1} \leqq n.$$

(iii) *If $Q_{i,\ell}$ is an embedded point, then the value δ of Definition 1.5 is given as follows:*

$$\delta_{Q_{i,\ell}}(\tilde{X}''_n) = d_i - \lceil \frac{a_i}{n} \rceil \quad \text{and} \quad \delta(\tilde{X}''_n) = \sum_{i=1}^{\bar{r}} (d_i - \lceil \frac{a_i}{n} \rceil) s_i.$$

(iv) $\mathrm{mult}(X_n, o) = M_E \cdot (h\circ\sigma)_{non-exc} = \sum_{i=1}^{\bar{r}} d_i s_i$.

Proof. We use the notations in the proof of Theorem 3.4. Let $C'' := (\psi'')^{-1}C$. Let U' be a small open neighborhood of $E_i \cap C_{i,\ell}$ as in Theorem 3.4; also let $W' := (\psi'')^{-1}(U')$. Then W' is a connected open subset in $\tilde{X}''_n$ because h is a reduced element.

First, let us consider the case of $n \mid a_i$. Then, W' is a non-singular surface and $Q_{i,\ell} = E''_i \cap C''_{i,\ell}$ from Lemma 2.1. Hence, $\mathrm{Coeff}_{E''_i}(z\circ\phi)_{E(n)''} = \frac{a_i}{n} \leqq \frac{d_i n}{\gcd(n, a_i)} = \mathrm{Coeff}_{E''_i}(g\circ\sigma\circ\psi'')_{E(n)''}$ from $n \geqq N_1(h)$. Then we have the following:

$$\mathrm{Coeff}_{E''_i} M_{E(n)''} = \frac{a_i}{n}. \tag{3.7}$$

Second, let us consider the case of $n \nmid a_i$. Put $\lambda := \gcd(a_i, n)$, $\bar{n} := \frac{n}{\lambda}$ and $\bar{a}_i := \frac{a_i}{\lambda}$. By Lemma 2.1, W' is the minimal resolution space of a cyclic quotient singularity of type $C_{\bar{n},q_1}$, where q_1 is an integer with $\bar{a}_i q_1 + 1 \equiv 0$

(mod $\bar{n}$) and $0 < q_1 < \bar{n}$. We have $Q_{i,\ell} = F''_{i,\ell,t_1} \cap C''_{i,\ell}$ in the figure (3.5). Since $\mathrm{Coeff}_{E_i}(g \circ \sigma)_E = d_i$, we have $\mathrm{Coeff}_{E''_i}(g \circ \sigma \circ \psi'')_{\tilde{X}''_n} = \bar{n}d_i$ from Lemma 3.2 (ii). Also, $\mathrm{Coeff}_{C''_{i,\ell}}(g \circ \sigma \circ \psi'')_{\tilde{X}''_n} = 0$ from the assumption of $\mathrm{supp}(g \circ \sigma)_{non-exc} \cap \mathrm{supp}(h \circ \sigma)_{non-exc} = \emptyset$. Put $e_0 := \mathrm{Coeff}_{E''_i}(g \circ \sigma \circ \psi'')_{\tilde{X}''_n} = \bar{n}d_i$, $e_k := \mathrm{Coeff}_{F''_{i,\ell,k}}(g \circ \sigma \circ \psi'')_{\tilde{X}''_n}$ for $k = 1, \cdots, t_1$ and $e_{t_1+1} := 0$. Then, $e_{k-1} - e_k\beta_k + e_{k+1} = F''_{i,\ell,k} \cdot (g \circ \sigma \circ \psi'')_{\tilde{X}''_n} = 0$ $(k = 1, \cdots, t_1)$; hence we can easily see that $\mathrm{Coeff}_{F''_{i,\ell,t_1}}(g \circ \sigma \circ \psi'')_{\tilde{X}''_n} = d_i$ from $\dfrac{\bar{n}}{q_1} = [[\beta_1, \cdots, \beta_{t_1}]]$ and (3.5). From Lemma 3.2 (i), we have $\mathrm{Coeff}_{E''_i}(z \circ \phi)_{\tilde{X}''_n} = \bar{a}_i$ and $\mathrm{Coeff}_{C''_{i,\ell}}(z \circ \phi)_{\tilde{X}''_n} = 1$. Then we have $\mathrm{Coeff}_{F''_{i,\ell,t_1}}(z \circ \phi)_{\tilde{X}''_n} = \dfrac{q'_1 + \bar{a}_i}{\bar{n}}$ from Lemma 2.2, where $q_1 q'_1 \equiv 1$ (mod $\bar{n}$) and $0 < q'_1 < \bar{n}$. Then $\dfrac{q'_1 + \bar{a}_i}{\bar{n}} = \dfrac{\lambda q'_1 + a_i}{n} = \lceil \dfrac{a_i}{n} \rceil$ because of $0 < \lambda q'_1 < n$. From the assumption $N_1(h) \leqq n$, we have $\lceil \dfrac{a_i}{d_i} \rceil \leqq n$ and so $\lceil \dfrac{a_i}{n} \rceil \leqq d_i$. Therefore, we have $\mathrm{Coeff}_{F''_{i,\ell,t_1}}(z \circ \phi)_{\tilde{X}''_n} = \lceil \dfrac{a_i}{n} \rceil \leqq d_i = \mathrm{Coeff}_{F''_{i,\ell,t_1}}(g \circ \sigma \circ \psi'')_{\tilde{X}''_n}$; then

$$\mathrm{Coeff}_{F''_{i,\ell,t_1}} M_{E(n)''} = \lceil \frac{a_i}{n} \rceil. \tag{3.8}$$

From (3.7) and (3.8), (i) is proven.

From now on, we prove (ii) and (iii). We remark that any embedded point of $\mathfrak{m}_{X_n,o}\mathcal{O}_{\tilde{X}''_n}$ is included in $E(n)'' \cap C''$. Therefore we check whether $(\mathfrak{m}_{X_n,o}\mathcal{O}_{\tilde{X}''_n})_{Q_{i,\ell}}$ is a principal ideal or not. Assume that $n \nmid a_i$. Let $U_{i,\ell}$ be a small open neighborhood of $Q_{i,\ell}$ in $\tilde{X}''_n$. Let (u, v) be a local coordinate of $U_{i,\ell}$ such that $F''_{i,\ell,t_1} \cap U_{i,\ell} = \{u = 0\}$ and $C''_{i,\ell} \cap U_{i,\ell} = \{v = 0\}$. Since $\tilde{X}''_n$ is a non-singular surface, $I := (\mathfrak{m}_{X_n,o}\mathcal{O}_{\tilde{X}''_n})_{Q_{i,\ell}}$ is an ideal of the convergent power series ring $\mathbb{C}\{u, v\}$. If $\lceil \dfrac{a_i}{n} \rceil \geqq d_i$, then $I = (u^{d_i}, u^{\lceil \frac{a_i}{n} \rceil} v) = (u^{d_i})$ from (3.8) and so $Q_{i,\ell}$ is not an embedded point. Assume that $\lceil \dfrac{a_i}{n} \rceil < d_i$. Then $I = (u^{d_i}, u^{\lceil \frac{a_i}{n} \rceil} v)$ is not a principal ideal. Further, we can easily see that $Q_{i,\ell}$ is an embedded point with $\delta_{Q_{i,\ell}}(\tilde{X}''_n) = d_i - \lceil \dfrac{a_i}{n} \rceil$ (see [KN; Section 3]). For the case of $n \mid a_i$, we can get the same equality above by putting $F''_{i,\ell,0} := E''_i$ (i.e., $t_1 = 0$). Therefore, (ii) and (iii) are proven since $\lceil \dfrac{a_i}{n} \rceil < d_i$ if and only if $\dfrac{a_i}{d_i - 1} \leqq n$.

Next, let us prove (iv). From Theorem 3.4, we have $M_{E(n)''} = (z \circ \phi)_{E(n)''}$ and $(z \circ \phi)_{\tilde{X}''_n} \sim 0$; thus $0 = M_{E(n)''} \cdot (z \circ \phi)_{\tilde{X}''_n} =$

$M^2_{E(n)''} + M_{E(n)''}C''$. From $\mathrm{Coeff}_{C''_{i,\ell}}(z \circ \phi)_{\tilde{X}''_n} = 1$, we have $-M^2_{E(n)''} = M_{E(n)''}C'' = \sum_{i=1}^{\bar{r}} \lceil \frac{a_i}{n} \rceil s_i$ by (3.7) and (3.8). From (iii) and Theorem 1.3, we have $\mathrm{mult}(X_n, o) = -M^2_{E(n)''} + \delta(\tilde{X}''_n) = \sum_{i=1}^{\bar{r}} d_i s_i$; thus (iv) is proven. □

Remark 3.7. In Theorem 3.6, the assumption "$\mathrm{supp}(g \circ \sigma)_{non-exc} \cap \mathrm{supp}(h \circ \sigma)_{non-exc} = \emptyset$" seems to be necessary in general. However, if (X, o) is a rational singularity, then the assumption is unnecessary. In fact, when we give a reduced element h, then we can always find $g \in \mathfrak{m}_{X,o}$ satisfying $M_E = (g \circ \sigma)_E$ and $\mathrm{supp}(g \circ \sigma)_{non-exc} \cap \mathrm{supp}(h \circ \sigma)_{non-exc} = \emptyset$ (see [Li, Theorem 12.1] and [T2, Lemma 2.1]).

Example 3.8. Let (X, o) be a rational singularity such that the w.d. graph and the fundamental cycle associated to the minimal resolution $\sigma\colon (\tilde{X}, E) \to (X, o)$ is given as follows:

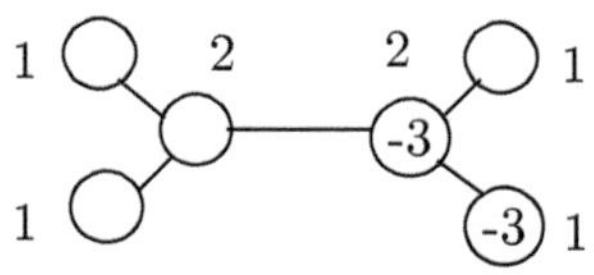

Then we have $\mathrm{mult}(X, o) = -Z^2 = 5$ by Theorem 1.2. Let g and h be two elements of $\mathfrak{m}_{X,o}$ such that the divisors $(h \circ \sigma)_{\tilde{X}}$ and $(g \circ \sigma)_{\tilde{X}}$ are given as follows:

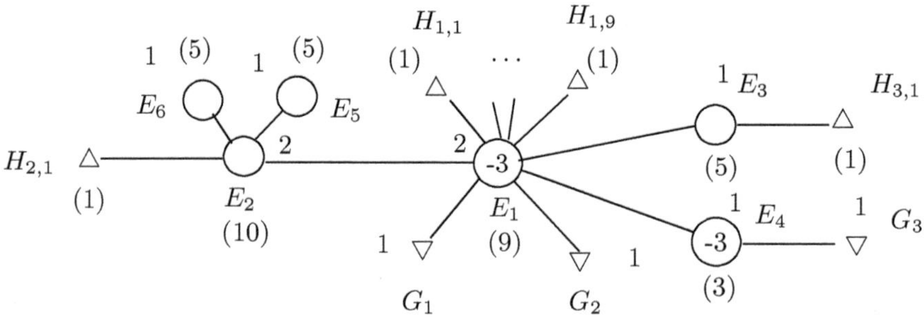

where $H_{i,j}$ (resp. $G_{k,\ell}$) is an irreducible non-exceptional curve of $(h \circ \sigma)_{\tilde{X}}$ (resp. $(g \circ \sigma)_{\tilde{X}}$) and it is represented by △ (resp. ▽). Moreover, for each irreducible component in the figure above, the integers (a)'s mean the coefficients of $(h \circ \sigma)_{\tilde{X}}$ and other integers mean the coefficients of $(g \circ \sigma)_{\tilde{X}}$ (for example, $\mathrm{Coeff}_{E_1}(h \circ \sigma)_{\tilde{X}} = 9$ and $\mathrm{Coeff}_{E_1}(g \circ \sigma)_{\tilde{X}} = 2$). Since (X, o) is a rational singularity, the reduced element h satisfying $\mathrm{supp}(g \circ \sigma)_{non-exc} \cap$

$\mathrm{supp}(h \circ \sigma)_{non-exc} = \emptyset$ exists always from Remark 3.7. We have $\bar{r} = 3$, $s_1 = 9$, $s_2 = 1$ and $s_3 = 1$; also $N_0(h) = 10$ and $N_1(h) = 5$. Therefore, if $n \geqq 5$, we can see that $\mathrm{mult}(X, o) = 21$ for any n from Theorem 3.6 (iv). However, the values of $M^2_{E(n)''}$ and $\delta(\tilde{X}''_n)$ vary according to the value of n as follows:

n	2	3	4	5	6	7	8	9	$n \geqq 10$
$-M^2_{E(n)''}$	10	15	19	21	21	21	21	12	11
$\delta(\tilde{X}''_n)$	0	0	0	0	0	0	0	9	10
$\mathrm{mult}(X_n, o)$	10	15	19	21	21	21	21	21	21

The values of $M^2_{E(n)''}$ and $\delta(\tilde{X}''_n)$ change delicately from $n = 8$ to 10. For each case, we can see that the z-minimal resolution coincides the minimal resolution. Also, $M^2_{E(n)''} = M^2_{E(n)}$ and $\delta(\tilde{X}''_n) = \delta(\tilde{X}_n)$. Hence we describe the configuration of $E(n)$ in the resolution space $\tilde{X}_n$ in the following.

[The case of $n = 8$]

The w.d. graph of the minimal resolution of (X_8, o) is given as follows:

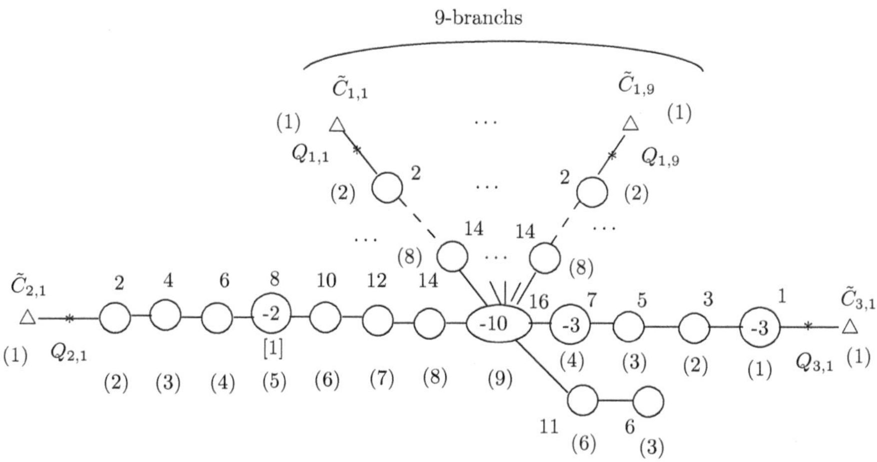

where (a) indicates the coefficient of the divisor $(z \circ \pi_8)_{\tilde{X}_8}$ and another integer indicates the coefficient of the divisor $(g \circ \psi \circ \pi_8)_{\tilde{X}_8}$ for each component.

We can see that $Z_{E(8)} = M_{E(8)} = (z \circ \pi_8)_{E(8)}$ and any $Q_{i,j}$ above is not an embedded point; hence $\delta(\tilde{X}_8) = 0$. Thus $\mathrm{mult}(X, o) = -M^2_{E(8)} = 21$.

[The case of $n = 9$]

The w.d. graph of the minimal resolution of (X_9, o) is given as follows:

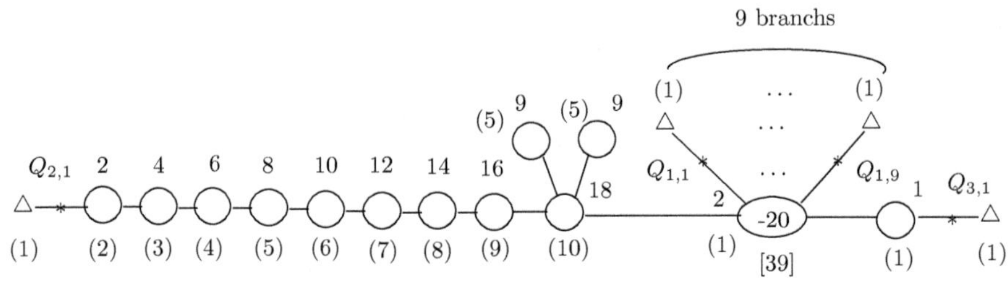

Around $Q_{1,\ell}$ $(1 \leqq \ell \leqq 9)$, $(\mathfrak{m}_{X_9,o}\mathcal{O}_{\tilde{X}_9})_{Q_{1,\ell}}$ is represented by the ideal (u^2, uv) in $\mathbb{C}\{u, v\}$. This is not a locally principal ideal and so $Q_{1,\ell}$ is an embedded point with $\delta(Q_{1,\ell}) = 1$. However, $Q_{2,1}$ and $Q_{3,1}$ are not embedded points because the ideal above are given by $(uv^2, v^2) = (v^2)$ and $(uv, v) = (v)$ respectively. Also, $Z_{E(9)} = M_{E(9)} = (z \circ \pi_9)_{E(9)}$, $-M^2_{E(9)} = 12$ and $\delta(\tilde{X}_9) = 9$. Thus $\mathrm{mult}(X_9, o) = 12 + 9 = 21$.

[The case of $n = 10$]

The w.d. graph of the minimal resolution of (X_{10}, o) is given as follows:

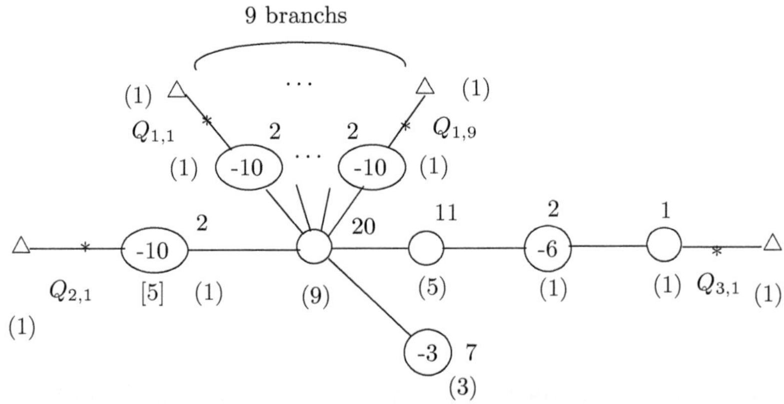

By the similar computation as the case of $n = 9$, we can see that $Q_{1,1}, \cdots, Q_{1,9}$ and $Q_{2,1}$ are embedded points with $\delta(Q_{i,j}) = 1$ and then $\delta(\tilde{X}_{10}) = 10$. Thus we have $-M_{E(10)}^2 = 11$ and $\mathrm{mult}(X_{10}, o) = 21$. Since $n = N_0(h) = 10$, $(X_{10}, 0)$ is a Kodaira singularity; hence $M_{E(10)} = Z_{E(10)}$.

Acknowledgments

The author would like to thank the referee for his careful reading and many useful comments.

References

[A] M. Artin, *On isolated rational singularities of surfaces*, Amer. J. Math. **88** (1963), 1129–138.

[C] S. D. Cutkosky, Resolution of Singularities, Graduate Studies in Math., 63, Amer. Math. Soc., (2004).

[K] U. Karras, *On pencils of curves and deformations of minimally elliptic singularities*, Math. Ann. **247** (1980), 43–65.

[KN] K. Konno and D. Nagashima, *Maximal ideal cycles over normal surface singularities of Brieskorn type*, Osaka J. Math. **49** (2012), no. 1, 225–245.

[La] H. Laufer, *On minimally elliptic singularities*, Amer. J. Math. **99** (1977), no. 6, 1257–1295.

[Li] J. Lipman, *Rational singularities, with applications to algebraic surfaces and unique factorization*, Inst. Hautes Etudes Sci. Publ. Math. No. **36** (1969), 195–279.

[MO] F. N. Meng and T. Okuma, *The maximal ideal cycles over Complete intersection surface singularities of Brieskorn type*, Kyushu J. Math. **68** (2014), no. 1, 121–137.

[N] A. Némethi, *The cohomology of line bundles of splice-quotient singularities*, Adv. Math. **229** (2012), no. 4, 2503–2524.

[O] T. Okuma, *The multiplicity of abelian covers of splice quotient singularities*, Math. Nachr. **288** (2015), no. 2-3, 343–352.

[TT] M. Tomari and T. Tomaru, *The maximal ideal cycles over normal surface singularities with $\mathbb{C}^*$-action*, Tohoku Math. J. (2) **69** (2017), no. 3, 415–430.

[TW] M. Tomari and K.-i. Watanabe, *Cyclic covers of normal graded rings*, Kōdai Math. J. **24** (2001), 436–457.

[T1] T. Tomaru, *On Kodaira singularities defined by $z^n = f(x, y)$*, Math. Z. **236** (2001), 133–149.

[T2] T. Tomaru, *Pencil genus for normal surface singularities*, J. Math. Soc. Japan **59** (2007), no. 1, 35–80.

[T3] T. Tomaru, *$\mathbb{C}^*$-equivariant degenerations of curves and normal surface singularities with $\mathbb{C}^*$-action*, J. Math. Soc. Japan **65** (2013), no. 3, 829–885.

[W] P. Wagreich, *Elliptic singularities of surfaces*, Amer. J. Math. **92** (1970), 421–454.
[Y] S. S.-T. Yau, *On maximally elliptic singularities*, Trans. Amer. Math. Soc. **257** (1980), no. 2, 269–329.

Notes on the motivic McKay correspondence for the group scheme α_p

Fabio Tonini

Università degli Studi di Firenze,
Dipartimento di Matematica e Informatica 'Ulisse Dini',
Viale Morgagni, 67/a - 50134 Firenze, Italy
E-mail: jacobbb84@gmail.com

Takehiko Yasuda

Department of Mathematics, Graduate School of Science,
Osaka University, Toyonaka, Osaka 560-0043, Japan
E-mail: takehikoyasuda@math.sci.osaka-u.ac.jp

We formulate a conjecture on the motivic McKay correspondence for the group scheme α_p in characteristic $p > 0$ and give a few evidences. The conjecture especially claims that there would be a close relation between quotient varieties by α_p and ones by the cyclic group of order p.

Keywords: the McKay correspondence, motivic integration, quotient singularities, finite group schemes

2010 MSC: 14E16, 14E18, 14L30, 14G17

1. Introduction

The motivic McKay correspondence was established by Batyrev [1] and Denef–Loeser [3] in characteristic zero. A version of this theory says that given a linear action of a finite group G on an affine space $\mathbb{A}_k^d$ without pseudo-reflection, we can express the motivic stringy invariant $M_{\mathrm{st}}(\mathbb{A}_k^d/G)$ of the quotient variety $\mathbb{A}_k^d/G$ as a sum of the form $\sum_{g\in\mathrm{Conj}(G)}\mathbb{L}^{a(g)}$, where $\mathrm{Conj}(G)$ is the set of conjugacy classes of G and a is a function on $\mathrm{Conj}(G)$ with values in $\frac{1}{\sharp G}\mathbb{Z}$. This can be generalized to the tame case in characteristic $p > 0$ (the case $p \nmid \sharp G$) without essential change (see [9]). After studying the case of the cyclic group of order p, the second author formulated a conjectural generalization to the wild case (the case $p \mid \sharp G$) (see [10, 11]). In this conjecture, the sum $\sum_{g\in\mathrm{Conj}(G)}\mathbb{L}^{a(g)}$ is replaced with an integral of the form $\int_{\Delta_G}\mathbb{L}^{a(g)}$, where Δ_G is the moduli space of G-torsors over the punctured formal disk $\operatorname{Spec} k((t))$ and a is a $\frac{1}{\sharp G}\mathbb{Z}$-valued function on it.

The aim of this paper is to discuss the case where G is the group scheme α_p rather than a genuine finite group, as a first step towards further generalization to arbitrary finite group schemes. We will formulate a conjectural expression (Conjecture 4.3) for $M_{\mathrm{st}}(\mathbb{A}_k^d/G)$ again of the form $\int_{\Delta_G} \mathbb{L}^{a(g)}$ under the condition $D_{\mathbf{d}} \geq 2$ (for the definition of $D_{\mathbf{d}}$, see Section 4). But here we have a new phenomenon: the moduli space Δ_G in this case is an ind-pro-limit of finite dimensional spaces rather than an ind-limit as in the case of genuine finite groups. Therefore we need to define a motivic measure on Δ_G in terms of truncation maps as we do for the arc space. Our conjecture also indicates a close relation between the case of $G = \alpha_p$ and the case of the cyclic group $H := \mathbb{Z}/p\mathbb{Z}$ of order p. There exists a one-to-one correspondence between G-representations and H-representations. The conjecture says that if $\mathbb{A}_k^d/G$ and $\mathbb{A}_k^d/H$ are quotient varieties associated to G and H-representations corresponding to each other, then they have equal motivic stringy invariant. We will give a few evidences for this conjecture. Note that Hiroyuki Ito [6] earlier observed a similarity between surface quotient singularities by (non-linear) G-actions and H-actions. This was an inspiration for our conjecture.

The paper is organized as follows. In Section 2, we describe the moduli spaces Δ_G and Δ_H. In Section 3, we define motivic measures on these spaces. In Section 4, we formulate our main conjecture. In Section 5, we give two examples supporting the equality of stringy invariants of $\mathbb{A}_k^d/G$ and $\mathbb{A}_k^d/H$. In Section 6, we discuss the case where G acts on $\mathbb{A}_k^2$, that is, the case $D_{\mathbf{d}} = 1$, as a toy model and show the change of variables formula for the quotient map $\mathbb{A}_k^2 \to \mathbb{A}_k^2/G$. This would be viewed as a supporting evidence for our conjecture that $M_{\mathrm{st}}(\mathbb{A}_k^d/G)$ is expressed as $\int_{\Delta_G} \mathbb{L}^{a(g)}$ when $D_{\mathbf{d}} \geq 2$. In Appendix, we briefly recall the representation theory of α_p in terms of relations to nilpotent endomorphisms and derivations.

In what follows, we work over an algebraically closed field k of characteristic $p > 0$. We always denote by G the group scheme α_p and by H the cyclic group of order p. We write the coordinate ring of α_p as $k[\epsilon] = k[x]/(x^p)$ with ϵ the image of x in this quotient ring.

The authors would like to thank Hiroyuki Ito for helpful discussion. The second author was supported by JSPS KAKENHI Grant Numbers JP15K17510 and JP16H06337.

2. Moduli of G and H-torsors over the punctured formal disk

The group scheme G fits into the exact sequence

$$0 \to G \to \mathbb{G}_a \xrightarrow{F} \mathbb{G}_a \to 0,$$

where F is the Frobenius map. Therefore the G-torsors over $\operatorname{Spec} k((t))$ are parameterized by $k((t))/F(k((t)))$. Let

$$\Delta_G := \left\{ \sum_{i \in \mathbb{Z};\, p \nmid i} c_i t^i \mid c_i \in k \right\} \subset k((t)),$$

the set of Laurent power series having only terms of degree prime to p. The inclusion map $\Delta_G \to k((t))$ induces a bijection $\Delta_G \to k((t))/F(k((t)))$. Thus we regard Δ_G as the "moduli space" of G-torsors over $\operatorname{Spec} k((t))$. The G-torsor corresponding to a Laurent power series $f \in \Delta_G$ is $\operatorname{Spec} k((t))[z]/(z^p - f)$ and the action of α_p is defined so that the associated coaction is the $k((t))$-algebra homomorphism

$$\frac{k((t))[z]}{(z^p - f)} \to \frac{k((t))[z]}{(z^p - f)}[\epsilon],\ z \mapsto z + \epsilon.$$

We can make a similar construction for H-torsors. Let $\wp\colon k((t)) \to k((t))$ be the Artin-Schreier map given by $\wp(f) = f^p - f$ and let

$$\Delta_H := \left\{ \sum_{i \in \mathbb{Z};\, i<0,\, p \nmid i} c_i t^i \mid c_i \in k \right\} \subset k((t)),$$

the set of Laurent polynomials having only terms of *negative* degree prime to p. Then the H-torsors are parameterized by $k((t))/\wp(k((t)))$ and the composite map $\Delta_H \hookrightarrow k((t)) \twoheadrightarrow k((t))/\wp(k((t)))$ is bijective. Thus we regard Δ_H as the "moduli space" of H-torsors over $\operatorname{Spec} k((t))$. The H-torsor corresponding to $f \in \Delta_H$ is $\operatorname{Spec} k((t))[z]/(z^p - z - f)$ where a generator of H acts by $z \mapsto z + 1$.

Remark 2.1. Constructing the true moduli space (stack) which represents a relevant moduli functor is a more difficult problem. However, the above ad hoc version would be sufficient for our application to motivic integration, since we work over an algebraically closed field. When k is algebraically closed, the coarse moduli space for Δ_H was constructed by Harbater [4]. The fine moduli stack for Δ_H over an arbitrary field of characteristic $p > 0$ was constructed by the authors [7]. The moduli space Δ_G has not been seriously studied yet, as far as the authors know.

3. Motivic measures on Δ_G and Δ_H

For a positive integer j prime to p, the set $\Delta_H^{\geq -j} := \{f \in \Delta_H \mid \mathrm{ord}(f) \geq -j\}$ is the affine space $\mathbb{A}_k^{j-\lfloor j/p \rfloor}$. Thus Δ_H is the union of affine spaces $\Delta_H^{\geq -j}$. We say that a subset C of Δ_H is *constructible* if C is a constructible subset of some $\Delta_H^{\geq -j}$. We define the *motivic measure* μ_H on Δ_H by $\mu_H(C) := [C]$ say in $\hat{\mathcal{M}}'$, a version of the complete Grothendieck ring of varieties used in [10]. (In this note, we do not discuss what additional relation would be really necessary to put on the complete Grothendieck ring for the McKay correspondence in the case of the group scheme G. This should be clarified in a future study.)

For $n \in \mathbb{Z}$, let

$$\tau_n \colon \Delta_G \to \Delta_{G,n} := \frac{\Delta_G}{\Delta_G \cap t^{np}k[[t]]}$$

be the quotient map, which truncates the terms of degrees $\geq np$. We often identify $\Delta_{G,0}$ with Δ_H through the natural bijection $\Delta_H \hookrightarrow \Delta_G \to \Delta_{G,0}$ and τ_0 with a map $\Delta_G \to \Delta_H$. When $n \geq 0$, $\Delta_{G,n}$ is the product of Δ_H and an affine space $\mathbb{A}_k^{n(p-1)}$. We say that a subset $C \subset \Delta_G$ is a *cylinder of level* n if $\tau_n(C)$ is a constructible subset of $\Delta_{G,n}$ and $C = \tau_n^{-1}(\tau_n(C))$. For a cylinder $C \subset \Delta_G$ of level n, we define its measure as

$$\mu_G(C) := [\tau_n(C)]\mathbb{L}^{-n(p-1)} \in \hat{\mathcal{M}}'.$$

Since the natural map $\Delta_{G,n+1} \to \Delta_{G,n}$ is the trivial $\mathbb{A}_k^{p-1}$-fibration, the element $[\tau_n(C)]\mathbb{L}^{-n(p-1)}$ is independent of the choice of a sufficiently large n. For instance, $\Delta_G^{\geq 0} := \{f \in \Delta_G \mid \mathrm{ord}(f)\}$ is a cylinder of level zero such that $\tau_0(\Delta_G^{\geq 0})$ is a singleton. Note that this set contains $0 \in k((t))$, according to the convention $\mathrm{ord}(0) = +\infty$. The measure of $\Delta_G^{\geq 0}$ is

$$\mu_G(\Delta_G^{\geq 0}) = [1\,\mathrm{pt}] = 1.$$

4. A conjecture on the McKay correspondence for linear G-actions

To a sequence of integers, $\mathbf{d} = (d_1, d_2, \ldots, d_l)$ such that $1 \leq d_\lambda \leq p$ and $d_\lambda \geq d_{\lambda+1}$, we associate the linear representation W of H over k that have d_λ-dimensional indecomposables as direct summands. Namely a generator of H acts on the vector space $W = k^d$, $d := |\mathbf{d}| = \sum_{\lambda=1}^{l} d_\lambda$ by a matrix whose Jordan normal form has Jordan blocks of sizes d_λ with diagonal entries 1. The map $\mathbf{d} \mapsto W$ gives a one-to-one correspondence between sequences $\mathbf{d}$ of integers as above and isomorphism classes of H-representions over k.

Similarly, to a sequence $\mathbf{d}$ as above, we associate also the linear representation V of G over k that have d_λ-dimensional indecomposables as direct summands. If ρ denotes the nilpotent linear endomorphism of $k[x_1, \dots, x_d]$ defined by Jordan blocks of sizes d_λ with diagonal entries 0, then the map

$$k[x_1, \dots, x_d] \to k[x_1, \dots, x_d][\epsilon],\ f \mapsto \sum_{i=0}^{p-1} \frac{\rho^i(f)}{i!}\epsilon^i$$

defines the linear G-action on $\mathbb{A}_k^d$. For details, see Appendix. In particular, through sequences $\mathbf{d}$, we get a one-to-one correspondence between G-representations and H-representations. We expect that the quotient varieties V/G and W/H for the same $\mathbf{d}$ would be very similar in the sense we will make more precise.

Let us fix a sequence $\mathbf{d}$ as above. Following [10], for a positive integer j prime to p, we define

$$\mathrm{sht}(j) := \sum_{\lambda=1}^{l} \sum_{i=1}^{d_\lambda - 1} \left\lfloor \frac{ij}{p} \right\rfloor.$$

We define a function $\mathrm{sht}\colon \Delta_H \to \mathbb{Z}$ by

$$\mathrm{sht}(f) := \begin{cases} \mathrm{sht}(-\mathrm{ord}(f)) & (f \neq 0) \\ 0 & (f = 0). \end{cases}$$

We define another function sht' on Δ_H by

$$\mathrm{sht}'(f) := \begin{cases} \mathrm{sht}(f) - l & (f \neq 0) \\ -d & (f = 0). \end{cases}$$

Fibers of these functions are constructible subsets. In general, for a function $u\colon \Delta_H \to \mathbb{Z}$ with constructible fibers, we define the motivic integral

$$\int_{\Delta_H} \mathbb{L}^u \, d\mu_H := \sum_{i \in \mathbb{Z}} [u^{-1}(i)]\mathbb{L}^i \in \hat{\mathcal{M}}',$$

provided that the last infinite sum converges in $\hat{\mathcal{M}}'$. When it diverges, we formally put $\int_{\Delta_H} \mathbb{L}^f \, d\mu_H := \infty$.

We define a numerical invariant $D_\mathbf{d} := \sum_{\lambda=1}^{l} (d_\lambda - 1)d_\lambda/2$, which we also think of as invariants of representations $V_\mathbf{d}$ and $W_\mathbf{d}$. Integrals

$$\int_{\Delta_H} \mathbb{L}^{-\mathrm{sht}} \, d\mu_H, \quad \int_{\Delta_H} \mathbb{L}^{-\mathrm{sht}'} \, d\mu_H$$

converge exactly when $D_{\mathbf{d}} \geq p$ (for details of computation, see [10]). If $D_{\mathbf{d}} = 0$, then the corresponding H-action is trivial. If $D_{\mathbf{d}} = 1$, then $\mathbf{d} = (2, 1, \dots, 1)$, $W/H \cong \mathbb{A}^d_k$ and the quotient map $W \to W/H$ has ramification locus of codimension one. Therefore the case $D_{\mathbf{d}} \geq 2$ is of our main interest, although the case $D_{\mathbf{d}} = 1$ will be discussed in Section 6 as a toy case.

The quotient variety W/H is factorial [2, Th. 3.8.1], in particular, has the invertible canonical sheaf $\omega_{W/H}$. The *ω-Jacobian ideal sheaf* $\mathcal{J} \subset \mathcal{O}_{W/H}$ is defined by $\mathcal{J}\omega_{W/H} := \mathrm{Im}(\bigwedge^{|\mathbf{d}|} \Omega_{W/H} \to \omega_{W/H})$. Let $o \in W/H$ be the image of the origin of W, $J_\infty(W/H)$ the arc space of W/H and $J_\infty(W/H)_o$ the preimage of o by the natural map $J_\infty(W/H) \to W/H$. The *motivic stringy invariant* (resp. the *motivic stringy invariant at* o) of W/H is defined to be

$$M_{\mathrm{st}}(W/H) := \int_{J_\infty(W/H)} \mathbb{L}^{\mathrm{ord}\mathcal{J}}\, d\mu_{W/H}$$

$$\left(\text{resp. } M_{\mathrm{st}}(W/H)_o := \int_{J_\infty(W/H)_o} \mathbb{L}^{\mathrm{ord}\mathcal{J}}\, d\mu_{W/H}\right).$$

If there exists a crepant resolution $\phi \colon Y \to W/H$, then we have $M_{\mathrm{st}}(W/H) = [Y]$ and $M_{\mathrm{st}}(W/H)_o = [\phi^{-1}(o)]$.

The following theorem proved in [10] can be considered as the motivic McKay correspondence for linear H-actions:

Theorem 4.1. *If $D_{\mathbf{d}} \geq 2$, we have the following equalities in $\hat{\mathcal{M}}' \cup \{\infty\}$,*

$$M_{\mathrm{st}}(W/H)_o = \int_{\Delta_H} \mathbb{L}^{-\mathrm{sht}}\, d\mu_H, \quad M_{\mathrm{st}}(W/H) = \int_{\Delta_H} \mathbb{L}^{-\mathrm{sht}'}\, d\mu_H.$$

Since the convergence of $M_{\mathrm{st}}(W/H)$ (or $M_{\mathrm{st}}(M/H)_o$) is equivalent to that W/H has only canonical singularities, this theorem in particular implies that W/H has canonical singularities if and only if $D_{\mathbf{d}} \geq p$ (see [8]).

In the rest of this section, we will formulate a conjecture for V/G similar to this theorem. For this purpose, we extend functions $\mathrm{sht}, \mathrm{sht}' \colon \Delta_H \to \mathbb{Z}$ to functions on Δ_G by

$$\mathrm{sht}(f) = \mathrm{sht}(\tau_0(f)) = \begin{cases} \mathrm{sht}(-\mathrm{ord}(f)) & (\mathrm{ord}(f) < 0) \\ \mathrm{sht}(0) & (\mathrm{ord}(f) \geq 0) \end{cases},$$

$$\mathrm{sht}'(f) := \mathrm{sht}'(\tau_0(f)) = \begin{cases} \mathrm{sht}'(-\mathrm{ord}(f)) & (\mathrm{ord}(f) < 0) \\ \mathrm{sht}'(0) & (\mathrm{ord}(f) \geq 0) \end{cases}.$$

Each fiber of the extended functions sht or sht′ is a cylinder of level zero and

$$\tau_0(\mathrm{sht}^{-1}(i)) = (\mathrm{sht}|_{\Delta_H})^{-1}(i), \quad \tau_0((\mathrm{sht}')^{-1}(i)) = \left(\mathrm{sht}'|_{\Delta_H}\right)^{-1}(i). \quad (4.1)$$

For a function $u\colon \Delta_G \to \mathbb{Z}$ whose fibers are cylinders, we define

$$\int_{\Delta_G} \mathbb{L}^u \, d\mu_G := \sum_{i\in\mathbb{Z}} \mu_G(u^{-1}(i))\mathbb{L}^i.$$

From (4.1), we have

$$\int_{\Delta_G} \mathbb{L}^{-\mathrm{sht}} \, d\mu_G = \int_{\Delta_H} \mathbb{L}^{-\mathrm{sht}} \, d\mu_H, \quad \int_{\Delta_G} \mathbb{L}^{-\mathrm{sht}'} \, d\mu_H = \int_{\Delta_H} \mathbb{L}^{-\mathrm{sht}'} \, d\mu_H.$$

Lemma 4.2. *The scheme V/G is of finite type over k and factorial.*

Proof. Let $k[V]$ and $k[V/G]$ be the coordinate rings of V and V/G respectively, so that $k[V/G] = k[V]^D$. In particular $k[V]^p \subseteq k[V/G]$, which easily implies that $V \longrightarrow V/G$ is a homeomorphism. Moreover $V/G \longrightarrow V$ is finite, which implies that V/G is of finite type over k because k is perfect.

Let P be a prime of height one of $k[V/G]$. We must show that P is principal. We have that P is the restriction of a height one prime ideal of $k[V]$, which is therefore generated by an irreducible polynomial f. If $f \mid D(f)$ then, since $\deg(D(f)) \le \deg(f)$, we have $D(f) = cf$ for some $c \in k$. In particular $0 = D^p(f) = c^p f$ so that $c = 0$ and $D(f) = 0$. In this case it follows easily that $P = k[V/G] \cap (fk[V]) = fk[V/G]$.

So assume that $f \nmid D(f)$. We claim that $P = f^p k[V/G]$. Let $x \in P - \{0\}$, that is $x = hf^l$ with $l > 0$, h coprime with f and $D(x) = 0$. We have

$$0 = D(x) = D(h)f^l + hlf^{l-1}D(f) \implies p \mid l \text{ and } D(h) = 0$$

So $x = h(f^p)^{l/p} \in f^p k[V/G]$. □

Thanks to this lemma, we can define the ω-Jacobian ideal on V/G similarly to the case of W/H. In turn, we can define $M_{\mathrm{st}}(V/G), M_{\mathrm{st}}(V/G)_o$. The following is our main conjecture.

Conjecture 4.3. *If $D_{\mathrm{d}} \ge 2$, we have the following equalities in $\hat{\mathcal{M}}' \cup \{\infty\}$,*

$$M_{\mathrm{st}}(V/G)_o = \int_{\Delta_G} \mathbb{L}^{-\mathrm{sht}_{\mathrm{d}}} \, d\mu_G \left(= M_{\mathrm{st}}(W/H)_o\right),$$

$$M_{\mathrm{st}}(V/G) = \int_{\Delta_G} \mathbb{L}^{-\mathrm{sht}'_{\mathrm{d}}} \, d\mu_G \left(= M_{\mathrm{st}}(W/H)\right).$$

5. Two Examples

In this section, we see two examples supporting the conjecture that $M_{\text{st}}(V/G) = M_{\text{st}}(W/H)$ and $M_{\text{st}}(V/G)_o = M_{\text{st}}(W/H)_o$.

5.1.

We first consider the case $\mathbf{d} = (3)$, supposing $p \geq 3$. If $p = 3$, then W/H has a crepant resolution $\phi\colon U \to W/H$ such that

$$M_{\text{st}}(W/H) = [U] = \mathbb{L}^3 + 2\mathbb{L}^2,$$
$$M_{\text{st}}(W/H)_o = [\phi^{-1}(o)] = 2\mathbb{L} + 1.$$

If $p > 3$, then W/H is not log canonical, in particular, $M_{\text{st}}(W/H) = M_{\text{st}}(W/H)_o = \infty$. See [10, Example 6.23].

As for the G-action on V, the corresponding derivation D on the coordinate ring $k[x, y, z]$ is given by $D(x) = 0$, $D(y) = x$, $D(z) = y$. We can compute

$$k[V/G] = k[x, y^p, z^p, y^2 - 2xz] \cong k[X, Y, Z, W]/(Y^2 - W^p - X^p Z),$$

where X, Y, Z, W correspond to $2x, y^p, z^p, y^2 - 2xz$ respectively. By simple computation of blowups, we can easily see that if $p = 3$, then V/G has a crepant resolution. Using this resolution, we see that $M_{\text{st}}(V/G) = \mathbb{L}^3 + 2\mathbb{L}^2$ and $M_{\text{st}}(V/G)_o = 2\mathbb{L} + 1$. We also see that if $p > 3$, then V/G is not canonical and $M_{\text{st}}(V/G) = \infty$. More details of computation are as follows.

Let us compute a (partial) resolution of $U_0 := V/G$. The singular locus $U_{0,\text{sing}}$ of U_0 is the affine line defined by $X = Y = W = 0$. Let $U_1 \to U_0$ be the blowup of U_0 along $U_{0,\text{sing}}$. This is a crepant morphism and the exceptional locus is the trivial $\mathbb{P}^1$-bundle over $U_{0,\text{sing}}$.

When $p = 3$, then $U_{1,\text{sing}}$ is again an affine line and $U_{1,\text{sing}}$ has an affine open neighborhood isomorphic to $\operatorname{Spec} k[X, Y, Z]/(Y^2 + XZ) \times \mathbb{A}^1_k$. Therefore the blowup $U_2 \to U_1$ along $U_{1,\text{sing}}$ is a crepant resolution and its exceptional locus is again the trivial $\mathbb{P}^1$-bundle over $U_{1,\text{sing}}$. Let $\phi\colon U_2 \to U_0$ be the natural morphism, which is a crepant resolution. Then, the above computation shows that

$$M_{\text{st}}(V/G) = [U_2] = \mathbb{L}^3 + 2\mathbb{L}^2,$$
$$M_{\text{st}}(V/G)_o = [\phi^{-1}(o)] = 2\mathbb{L} + 1.$$

When $p > 3$, we claim that U_1 is not log canonical, and neither is $U_0 = V/G$, since $U_1 \to U_0$ is crepant. By an explicit computation, we see that U_1 has an affine open subset U_1' which is isomorphic to a hypersurface defined by $Y^2 - X^{p-2}W^p - X^{p-2}Z = 0$. Its singular locus $U'_{1,\mathrm{sing}}$ is defined by $X = Y = 0$. Let $U_2 \to U'_1$ be the blowup along $U'_{1,\mathrm{sing}}$ and $B \to \mathbb{A}^4_k$ the blowup of the ambient affine space along the same locus $U'_{1,\mathrm{sing}}$ so that U_2 is a closed subset of B. Let $E \subset B$ be the exceptional divisor of $B \to \mathbb{A}^4_k$. We see

$$K_{U_2/U'_1} = -E|_{U_2}.$$

If $p = 5$, then U_2 is normal and $-E|_{U_2} = -2E'$ for a prime divisor E'. Thus E' has discrepancy -2 over U_1, hence U_1 is not log canonical. If $p > 5$, then U_2 is not normal. We similarly take the blowup $\psi\colon U_3 \to U'_2$ of an affine open subset $U'_2 \subset U_2$ such that $K_{U_3/U'_2} = -F|_{U_3}$ and $\psi^*(E|_{U_2}) = F|_{U_3}$, where F is the exceptional divisor of the blowup of the ambient affine space. We conclude that $K_{U_3/U'_1} = -2F|_{U_3}$. This shows that U_1 is not log canonical. Thus V/G is not log canonical and we have $M_{\mathrm{st}}(V/G) = \infty$. Let $T \to V/G$ be a log resolution on which the above exceptional divisor with discrepancy < -1 appears. This exceptional divisor on T surjects onto the singular locus of V/G. This shows that $M_{\mathrm{st}}(V/G)_o = \infty$.

Remark 5.1. When $p = 3$, the quotient singularities above by H and G are the same as the two hypersurface singularities in characteristic three in [5, Th. 3] up to suitable coordinate transforms.

5.2.

Next we consider the case $p = 2$ and $\mathbf{d} = (2, 2)$. Then W/H is the symmetric product of two copies of $\mathbb{A}^2_k$. It has a crepant resolution $\phi\colon U \to W/H$ constructed simply by blowing up the singular locus once. This resolution coincides with the Hilbert scheme of two points of $\mathbb{A}^2_k$, usually denoted by $\mathrm{Hilb}^2(\mathbb{A}^2_k)$. From an explicit description, we know that

$$M_{\mathrm{st}}(W/H) = [U] = \mathbb{L}^4 + \mathbb{L}^3,$$
$$M_{\mathrm{st}}(W/H)_o = [\phi^{-1}(o)] = \mathbb{L} + 1.$$

The corresponding derivation on the polynomial ring $k[x_0, y_0, x_1, y_1]$ is given by $D(x_i) = 0$, $D(y_i) = x_i$. The coordinate ring of V/G is

$$k[x_0, y_0^2, x_1, y_1^2, x_0y_1 + x_1y_0] \cong k[V, W, X, Y, Z]/(Z^2 + V^2Y + X^2W).$$

The singular locus of V/G is defined by $X = V = Z = 0$ and isomorphic to an affine plane. Let $\phi\colon U \to V/G$ be the blowup along the singular locus. This is a crepant resolution such that the exceptional locus is the trivial $\mathbb{P}^1$-bundle over $(V/G)_{\mathrm{sing}} \cong \mathbb{A}^2_k$. Thus $[U] = \mathbb{L}^4 + \mathbb{L}^3$ and $[\phi^{-1}(o)] = \mathbb{L} + 1$.

6. The change of variables in dimension two

In this section, we present some computation supporting the equalities $M_{\mathrm{st}}(V/G)_o = \int_{\Delta_G} \mathbb{L}^{-\mathrm{sht}_{\mathbf{d}}} d\mu_G$ and $M_{\mathrm{st}}(V/G) = \int_{\Delta_G} \mathbb{L}^{-\mathrm{sht}'_{\mathbf{d}}} d\mu_G$ in Conjecture 4.3. However, when $D_{\mathbf{d}} \geq 2$, V/G and W/H have singularities, which make analysis more difficult. Therefore we consider the case $D_{\mathbf{d}} = 1$ as a toy model. We will use jet schemes and the theory of integration above those spaces. For generalities see [10, Section 4].

Then $\mathbf{d}$ is of the form $(2, 1, \ldots, 1)$, but it is enough to consider the special case $\mathbf{d} = (2)$, because there is no essential difference in the general case. When $\mathbf{d} = (2)$, if we write $W = \operatorname{Spec} k[x, y]$ and let a generator of H act on it by $y \mapsto y$, $x \mapsto x + y$, then $W/H = \operatorname{Spec} k[x^p - xy^{p-1}, y] \cong \mathbb{A}^2_k$. The quotient map $W \to W/H$ is ramified along the divisor y, in particular, the map is not crepant. Because of this, we do not have the equalities in Conjecture 4.3 in this case. Instead we have

$$\mathbb{L}^2 = M_{\mathrm{st}}(W/H) = \int_{J_\infty(W/H)} 1 \, d\mu_{W/G} = \int_{\mathcal{J}_\infty \mathcal{W}} \mathbb{L}^{-\mathrm{ord}(y^{p-1}) - \mathfrak{s}} \, d\mu_{\mathcal{W}},$$

$$1 = M_{\mathrm{st}}(W/H)_o = \int_{(J_\infty(W/H))_o} 1 \, d\mu_{W/G} = \int_{(\mathcal{J}_\infty \mathcal{W})_o} \mathbb{L}^{-\mathrm{ord}(y^{p-1}) - \mathfrak{s}} \, d\mu_{\mathcal{W}}$$

the last integral of which we will explain now. The domain of integral, $\mathcal{J}_\infty \mathcal{W}$, is the space of twisted arcs of the quotient stack $\mathcal{W} = [W/H]$. The use of the stack $\mathcal{W}$ is only for this conventional notation and not really necessary. We can describe this space as

$$\mathcal{J}_\infty \mathcal{W} := \bigsqcup_{f \in \Delta_H} \mathrm{Hom}^H(\operatorname{Spec} \mathcal{O}_f, W)/H,$$

where $\operatorname{Spec} \mathcal{O}_f$ is the normalization of $\operatorname{Spec} k[[t]]$ in the H-torsor over $\operatorname{Spec} k((t))$ corresponding to f, $\mathrm{Hom}^H(-,-)$ is the set of H-equivariant morphisms. The $(\mathcal{J}_\infty \mathcal{W})_o$ is the subset of $\mathcal{J}_\infty \mathcal{W}$ consisting of H-orbits of H-equivariant maps $\operatorname{Spec} \mathcal{O}_f \to W$ sending the closed point(s) onto the origin of W, and $(J_\infty(W/H))_o$ is the set of arcs $\operatorname{Spec} k[[t]] \to W/G$ sending the closed point to o. We can define a motivic measure on $\mathcal{J}_\infty \mathcal{W}$ and there exists a natural map $\mathcal{J}_\infty \mathcal{W} \to J_\infty(W/H)$ which is almost bijective (bijective outside measure zero subsets) and induces an almost bijection $(\mathcal{J}_\infty \mathcal{W})_o \to (J_\infty(W/H))_o$. The $\mathrm{ord}(y^{p-1})$ is the function on $\mathcal{J}_\infty \mathcal{W}$ assigning orders of y^{p-1} along twisted arcs. Note that the ideal $(y^{p-1}) \subset k[x, y]$ is the Jacobian ideal of the map $W \to W/H$. Finally $\mathfrak{s}$ is the composition of the natural map $\mathcal{J}_\infty \mathcal{W} \to \Delta_H$ and $\mathrm{sht}\colon \Delta_H \to \mathbb{Z}$. The equality $\int_{J_\infty(W/H)} 1 \, d\mu_{W/H} = \int_{\mathcal{J}_\infty \mathcal{W}} \mathbb{L}^{-\mathrm{ord}(y^{p-1}) - \mathfrak{s}} \, d\mu_{\mathcal{W}}$ can be thought of as the change of variables formula for the map $\mathcal{J}_\infty \mathcal{W} \to J_\infty(W/H)$. More generally, for a measurable function $F: C \to \mathbb{Z}$ on a subset $C \subset J_\infty(W/G)$, if $\phi: \mathcal{J}_\infty \mathcal{W} \to J_\infty(W/G)$ denotes the natural map, then

$$\int_C \mathbb{L}^F \, d\mu_{W/H} = \int_{\phi^{-1}(C)} \mathbb{L}^{F\circ\phi - \mathrm{ord}(y^{p-1}) - \mathfrak{s}} \, d\mu_{\mathcal{W}}. \tag{6.1}$$

The term $-\mathrm{ord}(y^{p-1})$ corresponds to the ramification divisor of $W \to W/H$, the divisor defined by the Jacobian ideal, or to the relative canonical divisor of the proper birational map $\mathcal{W} \to W/H$. If $D_{\mathbf{d}} \geq 2$, then $W \to W/H$ is étale in codimension one and has no ramification divisor, but W/H acquires singularities as compensation. Therefore the corresponding formula in that case is

$$\int_C \mathbb{L}^{F+\mathcal{J}} \, d\mu_{W/H} = \int_{\phi^{-1}(C)} \mathbb{L}^{F\circ\phi - \mathfrak{s}} \, d\mu_{\mathcal{W}}. \tag{6.2}$$

When $C = J_\infty(W/H)$ and $F \equiv 0$, then the right hand side becomes

$$\int_{\mathcal{J}_\infty \mathcal{W}} \mathbb{L}^{-\mathfrak{s}} \, d\mu_{\mathcal{W}} = \int_{\Delta_H} \mathbb{L}^{-\mathrm{sht}'} \, d\mu_H.$$

We show a similar formula in the case of G. We first introduce a counterpart of $\mathcal{J}_\infty \mathcal{W}$. For $f \in \Delta_G$, let $\operatorname{Spec} K_f \to \operatorname{Spec} k((t))$ be the corresponding G-torsor. The underlying topological space of $\operatorname{Spec} K_f$ is always a singleton.

Lemma 6.1. *If $f \neq 0$, then K_f is reduced, equivalently, K_f is a field.*

Proof. The K_f is a finite extension of $k((t))$ of degree p. If K_f is non-reduced, then the associated reduced ring $(K_f)_{\mathrm{red}}$ is an extension of degree one, hence the natural map $k((t)) \to (K_f)_{\mathrm{red}}$ is an isomorphism. This means that the G-torsor $\operatorname{Spec} K_f \to \operatorname{Spec} k((t))$ admits a section, hence it is a trivial torsor and $f = 0$. □

For $f \neq 0$, let $\operatorname{Spec} \mathcal{O}_f$ be the normalization of $\operatorname{Spec} k[[t]]$ in $\operatorname{Spec} K_f$. For $f = 0$, we define $\mathcal{O}_f := k[[t]][z]/(z^p)$. We say that a morphism $\operatorname{Spec} \mathcal{O}_f \to V$ is G-equivariant if the composition map $\operatorname{Spec} K_f \to \operatorname{Spec} \mathcal{O}_f \to V$ is G-equivariant. Note that the G-action on $\operatorname{Spec} K_f$ does not generally extend to $\operatorname{Spec} \mathcal{O}_f$, which is the reason that we define G-equivariant morphisms $\operatorname{Spec} \mathcal{O}_f \to V$ in this way. Let $\mathrm{Hom}^G(\operatorname{Spec} K_f, V)$ and $\mathrm{Hom}^G(\operatorname{Spec} \mathcal{O}_f, V)$ be the set of G-equivariant morphisms $\operatorname{Spec} K_f \to V$ and $\operatorname{Spec} \mathcal{O}_f \to V$ respectively. Regarding $\mathrm{Hom}(\operatorname{Spec} \mathcal{O}_f, V)$ as a subset of $\mathrm{Hom}(\operatorname{Spec} K_f, V)$, we have

$$\mathrm{Hom}^G(\operatorname{Spec} \mathcal{O}_f, V) = \mathrm{Hom}(\operatorname{Spec} \mathcal{O}_f, V) \cap \mathrm{Hom}^G(\operatorname{Spec} K_f, V).$$

We then put

$$\mathcal{J}_\infty \mathcal{V} := \bigsqcup_{f \in \Delta_G} \mathrm{Hom}^G(\mathrm{Spec}\, \mathcal{O}_f, V).$$

We now give an explicit description of this set. Let D denote the derivation on $k[x,y]$ given by $D(y) = x$, $D(x) = 0$, which corresponds to the G-action on $V = \mathrm{Spec}\, k[x,y]$. The G-action on V is given by the coaction:

$$\begin{aligned} \theta \colon k[x,y] &\to k[x,y][\epsilon] \\ x &\mapsto x \\ y &\mapsto y + x\epsilon \end{aligned}$$

The G-action on $\mathrm{Spec}\, K_f$ with $K_f = k((t))[z]/(z^p - f)$ is given by:

$$\begin{aligned} \psi_f \colon K_f &\to K_f[\epsilon] \\ z &\mapsto z + \epsilon \end{aligned}$$

An element of K_f is uniquely written as $\sum_{i=0}^{p-1} a_i z^i$, $a_i \in k((t))$ and a map $\gamma\colon \mathrm{Spec}\, K_f \to V$ is uniquely determined by two elements $\gamma^*(x) = \sum_{i=0}^{p-1} a_i z^i$ and $\gamma^*(y) = \sum_{i=0}^{p-1} b_i z^i$ of K_f. The map γ is G-equivariant if and only if

$$\psi_f(\gamma^*(x)) = (\gamma^* \otimes \mathrm{id}_{k[\epsilon]})(\theta(x)),\ \psi_f(\gamma^*(y)) = (\gamma^* \otimes \mathrm{id}_{k[\epsilon]})(\theta(y)).$$

The left equality is explicitly written as

$$\sum_i a_i (z+\epsilon)^i = \sum_i a_i z^i,$$

which is equivalent to saying that $a_i = 0$ for $i > 0$. The right equality then says that

$$\sum_i b_i (z+\epsilon)^i = \sum_i b_i z^i + \sum_i a_i z^i \epsilon.$$

This is equivalent to requiring $b_1 = a_0$ and $b_i = 0$ for $i > 1$. As a consequence, we can identify $\mathrm{Hom}^G(\mathrm{Spec}\, K_f, V)$ with

$$\{(a, b + az) \in K_f^2 \mid a, b \in k((t))\} \cong k((t))^2.$$

For $f \neq 0$, if we extend the order function on $k((t))$ to K_f as a valuation, then $\mathrm{ord}(z) = \frac{\mathrm{ord}(f)}{p}$. Therefore, with the above identification, $(a, b+az) \in \mathrm{Hom}^G(\mathrm{Spec}\, K_f, V)$ lies in $\mathrm{Hom}^G(\mathrm{Spec}\, \mathcal{O}_f, V)$ if and only if $\mathrm{ord}(a) \geq 0$ and $\mathrm{ord}(b + az) \geq 0$. Since $\mathrm{ord}(f)$ is coprime with p, we have $\mathbb{Z} \ni \mathrm{ord}(b) \neq \mathrm{ord}(az) \notin \mathbb{Z}$ and therefore

$$\mathrm{ord}(b + az) = \min\{\mathrm{ord}(b), \mathrm{ord}(az)\}$$

Thus the two conditions translate into $\operatorname{ord}(a) \geq \max\{0, \lceil -\operatorname{ord}(f)/p \rceil\}$ and $\operatorname{ord}(b) \geq 0$. In conclusion, for every f, the set $\operatorname{Hom}^G(\operatorname{Spec} \mathcal{O}_f, V)$ is identified with the following subset of $\mathcal{O}_f^2$,

$$\{(a, b + az) \in \mathcal{O}_f^2 \mid a \in t^{s_f} \cdot k[[t]],\ b \in k[[t]]\}$$
$$(s_f := \max\{0, \lceil -\operatorname{ord}(f)/p \rceil\} = \operatorname{sht}'(f) + 2).$$

We then identify $\mathcal{J}_\infty \mathcal{V}$ with

$$\bigsqcup_{f \in \Delta_G} t^{s_f} k[[t]] \oplus k[[t]]$$

and write its elements as triples (f, a, b). For $m \in \mathbb{Z}_{\geq 0}$, the image of $\operatorname{Hom}^G(\operatorname{Spec} \mathcal{O}_f, V)$ under the natural map $\mathcal{O}_f^2 \to \mathcal{O}_f^2 / t^{m+1} \mathcal{O}_f^2$, which we denote by $\operatorname{Hom}^G(\operatorname{Spec} \mathcal{O}_f, V)_m$, coincides with the image of the injective map

$$\frac{t^{s_f} k[[t]]}{t^{s_f+m+1} k[[t]]} \oplus \frac{k[[t]]}{t^{m+1} k[[t]]} \longrightarrow \mathcal{O}_f^2 / t^{m+1} \mathcal{O}_f^2,\ (a, b) \mapsto (a, b + az).$$

We define a motivic measure $\mu_\mathcal{V}$ on $\mathcal{J}_\infty \mathcal{V}$ as follows. For $m, n \in \mathbb{Z}_{\geq 0}$, let

$$\mathcal{J}_{m,n} \mathcal{V} := \bigsqcup_{f \in \Delta_{G,n}} \frac{t^{s_f} k[[t]]}{t^{s_f+m+1} k[[t]]} \oplus \frac{k[[t]]}{t^{m+1} k[[t]]}.$$

Here $s_f = s_g$ where $g \in \Delta_G$ is any lift of $f \in \Delta_{G,n}$. For integers n, j with $np \geq j$, let $\Delta_{G,n}^{\geq j} \subset \Delta_{G,n}$ be the subspace of $f \in \Delta_{G,n}$ with $\operatorname{ord}(f) \geq j$. This is an affine space of finite dimension and $\Delta_{G,n}$ is the union of $\Delta_{G,n}^{\geq j}$, $j \leq np$. This filtration also allows to write $\mathcal{J}_{m,n} \mathcal{V}$ as an increasing union of affine spaces, so that the notion of constructible subsets and their measure is well defined. For $m' \geq m$ and $n' \geq n$, the natural map $\mathcal{J}_{m',n'} \mathcal{V} \to \mathcal{J}_{m,n} \mathcal{V}$ is a trivial $\mathbb{A}_k^{2(m'-m)+(p-1)(n'-n)}$-bundle. Let $\pi_{m,n} \colon \mathcal{J}_\infty \mathcal{V} \to \mathcal{J}_{m,n} \mathcal{V}$ be the natural map. We say that a subset $C \subset \mathcal{J}_\infty \mathcal{V}$ is a *cylinder of level* (m, n) if $C = \pi_{m,n}^{-1} \pi_{m,n}(C)$ and $\pi_{m,n}(C)$ is a constructible subset of $\mathcal{J}_{m,n} \mathcal{V}$. Then we define the measure $\mu_\mathcal{V}(C)$ of C as

$$\mu_\mathcal{V}(C) := [\pi_{m,n}(C)] \mathbb{L}^{-2m-(p-1)n}.$$

We can further extend this measure $\mu_\mathcal{V}$ to measurable subsets, following [3, Appendix]. A subset $C \subset \mathcal{J}_\infty \mathcal{V}$ is *measurable* if there exists a sequence of cylinders $C_1, C_2, \ldots$ approximating C (which means that there exists another sequence $B_1, B_2, \ldots$ of cylinders such that $\lim_{i \to \infty} \mu_\mathcal{V}(B_i) = 0$ and for each i, the symmetric difference $C \triangle C_i = (C \cup C_i) \setminus (C \cap C_i)$ is contained in B_i). For a measurable subset C, we define $\mu_{\mathcal{V}}(C) := \lim_{i\to\infty} \mu_{\mathcal{V}}(C_i)$. A function $f\colon C \to \mathbb{Z}$ on a subset $C \subset \mathcal{J}_\infty\mathcal{V}$ is said to be *measurable* if all fibers $f^{-1}(n)$ are measurable. The integral $\int_C \mathbb{L}^f \, d\mu_{\mathcal{V}}$ is then defined to be $\sum_{n\in\mathbb{Z}}[f^{-1}(n)]\mathbb{L}^n$ in $\hat{\mathcal{M}}'$, provided that this infinite sum converges.

The quotient variety V/G has the coordinate ring $k[x, y^p]$. The arc space $J_\infty(V/G)$ of V/G is identified with $k[[t]]^2$ by looking at the images of x and y^p in $k[[t]]$. Similarly the m-th jet scheme $J_m(V/G)$ is identified with $\left(k[[t]]/(t^{m+1})\right)^2$. Given an element of $\mathcal{J}_\infty\mathcal{V}$ regarded as a G-equivariant morphism $\operatorname{Spec}\mathcal{O}_f \to V$, taking the G-quotient of the induced morphism $\operatorname{Spec}K_f \to V$, we obtain a morphism $\operatorname{Spec}k((t)) \to V/G$. We easily see that this morphism extends to a morphism $\operatorname{Spec}k[[t]] \to V/G$. Thus we obtain a map $\psi\colon \mathcal{J}_\infty\mathcal{V} \to J_\infty(V/G)$. In concrete terms, the map sends (f, a, b) to $(a, b^p + fa^p)$. For $n \geq m \geq 0$, the map ψ induces map

$$\psi_{m,n}\colon \mathcal{J}_{m,n}\mathcal{V} \to J_m(V/G),\ (f, a, b) \mapsto (a, b^p + fa^p).$$

Let us take an element

$$(\alpha, \beta) \in \left(\frac{k[[t]]}{t^{m+1}k[[t]]}\right)^{\oplus 2} = J_m(V/G).$$

We will describe the fiber $\psi_{m,n}^{-1}((\alpha, \beta))$. Namely we will describe the set of triples (f, a, b) with $f \in \Delta_{G,n}$, $a \in t^{s_f}k[[t]]/t^{s_f+m+1}k[[t]]$, $b \in k[[t]]/t^{m+1}k[[t]]$ such that $a = \alpha$ and $b^p + fa^p = \beta$ in $k[[t]]/t^{m+1}k[[t]]$. Let us write $a = \sum_{i\leq s_f+m} a_i t^i$, $b = \sum_{i\leq m} b_i t^i$, $f = \sum_{i\leq np-1} f_i t^i$. The equality $a = \alpha$ determines a_i, $i \leq m$, requires that $s_f \leq \operatorname{ord}(\alpha)$ and put no other constraint on a_i, $i > m$, b_i or f_i. For the equality $b^p + fa^p = \beta$, we note that b^p (resp. fa^p) has only terms of degrees divisible (resp. not divisible) by p. Therefore this equality determines b_i, $i \leq \lfloor m/p \rfloor$. If a is fixed and $m \geq p\cdot\operatorname{ord}(a) = p\cdot\operatorname{ord}(\alpha)$, then the same equation determines f_i, $i \leq m - p\cdot\operatorname{ord}(a)$, but put no more constraint on a_i, b_i, f_i. In this case, since the resulting f is such that fa^p has neither term of negative degree nor term of degree divisible by p, it does follow that $f \in \Delta_{G,n}$ and that $s_f \leq \operatorname{ord}(\alpha)$. Notice moreover that, if β' is the subsum of β of degree coprime with p, then s_f is a function of $\operatorname{ord}(\beta')$, so that, in particular, the number s_f does not depend on the choice of (f, a, b) over (α, β). For simplicity, suppose $m = m'p$ for some $m' \in \mathbb{N}$. As a consequence of the above computation, if $\alpha \neq 0$ and $m' \geq \operatorname{ord}(\alpha)$, then $\psi_{m,n}^{-1}((\alpha, \beta))$ is the affine space of dimension

$$\begin{aligned}&\overbrace{\{(s_f+m)-m\}}^{\text{no. of free } a_i}+\overbrace{(m-\lfloor m/p\rfloor)}^{\text{no. of free } b_i}+\overbrace{\{n-(m'-\mathrm{ord}(a))\}(p-1)}^{\text{no. of free } f_i}\\&=s_f+(p-1)n+(p-1)\mathrm{ord}(a).\end{aligned}$$

We define functions

$$\begin{aligned}s\colon \mathcal{J}_\infty\mathcal{V}\to\mathbb{Z},\ (f,a,b)\mapsto s_f=\mathrm{sht}'(f)+2,\\ \mathrm{ord}(x)\colon \mathcal{J}_\infty\mathcal{V}\to\mathbb{Z}\sqcup\{\infty\},\ (f,a,b)\mapsto\mathrm{ord}(a).\end{aligned}$$

From the above argument, s is the composition of $\psi\colon \mathcal{J}_\infty\mathcal{V}\to J_\infty(V/G)$ and a function $s'\colon J_\infty(V/G)\to\mathbb{Z}$, the latter having cylindrical fibers. The function $\mathrm{ord}(x)$ on $\mathcal{J}_\infty\mathcal{V}$ also factors through $\mathrm{ord}(x)\colon J_\infty(V/G)\to\mathbb{Z}\cup\{\infty\}$, whose fibers are also cylinders except that $\mathrm{ord}(x)^{-1}(\infty)$ is a measurable subset of measure zero. Let $C\subset J_\infty(V/G)$ be a cylinder of level m. The inverse image $\psi^{-1}(C)$ is a cylinder of level (m,m) and

$$\pi_{m,m}(\psi^{-1}(C))=\psi_{m,m}^{-1}(\pi_m(C)).$$

If s and $\mathrm{ord}(x)$ take constant values s_0 and r on $\psi^{-1}(C)$, then

$$\mu_{V/G}(C)=\mathbb{L}^{-s_0-(p-1)r}\mu_{\mathcal{V}}(\psi^{-1}(C)).$$

By a standard formal argument on measurable subsets (for instance, see [10, Proof of Th. 5.20]), this equality is valid also when C is a measurable subset. Subdividing a given measurable subset $C\subset J_\infty(V/G)$, we can reduce to the case where s and $\mathrm{ord}(x)$ are constant. These arguments lead to the change of variables formula:

Theorem 6.2. *Let $C\subset J_\infty(V/G)$ be a subset and $F\colon C\longrightarrow\mathbb{Z}$ be a measurable function. Then*

$$\int_C \mathbb{L}^F\,d\mu_{V/G}=\int_{\psi^{-1}(C)}\mathbb{L}^{F\circ\psi-s-(p-1)\mathrm{ord}(x)}\,d\mu_{\mathcal{V}}.$$

In particular,

$$\mathbb{L}^2=M_{\mathrm{st}}(V/G)=\int_{J_\infty(V/G)}1\,d\mu_{V/G}=\int_{\mathcal{J}_\infty\mathcal{V}}\mathbb{L}^{-s-(p-1)\mathrm{ord}(x)}\,d\mu_{\mathcal{V}}.$$

In the last line $M_{\mathrm{st}}(V/G)=\int_{J_\infty(V/G)}1\,d\mu_{V/G}$ follows from definition and the fact that V/G is smooth, while $\int_{J_\infty(V/G)}1\,d\mu_{V/G}=\mathbb{L}^2$ from the fact that $J_0(V/G)=V/G=\mathbb{A}_k^2$.

We describe here an alternative way to check the equality

$$\int_{\mathcal{J}_\infty \mathcal{V}} \mathbb{L}^{-s-(p-1)\mathrm{ord}(x)}\, d\mu_{\mathcal{V}} = \mathbb{L}^2.$$

The set

$$C_{\geq 0,i} := \{(f,a,b) \in \mathcal{J}_\infty \mathcal{V} \mid \mathrm{ord}(f) \geq 0,\ \mathrm{ord}(a) = i\}$$

is a cylinder of level $(i,0)$ with $\pi_{i,0}(C_{\geq 0,i}) \cong \mathbb{G}_m \times \mathbb{A}_k^{i+1}$. Therefore

$$\mu_{\mathcal{V}}(C_{\geq 0,i}) = (\mathbb{L}-1)\mathbb{L}^{i+1}\mathbb{L}^{-2i} = (\mathbb{L}-1)\mathbb{L}^{-i+1}.$$

Their disjoint union is $C_{\geq 0} := \{(f,a,b) \mid \mathrm{ord}(f) \geq 0\}$ and

$$\begin{aligned}\int_{C_{\geq 0}} \mathbb{L}^{-s-(p-1)\mathrm{ord}(x)}\, d\mu_{\mathcal{V}} &= \sum_{i\geq 0} \mu_{\mathcal{V}}(C_{0,i})\mathbb{L}^{-(p-1)i} \\ &= \frac{\mathbb{L}^2-\mathbb{L}}{1-\mathbb{L}^{-p}}.\end{aligned}$$

For $j = -(pd+e) < 0$ with $d \in \mathbb{N}$ and $1 \leq e \leq p-1$ and for $i \in \mathbb{N}$, let $C_{j,i} := \{(f,a,b) \in \mathcal{J}_\infty \mathcal{V} \mid \mathrm{ord}(f) = j,\ \mathrm{ord}(a) = s_f + i\}$. This set is a cylinder of level $(i,0)$ such that

$$\pi_{i,0}(C_{j,i}) \cong \overbrace{\mathbb{G}_m \times \mathbb{A}_k^{d(p-1)+e-1}}^{f} \times \overbrace{\mathbb{G}_m}^{a} \times \overbrace{\mathbb{A}_k^{i+1}}^{b}.$$

Thus

$$\mu_{\mathcal{V}}(C_{j,i}) = (\mathbb{L}-1)^2\mathbb{L}^{-i+d(p-1)+e}.$$

The disjoint union of all the $C_{j,i}$ is $C_{<0} = \{(f,a,b) \mid \mathrm{ord}(f) < 0\}$. Then

$$\begin{aligned}\int_{C_{<0}} \mathbb{L}^{-s-(p-1)(x)}\, d\mu_{\mathcal{V}} &= \sum_{1\leq e\leq p-1} \sum_{d\in\mathbb{N}} \sum_{i\in\mathbb{N}} (\mathbb{L}-1)^2\mathbb{L}^{-i+d(p-1)+e} \\ &\qquad \times \mathbb{L}^{-(d+1)-(p-1)(d+1+i)} \\ &= \sum_{1\leq e\leq p-1} \sum_{d\in\mathbb{N}} \sum_{i\in\mathbb{N}} (\mathbb{L}-1)^2\mathbb{L}^{-pi+e-d-p} \\ &= (\mathbb{L}-1)^2\mathbb{L}^{-p}\frac{\mathbb{L}^1+\cdots+\mathbb{L}^{p-1}}{(1-\mathbb{L}^{-p})(1-\mathbb{L}^{-1})} \\ &= \frac{\mathbb{L}-\mathbb{L}^{2-p}}{1-\mathbb{L}^{-p}}.\end{aligned}$$

It follows that

$$\int_{\mathcal{J}_\infty \mathcal{V}} \mathbb{L}^{-s-(p-1)\mathrm{ord}(x)}\, d\mu_{\mathcal{V}} = \frac{\mathbb{L}^2-\mathbb{L}}{1-\mathbb{L}^{-p}} + \frac{\mathbb{L}-\mathbb{L}^{2-p}}{1-\mathbb{L}^{-p}} = \mathbb{L}^2.$$

Remark 6.3. It is natural to see the function $(p-1)\mathrm{ord}(x)$ in Theorem 6.2 as a counterpart of the function $\mathrm{ord}(y^{p-1}) = (p-1)\mathrm{ord}(y)$ in the change of variables formula (6.1) for the case of H. The latter is the order function associated to the Jacobian ideal $(y^{p-1}) \subset k[x,y]$ of the map $W \to W/H$. It is a natural problem, how to derive the ideal (x^{p-1}) as the "Jacobian ideal" of $V \to V/G$, a map not generically étale.

Appendix A. Representation theory of α_p

In this appendix we recall the representation theory of the group scheme α_p over $\mathbb{F}_p$.

Given an $\mathbb{F}_p$-algebra A we denote by $\mathrm{Mod}^{\alpha_p} A$ the category of A-modules with an action of $\alpha_p \times A$, or, equivalently, a coaction of the Hopf algebra $A[\alpha_p] = A[\varepsilon]$. We introduce also the category $\mathrm{Mod}^{nil} A$ of pairs (M, ξ) where M is an A-module and $\xi\colon M \longrightarrow M$ is an A-linear map which is p-nilpotent, that is $\xi^p = 0$. Given $(M,\xi) \in \mathrm{Mod}^{nil} A$ we define

$$\exp(\xi\varepsilon) = \sum_{i=0}^{p-1} \frac{\xi^i \varepsilon^i}{i!} \colon M \longrightarrow M \otimes \mathbb{F}_p[\varepsilon].$$

Proposition A.1. *The functor*

$$\mathrm{Mod}^{nil}(A) \longrightarrow \mathrm{Mod}^{\alpha_p}(A),\ (M,\xi) \longmapsto (M, \exp(\xi\varepsilon))$$

is well defined and an equivalence of categories. Moreover for (M,ξ), $(N,\eta) \in \mathrm{Mod}^{nil}(A)$ we have $M^{\alpha_p} = \mathrm{Ker}(\xi)$ and that

$$\xi \otimes \mathrm{id}_M + \mathrm{id}_N \otimes \eta \colon M \otimes N \longrightarrow M \otimes N$$

corresponds to the tensor product of α_p-modules.

If B is an A-algebra and $(B,\xi) \in \mathrm{Mod}^{nil}(A)$ then α_p acts on the A-algebra B, that is $A \longrightarrow B$ and $B \otimes_A B \longrightarrow B$ are α_p-equivariant, if and only if $\xi\colon B \longrightarrow B$ is an A-derivation.

Proof. Let M be an A-module and $\phi\colon M \longrightarrow M \otimes k[\varepsilon]$ be an A-linear map. The map ϕ can be written as

$$\phi = \sum_{i=0}^{p-1} \phi_i \varepsilon^i \text{ for } A\text{-linear maps } \phi_i\colon M \longrightarrow M.$$

The map ϕ must satisfy the following two conditions in order to be an α_p-action: $(\mathrm{id}_M \otimes z) \circ \phi = \mathrm{id}_M \colon M \longrightarrow M$, where $z\colon \mathbb{F}_p[\varepsilon] \longrightarrow \mathbb{F}_p$, $z(\varepsilon) = 0$ is the 0-section and $(\mathrm{id}_M \otimes \Delta) \circ \phi = (\phi \otimes \mathrm{id}_{\mathbb{F}_p[\epsilon]}) \circ \phi \colon M \longrightarrow M \otimes \mathbb{F}_p[\varepsilon] \otimes \mathbb{F}_p[\varepsilon]$, where $\Delta\colon \mathbb{F}_p[\varepsilon] \longrightarrow \mathbb{F}_p[\varepsilon]\otimes\mathbb{F}_p[\varepsilon]$, $\Delta(\varepsilon) = \varepsilon\otimes 1 + 1\otimes\varepsilon$ is the comultiplication. Those conditions translate into

$$\phi_0 = \mathrm{id}_M \text{ and } \phi_i\phi_j = \binom{i+j}{i}\phi_{i+j} \text{ for } 0 \le i, j < p$$

and into $\phi_1^p = 0$, $\phi = \exp(\phi_1\varepsilon)$. This easily proves the equivalence in the statement.

The trivial α_p-action on A corresponds to the nilpotent endomorphism $A \xrightarrow{0} A$. Thus

$$M^{\alpha_p} = \mathrm{Hom}^{\alpha_p}(A, M) = \mathrm{Hom}_{\mathrm{Mod}^{nil}(A)}((A, 0), (M, \xi)) = \mathrm{Ker}(\xi).$$

The claim about the tensor product follows from a direct check.

Consider now the last statement. The map $\iota\colon A \longrightarrow B$ is equivariant if and only if it is compatible with the nilpotent endomorphisms $0\colon A \longrightarrow A$ and $\xi\colon B \longrightarrow B$, that is $\xi|_A = 0$. From the assertion of the tensor product, the map $B \otimes_A B \longrightarrow B$ is α_p-equivariant if and only if this map is compatible with $\xi\colon B \longrightarrow B$ and $\xi \otimes \mathrm{id}_B + \mathrm{id}_B \otimes \xi\colon B \otimes_A B \to B \otimes_A B$ if and only if ξ satisfies the Leibniz rule. This ends the proof. □

Example A.2. Let $(M, \xi) \in \mathrm{Mod}^{nil} A$. Then α_p acts on the A-algebra $\mathrm{Sym}(M)$ and the corresponding p-nilpotent endomorphism $\xi_*\colon \mathrm{Sym}(M) \longrightarrow \mathrm{Sym}(M)$ is the unique A-derivation such that $(\xi_*)_{|M} = \xi$.

In particular the corresponding p-nilpotent endomorphism $\xi_n\colon \mathrm{Sym}^n M \longrightarrow \mathrm{Sym}^n M$ is given by

$$\xi_n(m_1 \cdots m_n) = \xi(m_1)m_2 \cdots m_n + \cdots + m_1 \cdots m_{n-1}\xi(m_n).$$

Example A.3. Assume that $A = k$ is a field. Then any p-nilpotent endomorphism $\xi\colon k^n \longrightarrow k^n$ can be put in Jordan form and, in this case, this just means that all blocks have 0 diagonal and have size at most p. It follows that, up to isomorphisms, the α_p-represention over k correspond bijectively to sequences $\mathbf{d} = (d_1, \ldots, d_l)$ with $1 \le d_i \le p$, $d_i \ge d_{i+1}$ and $l \in \mathbb{N}$.

References

[1] Victor V. Batyrev. Non-Archimedean integrals and stringy Euler numbers of log-terminal pairs. *J. Eur. Math. Soc. (JEMS)*, 1(1):5–33, 1999.

[2] H. E. A. Eddy Campbell and David L. Wehlau. *Modular invariant theory*, volume 139 of *Encyclopaedia of Mathematical Sciences*. Springer-Verlag, Berlin, 2011. Invariant Theory and Algebraic Transformation Groups, 8.

[3] Jan Denef and François Loeser. Motivic integration, quotient singularities and the McKay correspondence. *Compositio Math.*, 131(3):267–290, 2002.
[4] David Harbater. Moduli of p-covers of curves. *Comm. Algebra*, 8(12):1095–1122, 1980.
[5] Masayuki Hirokado, Hiroyuki Ito, and Natsuo Saito. Three dimensional canonical singularities in codimension two in positive characteristic. *J. Algebra*, 373:207–222, 2013.
[6] Hiroyuki Ito. Talk at Osaka, 2018.
[7] Fabio Tonini and Takehiko Yasuda. Moduli of formal torsors. arXiv: 1709.01705.
[8] Takehiko Yasuda. Discrepancies of p-cyclic quotient varieties. arXiv: 1710.06044.
[9] Takehiko Yasuda. Motivic integration over Deligne-Mumford stacks. *Adv. Math.*, 207(2):707–761, 2006.
[10] Takehiko Yasuda. The p-cyclic McKay correspondence via motivic integration. *Compos. Math.*, 150(7):1125–1168, 2014.
[11] Takehiko Yasuda. The wild McKay correspondence and p-adic measures. *J. Eur. Math. Soc. (JEMS)*, 19(12):3709–3734, 2017.

Singular patterns of generic maps of surfaces with boundary into the plane

Dominik J. Wrazidlo
Institute of Mathematics for Industry,
Kyushu University,
Motooka 744, Nishi-ku, Fukuoka 819-0395, Japan
E-mail: d-wrazidlo@imi.kyushu-u.ac.jp

For generic maps from compact surfaces with boundary into the plane we develop an explicit algorithm for minimizing both the number of cusps and the number of components of the singular locus. More precisely, we minimize among maps with fixed boundary conditions and prescribed *singular pattern*, by which we mean the combinatorial information of how the 1-dimensional singular locus meets the boundary. Each step of our algorithm modifies the given map only locally by either creating or eliminating a pair of cusps. We show that the number of cusps is an invariant modulo 2 and can be reduced to at most one, and we compute the minimal number of components of the singular locus in terms of the prescribed data. Applications include a discussion of pseudo-immersions as well as the computation of state sums in Banagl's positive topological field theory.

Keywords: Generic smooth map, elimination of cusps, swallowtail, winding number, Euler characteristic, pseudo-immersion

2010 MSC: 57R45, 58K15, 57R56

1. Introduction

By a classical result of Whitney [13], any smooth map $\mathbb{R}^2 \to \mathbb{R}^2$ ("smooth" always means differentiable of class C^∞) can be approximated arbitrarily well by a smooth map $F\colon \mathbb{R}^2 \to \mathbb{R}^2$ such that for every point $p \in \mathbb{R}^2$, the map germ of F at p is smoothly right-left equivalent to one of the following map germs $(\mathbb{R}^2, 0) \to (\mathbb{R}^2, 0)$:

$$(x, y) \mapsto \begin{cases} (x, y), & p \text{ is a regular point of } F, \\ (x, y^2), & p \text{ is a fold point of } F, \\ (x, xy + y^3), & p \text{ is a cusp of } F. \end{cases}$$

In general, a smooth map $F\colon W \to V$ between smooth 2-manifolds W and V is called *generic* if its singular locus

$$S(F) = \{p \in W;\ \text{the differential } dF_p\colon T_pW \to T_{F(p)}V \text{ has rank } < 2\}$$

consists only of fold points and cusps. If $F\colon W \to V$ is generic, then $S(F) \subset W$ is a 1-dimensional submanifold, and the cusps of F form a discrete subset of W. Moreover, a generic map $F\colon W \to V$ is called fold map if $S(F)$ contains no cusps. Note that any fold map $F\colon W \to V$ restricts to an immersion $S(F) \to V$.

The problem of classifying generic maps between surfaces in terms of suitable notions of complexity related to their singular locus has attracted the attention of several authors; see e.g. [10, 15] for counting cusps and components of the singular locus, [16] for the study of apparent contours of stable maps (compare Remark 8.1), and [9] for the combinatorics of certain graphs associated to stable maps.

Our paper contributes to this field of study by determining the possible number of cusps and of singular components (i.e., components of the singular locus) for generic maps of compact surfaces with boundary into the plane. For the generic maps under consideration, we impose reasonable boundary conditions, and also fix their *singular pattern* (see Definition 5.1), namely the combinatorial information of how the 1-dimensional singular locus runs into the boundary.

Initial motivation for our setting comes from the requirement that the number of singular components of a generic map should be detectable in a way that behaves nicely under gluing of cobordisms – a property that is desirable from the viewpoint of topological field theory (TFT). In fact, suppose that a 2-dimensional cobordism W is the result of gluing a cobordism W_1 from M to N and a cobordism W_2 from N to P along their common boundary part N, i.e., $W = W_1 \cup_N W_2$. Consider a generic map $F\colon W \to \mathbb{R}^2$ whose singular locus $S(F)$ is transverse to M, N and P. Then, the number of singular components of F cannot directly be computed from the number of singular components of $F|_{W_1}$ and $F|_{W_2}$. The remedy is to record also the additional information that is provided by singular patterns. It is exactly this gluing principle that has been exploited by Banagl [2, Section 10] in order to construct an explicit example of a so-called positive TFT which is based on certain fold maps of cobordisms into the plane (see Remark 9.5). The spirit of TFT is also visible in the proof of Proposition 4.2, where a relative version of the winding number for arcs in the plane is used.

Let us introduce some terminology to prepare the discussion of our main results. Suppose that W is a connected compact smooth 2-manifold with boundary $P = \partial W$, and fix a collar $[0, \infty) \times P \subset W$. A *boundary condition* is a fold map $f\colon (-\varepsilon, \varepsilon) \times P \to \mathbb{R}^2$ whose singular locus $S(f)$ is transverse to $\{0\} \times P$. By a *(singular) pattern* (see Definition 5.1) on W we mean a pair (f, φ) consisting of a boundary condition f, and a partition φ of the finite set $P_f := (\{0\} \times P) \cap S(f)$ into subsets of cardinality 2. Finally, a *realization* of a pattern (f, φ) is a generic extension $F\colon W \to \mathbb{R}^2$ of the germ of $f|_{[0,\varepsilon)\times P}$ at $\{0\} \times P$ (such F do always exist) such that a subset $A \subset P_f$ belongs to the partition φ if and only if A is the boundary of a component of $S(F)$. In Proposition 5.4 we will provide a combinatorial criterion for deciding whether a realization of a given pattern exists.

The proofs of our main results will provide an explicit algorithm for minimizing the number of cusps (Theorem 1.1) and the number of components (Theorem 1.2) of realizations of a given pattern on W. Our algorithm relies merely on two local modifications of generic maps which can be realized by homotopies supported in $W \setminus \partial W$. Namely, we use Levine's homotopy [11] for eliminating pairs of cusps of generic maps between surfaces, as well as the complementary process of creating a pair of cusps on a fold line by means of the swallow-tail homotopy. These two modifications, say (E) and (C), serve as elementary building blocks for more complicated moves (see Section 2) that will be used to build up our algorithm. Note that the moves (E) and (C) correspond to two out of 10 types of codimension one local strata of the discriminant hypersurface $\Gamma \subset C^\infty(W, \mathbb{R}^2)$ consisting of smooth maps $W \to \mathbb{R}^2$ that have non-generic apparent contours [1].

Theorem 1.1. *Let (f, φ) be a pattern on W.*

(a) *All realizations of (f, φ) have the same number of cusps modulo* 2.
(b) *Every realization of (f, φ) can be modified on $W \setminus \partial W$ by a finite sequence of* (E) *and* (C) *moves to obtain a realization with at most one cusp.*

In particular, part (a) yields a $\mathbb{Z}/2$-valued invariant of certain patterns on W. This invariant turns out to be independent of φ, and can be considered as a relative Thom polynomial for cusps of generic maps of surfaces into the plane. In Proposition 6.1 we provide a formula for computing this invariant in terms of f. It involves the *boundary turning invariant* of f, which will be defined in Section 3 by means of winding numbers of certain immersions of the circle into the plane. Our formula resembles the formula of Fukuda-Ishikawa [7] for maps between compact surfaces with boundary (see Remark 6.2). In the context of Proposition 6.1 we define an integer

Γ^σ that depends on an orientation σ of P. We will use Γ^σ in the definition of another integer Δ^σ (see Section 7) that will appear in our second main result concerning the minimal number of loops (i.e., singular components diffeomorphic to the circle) that can occur for realizations of a given pattern on W.

Theorem 1.2. *Let (f, φ) be a pattern on W such that $P_f \neq \emptyset$. Suppose that $F\colon W \to \mathbb{R}^2$ is a realization of (f, φ) without cusps.*

(a) *Let W be non-orientable. Then, given an integer $l \geq 0$, F can be modified on $W \setminus \partial W$ by a finite sequence of* (E) *and* (C) *moves to a realization of (f, φ) which has no cusps and l loops.*
(b) *Let W be orientable, and let σ denote an orientation of $P = \partial W$ which is induced by one of the two orientations of W. Then, the following statements are equivalent for any integer l:*
   (i) *F can be modified on $W \setminus \partial W$ by a finite sequence of* (E) *and* (C) *moves to a realization of (f, φ) which has no cusps and l loops.*
   (ii) $l \in \mathbb{N} \cap (\Delta^\sigma + 2\mathbb{N}) \cap (\Delta^{-\sigma} + 2\mathbb{N})$.

In Section 9 we use the gluing principle for singular patterns to generalize Theorem 1.2 to the case $P_f = \emptyset$ of pseudo-immersions (see Example 9.2), as well as to the case that F has cusps (see Proposition 9.4).

The paper is structured as follows. In Section 2 we present the local moves that will be essential for our approach. The concept of boundary turning invariant will be introduced in Section 3. The proofs of our main results will cover Section 4 to Section 8. Section 9 concludes with some applications.

Notation

Let $\mathbb{N} = \{0, 1, 2, \dots\}$ denote the set of non-negative integers. Let W denote a 2-dimensional connected compact smooth manifold. In case of a non-empty boundary $P = \partial W$ we also fix a collar neighborhood $[0, \infty) \times P \subset W$. Let $\chi(W)$ denote the Euler characteristic of W. The plane $\mathbb{R}^2$ is equipped with the standard orientation. We use the following convention for the induced orientation on the boundary of an oriented surface. If the first vector of an oriented frame points out of the surface at a boundary point, then the second vector defines the orientation of the boundary at that point.

Acknowledgements

The material presented in this paper is based on Chapter 5 of the author's PhD thesis [14]. The author is grateful to his advisor Professor Markus Banagl for constant encouragement and many fruitful discussions. The author would like to thank Professor Takahiro Yamamoto, Professor Minoru Yamamoto and Professor Osamu Saeki for discussions and invaluable comments on an early version of the manuscript. Also, the author thanks the anonymous referee for reading the paper very carefully, which led to several improvements.

The author is grateful to the German National Merit Foundation (Studienstiftung des deutschen Volkes) for financial support. Moreover, the author has been supported by JSPS KAKENHI Grant Number JP17H06128.

2. Local modifications

Throughout this section, let $F\colon W \to \mathbb{R}^2$ denote a realization of some pattern (f, φ) on W as defined in the introduction (see also Definition 5.1).

The discussion in this section will be supported by figures showing the singular locus of F in open neighborhoods $U \subset W \setminus \partial W$. These neighborhoods U will be colored in grey. The intersection $S(F) \cap U$ will be represented by massive black lines, whereas $S(F) \cap (W \setminus U)$ will be indicated by spotted lines in the figures. The cusps of F will be marked by discrete points on the massive lines. Every cusp c is furthermore equipped with a vector in $T_x W$ that points *downward* in the sense of Levine (see the definition on p. 284 of [11]). (A vector points downward at a cusp if its image in the plane points in the direction of the cusp. Intuitively, the vector itself points into the half plane into which the cusp can propagate.) The direction of a downward pointing tangent vector will be referred to as the direction into which the cusp *points*. Finally, *joining curves* (see Section (4.4) in [11, p. 285]) between cusps will be symbolized by dashed lines (see the left of Figure 1). (A joining curve between two cusps c_0 and c_1 is an embedding $\alpha\colon [0,1] \to W \setminus \partial W$ such that $\alpha^{-1}(S(F)) = \{0,1\}$, $\alpha(i) = c_i$, $i = 0, 1$, and $(-1)^i \cdot \alpha'(i) \in T_{c_i} W$ points downward at c_i.)

Our approach is based on the following two local modifications (see Figure 1).

(E) **Elimination of cusp pairs** (see Figure 1(a)). A pair of cusps of F can be eliminated if the cusps point into a common component of $W \setminus S(F)$ (see Fig. 3 in [11, p. 286]). In fact, it follows from the proofs of Lemma (1) and (2) in Section (4.4) of [11, p. 285] that such a pair of cusps can be connected by a joining curve. As W is a 2-dimensional cobordism, the pair of cusps of F is a matching pair in the sense of [11] and hence *removable* (as defined in Section (4.5) of [11, p. 285]). As explained in [11, p. 286], the pair of cusps can therefore be eliminated by a homotopy of F that modifies F only in a tubular neighborhood of the joining curve.

(C) **Creation of cusp pairs** (see Figure 1(b)). In a small neighborhood of a fold point of F a pair of two new cusps can be introduced on the fold line in such a way that the cusps point into opposite directions of the singular curve. This modification can be realized by a homotopy of F that modifies F only in this neighborhood and is based on the swallow-tail homotopy (for details, see for instance Section 4.7 in [14, p. 110]).

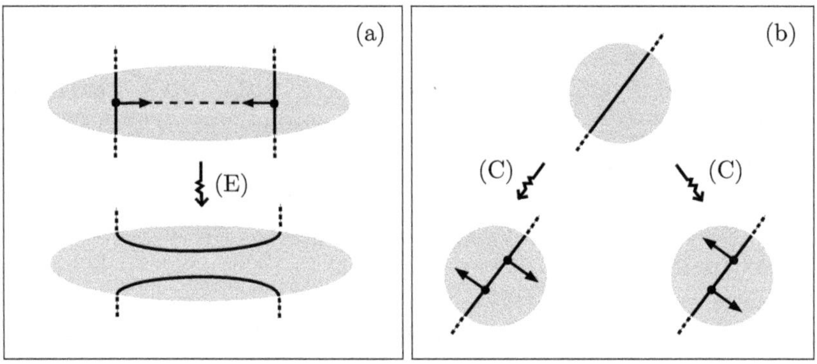

Fig. 1. (a): Levine's homotopy [11] for **eliminating pairs of cusps** of generic maps between surfaces. (b): The swallow-tail homotopy for **creating a pair of cusps** on a fold line.

Definition 2.1. By a *loop* of F we mean a component of $S(F)$ that is diffeomorphic to the circle. A loop C of F is called

- *pure* if C contains no cusps of F.
- *trivial* if the normal bundle of C in W is trivial.
- *contractible* if there is an open disc $D \subset W \setminus \partial W$ such that $D \cap S(F) = C$.

Proposition 2.2 (Loop generation).

(i) *Let* $p \in W \setminus \partial W$ *be a fold point of* F. *Given a neighborhood* $U \subset W$ *of* p, F *can be modified on* U *by two* (C) *moves followed by two* (E) *moves in such a way that the modified map* $\widetilde{F}\colon W \to \mathbb{R}^2$ *has two loops more than* F.

(ii) *Let* c_0, c_1 *be two cusps of* F. *Suppose that* $\alpha\colon [0,1] \to W \setminus \partial W$ *is a joining curve (see Section (4.4) in [11, p. 285]) between* $\alpha(0) = c_0$ *and* $\alpha(1) = c_1$. *Fix a neighborhood* V *of* $\alpha([0,1])$. *Let* $\gamma_i\colon [-1,1] \to S(F) \cap V$, $i = 0,1$, *be embeddings such that* $\gamma_i(0) = c_i$ *and* $\gamma_0([-1,1]) \cap \gamma_1([-1,1]) = \emptyset$. *Suppose that* $U \subset V$ *is a neighborhood of* $\alpha([0,1])$ *which does not contain the points* $\gamma_i(\pm 1)$, $i = 0,1$. *Then,* F *can be modified on* U *by one* (E) *move along the joining curve* α, *followed by two* (C) *moves, followed by two* (E) *moves in such a way that the modified map* $\widetilde{F}\colon W \to \mathbb{R}^2$ *has one loop more than* F *(namely, a contractible loop on* U*), and there exist embeddings* $\widetilde{\gamma}_i\colon [-1,1] \to S(\widetilde{F}) \cap V$, $i = 0,1$, *such that* $\widetilde{\gamma}_i(j) = \gamma_i(j)$ *for* $j = \pm 1$.

Definition 2.3. The *canonical orientation* of $S(F)$ is uniquely determined by requiring that, at every fold point of F, W is "folded to the left" into the plane, by which we mean the following. Recall that every fold point $p \in S(F)$ admits a neighborhood $U \subset W$ such that F restricts to an embedding $S(F) \cap U \to \mathbb{R}^2$, and $F(U) \subset \mathbb{R}^2$ is a codimension 0 submanifold with boundary $F(S(F) \cap U)$. Then the canonical orientation at $p \in S(F)$ is such that the induced orientation at $F(p) \in F(S(F) \cap U)$ coincides with the orientation of $F(S(F) \cap U)$ induced by $F(U) \subset \mathbb{R}^2$. Note that the resulting orientation of the fold locus of F extends unambiguously over the cusps of F.

Proposition 2.4 (General elimination of cusps). *Let* c_0, c_1 *be two cusps of* F. *Suppose that* $\alpha\colon [0,1] \to W \setminus \partial W$ *is an embedding such that* $\alpha(i) = c_i$, $i = 0,1$, $(-1)^i \cdot \alpha'(i)$ *is a downward pointing tangent vector of* c_i, $i = 0,1$, *and* $\alpha|_{(0,1)}$ *intersects* $S(F)$ *transversely and only in fold points of* F. *Fix a neighborhood* U *of* $\alpha([0,1])$ *and orientation preserving embeddings* $\gamma_0, \gamma_1\colon [0,1] \to S(F)$ *(where* $S(F)$ *is equipped with the canonical orientation of Definition 2.3) such that* $\gamma_0(0) \notin U$, $\gamma_0(1) = c_0$, *and* $\gamma_1(0) = c_1$, $\gamma_1(1) \notin U$. *Then,* F *can be modified on* U *by a finite sequence of* (E) *and* (C) *moves in such a way that the modified map* $\widetilde{F}\colon W \to \mathbb{R}^2$ *has two cusps less than* F, *and there exists an embedding* $\gamma\colon [0,1] \to S(\widetilde{F})$ *such that* $\gamma(i) = \gamma_i(i)$, $i = 0,1$.

Proof. We proceed by induction over the cardinality of $\alpha^{-1}(S(F)) =$

$\{t_0, \ldots, t_r\}$, where $0 = t_r < \cdots < t_0 = 1$. If $r = 1$, then the claim follows by applying (E) to the joining curve α, where note that the existence of γ is automatic. If $r > 1$, then apply one of the local modifications indicated in Figure 2 depending on the position of $\gamma_0(0)$, and modify α on $[0, t_{r-2})$ by reducing the number of cusps by one. □

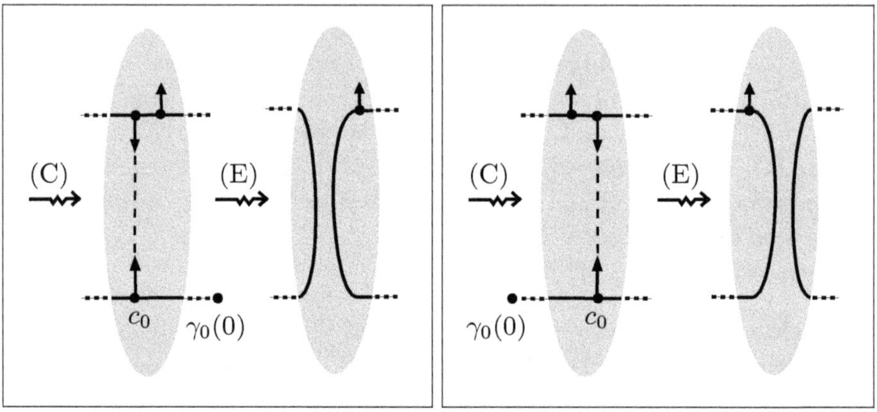

Fig. 2. Induction step for general elimination of cusps.

Remark 2.5. An explicit version of Proposition 2.4 (not claiming the existence of the embedding γ) is originally due to Kálmán (see Lemma 1.4 of [10, p. 308]).

Proposition 2.6 (Loop simplification, see Figure 3). *Let C be a loop of F. Suppose that $\alpha\colon [0,1] \to W \setminus \partial W$ is an embedding such that $\alpha^{-1}(S(F)) = \{0, 1\}$, where $\alpha(0) \in C$ and $\alpha(1) \notin C$ are fold points of F. Fix a neighborhood U of $\alpha([0,1])$. Then, F can be modified on U by a finite sequence of* (E) *and* (C) *moves in such a way that the modified map $\widetilde{F}\colon W \to \mathbb{R}^2$ has a contractible loop on U, and F and $\widetilde{F}$ have the same number of loops and the same number of cusps.*

Proposition 2.7 (Loop reduction, see Figure 4). *Let C_0, C_1 be two loops of F. Suppose that $\alpha\colon [0,1] \to W \setminus \partial W$ is an embedding such that $\alpha^{-1}(S(F)) = \{0, t, 1\}$ for some $t \in (0,1)$, where $\alpha(i) \in C_i$, $i = 0, 1$, are fold points of F, and α intersects $S(F)$ transversely at t in a fold point $\alpha(t) \notin C_0 \cup C_1$ of F. Fix a neighborhood U of $\alpha([0,1])$. Then, F can be modified by a finite sequence of* (E) *and* (C) *moves in such a way that the* *modified map* $\widetilde{F}\colon W \to \mathbb{R}^2$ *has two loops less than* F*, and* F *and* $\widetilde{F}$ *have the same number of cusps.*

Proposition 2.8 (Tunneling, see Figure 5). *Let* C *be a trivial loop of* F*. Suppose that* $\alpha\colon [0,1] \to W \backslash \partial W$ *is an embedding such that* $\alpha^{-1}(S(F)) = \{0, t_0, t_1\}$ *for some* $0 < t_0 < t_1 < 1$*, where* $\alpha(0) \in C$*, and* α *intersects* $S(F)$ *at* t_i*,* $i = 0, 1$*, transversely in fold points* $\alpha(t_i) \notin C$ *of* F*. Fix a neighborhood* V *of* $C \cup \alpha([0,1])$*. Let* $C' \subset V \setminus S(F)$ *be an embedded circle which is contained in a tubular neighborhood of* $C \subset W$ *such that* $\alpha^{-1}(C') = \{s\}$ *for some* $0 < s < t_0$*. Let* $\gamma_i\colon [-1,1] \to S(F) \cap V$*,* $i = 0, 1$*, be embeddings such that* $\gamma_i(0) = \alpha(t_i)$ *and* $\gamma_0([-1,1]) \cap \gamma_1([-1,1]) = \emptyset$*. Suppose that* $U \subset V$ *is a neighborhood of* $C' \cup \alpha([s, t_1])$ *which does neither contain* C *nor any of the points* $\alpha(1)$ *and* $\gamma_i(\pm 1)$*,* $i = 0, 1$*. Then,* F *can be modified on* U *by a finite sequence of* (E) *and* (C) *moves in such a way that the modified map* $\widetilde{F}\colon W \to \mathbb{R}^2$ *has the same number of loops and the same number of cusps as* F*, and there exists an embedding* $\widetilde{\alpha}\colon [0,1] \to V$ *such that* $\widetilde{\alpha}^{-1}(S(\widetilde{F})) = \{0\}$ *and* $\widetilde{\alpha}(i) = \alpha(i)$*,* $i = 0, 1$*, and embeddings* $\widetilde{\gamma}_i\colon [-1,1] \to S(\widetilde{F}) \cap V$*,* $i = 0, 1$*, such that* $\widetilde{\gamma}_i(j) = \gamma_i(j)$ *for* $j = 0, 1$*.*

Proposition 2.9 (Balancing, see Figure 6). *Let* C_0*,* C_1 *be two loops of* F*. Suppose that* $\alpha\colon [0,1] \to W \backslash \partial W$ *is an embedding such that* $\alpha^{-1}(S(F)) = \{0, t_0, t_1, 1\}$ *for some* $0 < t_0 < t_1 < 1$*, where* $\alpha(i) \in C_i$*,* $i = 0, 1$*, are fold points of* F*, and* α *intersects* $S(F)$ *at* t_i*,* $i = 0, 1$*, transversely in fold points* $\alpha(t_i) \notin C_0 \cup C_1$ *of* F*. Furthermore, suppose that* $H \subset W \setminus S(F)$ *is an embedded circle that intersects* α *transversely, and such that* $\alpha^{-1}(H) = \{s\}$ *for some* $s \in (t_0, t_1)$*. Fix a neighborhood* V *of* $H \cup \alpha([0,1])$*. Let* $\gamma_i\colon [-1,1] \to S(F) \cap V$*,* $i = 0, 1$*, be embeddings such that* $\gamma_i(0) = \alpha(t_i)$ *and* $\gamma_0([-1,1]) \cap \gamma_1([-1,1]) = \emptyset$*. Suppose that* $U \subset V$ *is a neighborhood of* $H \cup \alpha([0,1])$ *which does not contain the points* $\gamma_i(\pm 1)$*,* $i = 0, 1$*. Then,* F *can be modified on* U *by a finite sequence of* (E) *and* (C) *moves in such a way that the modified map* $\widetilde{F}\colon W \to \mathbb{R}^2$ *has two loops less than* F*, whereas* F *and* $\widetilde{F}$ *have the same number of cusps, and there exist embeddings* $\widetilde{\gamma}_i\colon [-1,1] \to S(\widetilde{F}) \cap V$*,* $i = 0, 1$*, such that* $\widetilde{\gamma}_i(j) = \gamma_i(j)$ *for* $j = 0, 1$*.*

3. Boundary Turning Invariant

Throughout this section let P denote a closed 1-dimensional manifold equipped with an orientation σ. Furthermore, let f be a boundary condition on P. That is, $f\colon (-\varepsilon, \varepsilon) \times P \to \mathbb{R}^2$, $\varepsilon > 0$, is a fold map whose

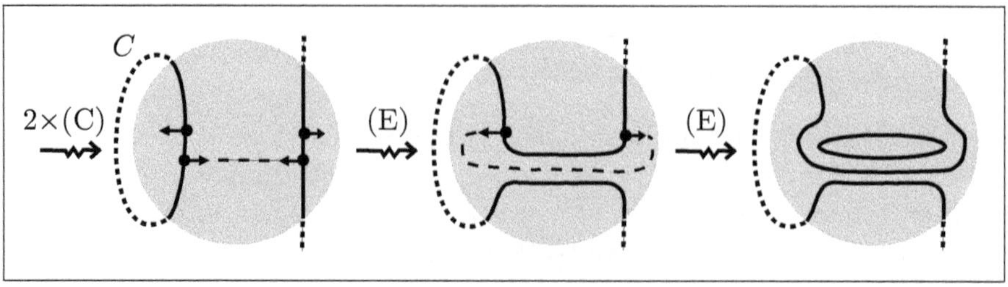

Fig. 3. Loop simplification. The modification is performed as indicated on an embedded open disc in $W \setminus \partial W$ containing $\alpha([0,1])$.

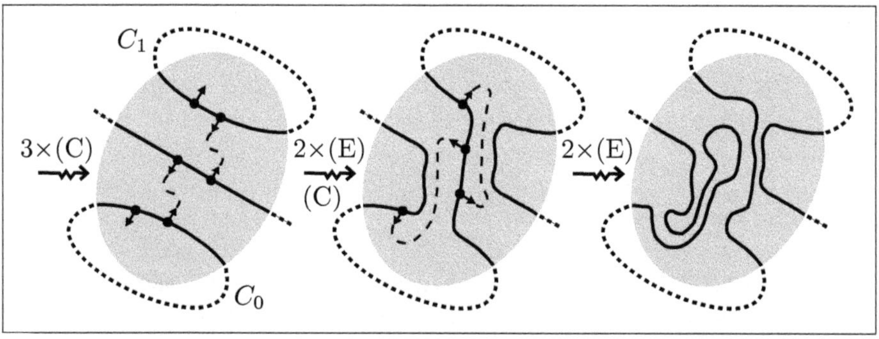

Fig. 4. Loop reduction. The modification is performed as indicated on an embedded open disc in $W \setminus \partial W$ containing $\alpha([0,1])$.

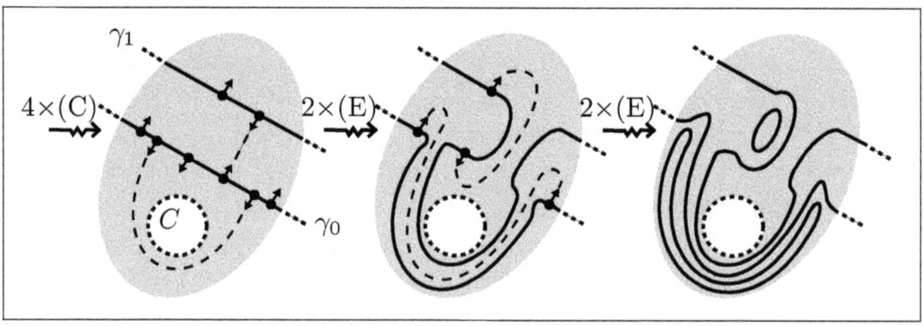

Fig. 5. Tunneling. The modification is performed as indicated on an embedded open annulus in U containing $C' \cup \alpha([s, t_1])$. The construction is completed by using loop reduction (Proposition 2.7) to eliminate the two contractible loops in the figure on the right.

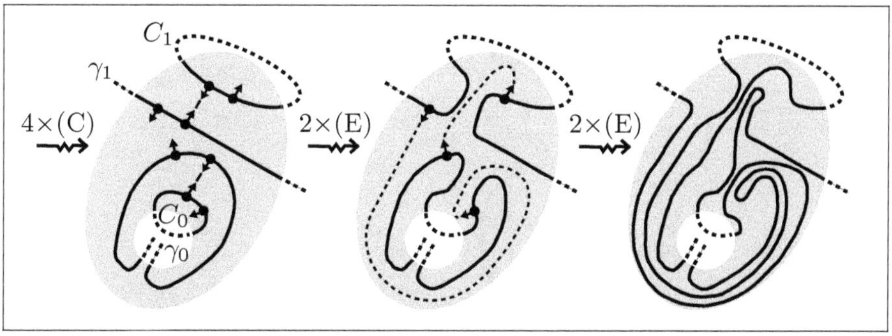

Fig. 6. Balancing. The modification is performed as indicated on an embedded open annulus in U containing $H \cup \alpha([0,1])$.

singular locus $S(f) \subset (-\varepsilon, \varepsilon) \times P$ is transverse to $\{0\} \times P$. In particular, the set $P_f := S(f) \cap \{0\} \times P$ is finite. To these data we will assign in Definition 3.5 an integer $\omega_\sigma(f)$ called the boundary turning invariant.

Lemma 3.1.

(a) *There exists $\varepsilon' \in (0, \varepsilon)$ such that the projection to the first factor $(-\varepsilon', \varepsilon') \times P \to (-\varepsilon', \varepsilon')$ restricts to a local diffeomorphism $S(f) \cap (-\varepsilon', \varepsilon') \times P \to (-\varepsilon', \varepsilon')$.*

(b) *For every component C of P, the (possibly empty) set $C_f := S(f) \cap \{0\} \times C$ is finite, and has even cardinality.*

Definition 3.2. Let C be a component of P. An embedding $\alpha\colon C \to (-\varepsilon, \varepsilon) \times C$ is called (σ, f)-*adapted* if the following properties are satisfied:

(i) There exists $\varepsilon' \in (0, \varepsilon)$ as in Lemma 3.1(a) such that $\alpha(C) \subset (-\varepsilon', \varepsilon') \times C$.

(ii) α is transverse to $S(f)$, and $\alpha(C) \cap S(f) = C_f$ (see Lemma 3.1(b)).

(iii) The composition of α with the projection to the second factor $(-\varepsilon, \varepsilon) \times C \to C$ has degree $+1$.

(iv) The composition $f \circ \alpha\colon C \to \mathbb{R}^2$ is an immersion.

Figure 7 illustrates the behavior of a (σ, f)-adapted embedding $\alpha\colon C \to (-\varepsilon, \varepsilon) \times C$ around a point $x \in C_f$. Let $u_x \in T_xS(f)$ denote a non-zero tangent vector which points into $(0, \varepsilon) \times C$. Let $v_x \in \ker d_xf$ be a non-zero vector such that the pair (u_x, v_x) gives the product orientation of $(-\varepsilon, \varepsilon) \times C$ at x. If $w_x \in T_x(\alpha(C))$ denotes a non-zero tangent vector that gives the orientation of $\alpha(C)$ induced by the orientation $\sigma|_C$, then properties (i), (ii) and (iii) of Definition 3.2 imply that the pair (u_x, w_x) is a basis of

$T_x((-\varepsilon,\varepsilon)\times C)$ which gives the product orientation of $(-\varepsilon,\varepsilon)\times C$. Moreover, property (iv) implies that the pair (v_x, w_x) is a basis of $T_x((-\varepsilon,\varepsilon)\times C)$.

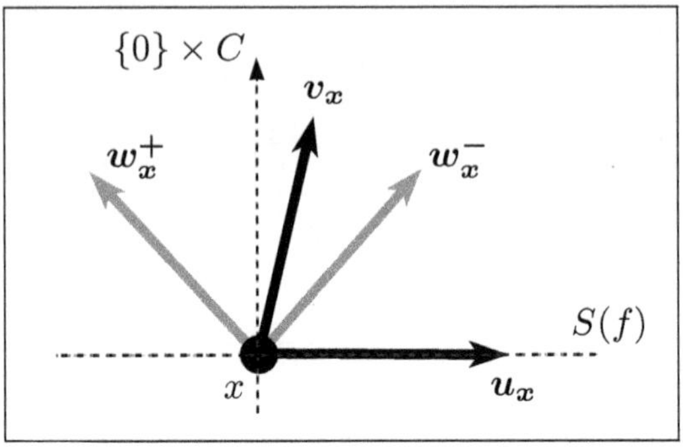

Fig. 7. A neighborhood in $(-\varepsilon,\varepsilon)\times C$ of a point $x \in C_f$. Given a (σ, f)-adapted embedding α, the vector w_x looks like w_x^+ or w_x^- depending on whether x is positive or negative with respect to α (compare Definition 3.3). There are no requirements about the positions of v_x and w_x relative to $\{0\}\times C$.

Definition 3.3. A point $x \in C_f$ is called *positive* or *negative* with respect to a (σ, f)-adapted embedding $\alpha\colon C \to (-\varepsilon,\varepsilon)\times C$ (compare Figure 7) depending on whether the pair (v_x, w_x) gives the product orientation of $(-\varepsilon,\varepsilon)\times C$ or its opposite orientation, respectively. Let C_f^α denote the sets of points of C_f which are negative with respect to α. A (σ, f)-adapted embedding $\alpha\colon C \to (-\varepsilon,\varepsilon)\times C$ is called *positive* if $C_f^\alpha = \emptyset$.

Lemma 3.4. *Let C be a component of P. There exists a positive (σ, f)-adapted embedding $\alpha\colon C \to (-\varepsilon,\varepsilon)\times C$. Moreover, any two such embeddings are smoothly isotopic in the space of all positive (σ, f)-adapted embeddings.*

Lemma 3.4 enables us to assign to every component C of P a well-defined integer $\omega_{\sigma|_C}(f|_{(-\varepsilon,\varepsilon)\times C})$ given by the winding number $W(f\circ\alpha)$ (i.e., the degree of the Gauss map, compare [12]) of the immersion $f\circ\alpha\colon C \to \mathbb{R}^2$ for any positive (σ, f)-adapted embedding $\alpha\colon C \to (-\varepsilon,\varepsilon)\times C$.

Definition 3.5. The *boundary turning invariant $\omega_\sigma(f)$ of f* is defined as

$$\omega_\sigma(f) := \sum_C \omega_{\sigma|_C}(f|_{(-\varepsilon,\varepsilon)\times C}) \qquad \in \mathbb{Z},$$

where the sum runs through the components C of P.

Example 3.6. The embedding $g\colon \mathbb{R} \times S^1 \to \mathbb{R}^2$, $g(r,s) = e^r \cdot s$, defines a tubular neighborhood of the circle $S^1 \subset \mathbb{R}^2$. Let f_{fold} denote the composition of g with the fold map $(x,y) \mapsto (x,y^2)$, and let f_{cusp} denote the composition of g with the stable Whitney cusp $(x,y) \mapsto (x, xy+y^3)$. Then, for any orientation σ on the circle S^1, one can show that $\omega_\sigma(f_{\text{fold}}) = 0$ and $\omega_\sigma(f_{\text{cusp}}) = \pm 1$.

Lemma 3.7. *Let C be a component of P. If $\alpha\colon C \to (-\varepsilon,\varepsilon) \times C$ is a (σ, f)-adapted embedding, then the winding number of the immersion $f \circ \alpha\colon C \to \mathbb{R}^2$ has the same parity as $\omega_{\sigma|_C}(f|_{(-\varepsilon,\varepsilon)\times C}) + |C_f^\alpha|$.*

Proof. Given $x \in C_f^\alpha$, we show how to modify α in a small neighborhood of $\alpha^{-1}(x)$ to obtain a (σ, f)-adapted embedding $\beta\colon C \to (-\varepsilon,\varepsilon) \times C$ such that $x \notin C_f^\beta$, and such that the winding numbers of the immersions $f \circ \alpha$ and $f \circ \beta$ differ by ± 1.

Let $\mathfrak{X}$ denote the space (equipped with the Whitney C^∞ topology) of all smooth maps $\gamma\colon C \to (-\varepsilon,\varepsilon) \times C$ for which the composition $f \circ \gamma$ is an immersion. Note that the winding number of $f \circ \gamma$ is invariant under homotopies of γ in $\mathfrak{X}$.

The given (σ, f)-adapted embedding α is homotopic in $\mathfrak{X}$ to the embedding $\alpha_1 : C \to (-\varepsilon,\varepsilon) \times C$ indicated in Figure 8(a) by means of a homotopy supported in a small neighborhood of x. Then, the winding numbers of the immersions $f \circ \alpha$ and $f \circ \alpha_1$ agree. Let $\alpha_2 \colon C \to (-\varepsilon,\varepsilon) \times C$ denote an immersion which arises from α_1 by inserting two loops based at x as indicated in Figure 8(b). Then, the winding numbers of the immersions $f \circ \alpha_1$ and $f \circ \alpha_2$ differ by ± 1. Finally, α_2 is homotopic in $\mathfrak{X}$ to the desired embedding $\beta\colon C \to (-\varepsilon,\varepsilon) \times C$ (see Figure 8(c)). $\square$

Proposition 3.8. *If ρ is any orientation of P, then $\omega_\sigma(f) \equiv \omega_\rho(f) \bmod 2$.*

Proof. It suffices to show that $\omega_{\sigma|_C}(f|_{(-\varepsilon,\varepsilon)\times C}) \equiv \omega_{-\sigma|_C}(f|_{(-\varepsilon,\varepsilon)\times C}) \bmod 2$ for every component C of P. For this purpose, note that if $\alpha\colon C \to (-\varepsilon,\varepsilon) \times C$ is a positive (σ, f)-adapted embedding, then $\beta := \alpha \circ \iota$ is a $(-\sigma, f)$-adapted embedding such that $C_f^\beta = C_f$, where $\iota\colon C \to C$ denotes an orientation reversing automorphism. Hence, the claim follows from Lemma 3.7 because C_f has even cardinality by Lemma 3.1(b). $\square$

4. Number of Cusps of Generic Extensions

Let f be a boundary condition on $P = \partial W$. By an *extension of f to W* we mean in the following a generic map $F\colon W \to \mathbb{R}^2$ such that $F|_{[0,\varepsilon')\times \partial W} =$

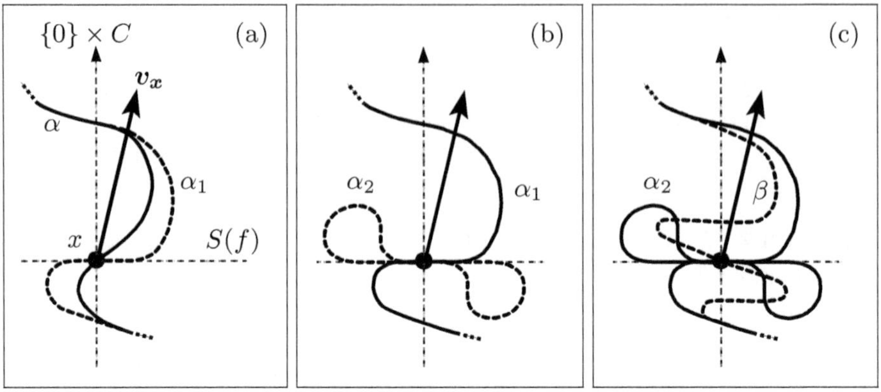

Fig. 8. Modification of α in a small neighborhood of $\alpha^{-1}(x)$.

$f|_{[0,\varepsilon')\times\partial W}$ for some $\varepsilon' \in (0,\varepsilon)$. Note that any f admits an extension, and any two extensions have the same parity of number of cusps. The purpose of this section is to prepare the ingredients needed to derive a formula for this parity in terms of the boundary turning invariant (see Proposition 6.1).

Definition 4.1. Suppose that W is oriented, and let σ be the induced orientation on the boundary $P = \partial W$. For any extension $F\colon W \to \mathbb{R}^2$ of f let W_F^σ denote the closure in W of the union of components U of $W \setminus S(F)$ for which the local diffeomorphism $F|_U : U \to \mathbb{R}^2$ is orientation preserving.

Under the assumptions of Definition 4.1, note that W_F^σ is a compact 2-dimensional submanifold of W with corners, and the set of corners is $P_f = S(f) \cap P$. Moreover, we have $W = W_F^\sigma \cup W_F^{-\sigma}$ and $\partial W_F^\sigma \cap \partial W_F^{-\sigma} = S(F) = W_F^\sigma \cap W_F^{-\sigma}$. Note that every component of $S(F)$ is contained in the boundary of exactly two components of $W \setminus S(F)$. One of these components belongs to W_F^σ, and the other component belongs to $W_F^{-\sigma}$.

The following proposition relates the boundary turning invariant as introduced in Section 3 to the Euler characteristic.

Proposition 4.2. *Suppose that W is oriented, and let σ be the induced orientation on $P = \partial W$. Then, for any extension $F\colon W \to \mathbb{R}^2$ of f without cusps, we have*

$$\chi(W_F^\sigma) - \chi(W_F^{-\sigma}) = \omega_\sigma(f).$$

Consequently, it follows from $\chi(W_F^\sigma) + \chi(W_F^{-\sigma}) = \chi(W) + |P_f|/2$ that

$$\chi(W_F^\sigma) = (\chi(W) + |P_f|/2 + \omega_\sigma(f))/2.$$

Proof. Let V be a component of W_F^{σ}. If $V \cap \partial W = \emptyset$, then $V \subset \operatorname{int} W$ is a submanifold with boundary $V \cap S(F)$, and we set $\widetilde{V} = V$. If, however, $V \cap \partial W \neq \emptyset$, then $V \subset W$ is a submanifold with corners, and the set of corners is given by $V \cap P_f$. In this case, we define a smooth submanifold with boundary $\widetilde{V} \subset V$ by smoothly cutting off the corners of V in sufficiently small neighborhoods of the corner points as indicated in Figure 9.

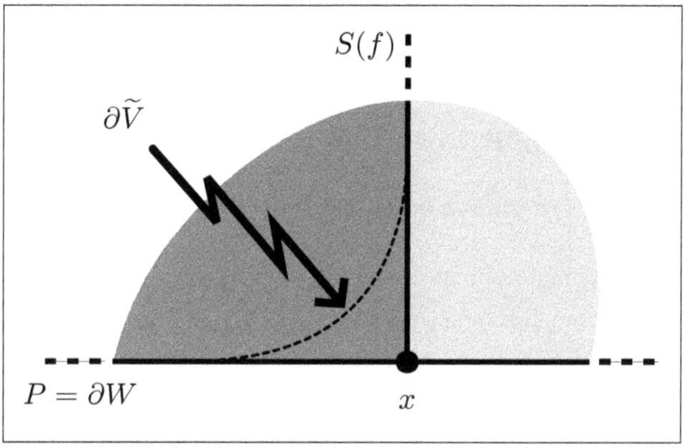

Fig. 9. A neighborhood (grey) in $[0, \varepsilon') \times P$ of a corner point $x \in V \cap P_f$ of a component V of W_F^{σ} (dark grey) for sufficiently small $\varepsilon' \in (0, \varepsilon)$. The boundary of $\widetilde{V}$ deviates from that of V as indicated only in small neighborhoods of the corners of V.

In any case, a theorem of Haefliger [8] implies that the winding number of the immersion $F|_{\partial\widetilde{V}} \colon \partial\widetilde{V} \to \mathbb{R}^2$ equals $\chi(V)$, where $\partial\widetilde{V}$ is equipped with the orientation induced by that of $\widetilde{V} \subset W$. Analogously, while keeping the same orientation on W, we can identify the Euler characteristic $\chi(V)$ of any component V of $W_F^{-\sigma}$ with the negative of the winding number of the immersion $F|_{\partial\widetilde{V}} \colon \partial\widetilde{V} \to \mathbb{R}^2$, where $\widetilde{V} \subset V$ is a suitably chosen smooth submanifold with boundary. Hence,

$$\chi(W_F^{\sigma}) - \chi(W_F^{-\sigma}) = \sum_{V \subset W_F^{\sigma} \sqcup W_F^{-\sigma}} \text{winding number of } \partial\widetilde{V},$$

where the sum runs through all components of $W_F^{\sigma} \sqcup W_F^{-\sigma}$. We observe that the contributions to the sum of winding numbers by adjacent components will cancel each other along common parts of their boundaries. Thus, only contributions of boundary parts contained in a neighborhood of $[0, \varepsilon') \times P$ for sufficiently small $\varepsilon' \in (0, \varepsilon)$ remain (compare Figure 9). (Our argument requires an extension of the concept of winding number to immersions of arcs into $\mathbb{R}^2$. This generalized winding number for immersions of arbitrary compact 1-manifolds turns out to behave additively under disjoint union and gluing of 1-manifolds along boundaries. By definition, the winding number of an immersion $\gamma\colon [a,b] \to \mathbb{R}^2$ is the real number $\lambda(b) - \lambda(a)$, where $\lambda\colon [a,b] \to \mathbb{R}$ denotes a lift of the Gauss map $[a,b] \to S^1$ of γ with respect to the universal cover $\mathbb{R} \to S^1$, $t \mapsto e^{2\pi t}$. We refer to Chapter 1 of [14] for details in the context of positive TFTs.) Finally, an argument in the spirit of the proof of Lemma 3.7 identifies the resulting sum of (generalized) winding numbers with $\omega_\sigma(f)$. □

5. Patterns and their Realizations

Definition 5.1. A *(singular) pattern on* W is a pattern (f, φ) consisting of

- a fold map $f\colon (-\varepsilon, \varepsilon) \times P \to \mathbb{R}^2$ for some $\varepsilon > 0$ whose singular locus $S(f) \subset (-\varepsilon, \varepsilon) \times P$ (which is well-known to be a 1-dimensional submanifold) is transverse to $\{0\} \times P$, and
- a fixed-point free involution $\varphi\colon P_f \to P_f$ of the finite set $P_f := S(f) \cap (\{0\} \times P)$. (Note that φ can equivalently be considered as a partition of P_f into subsets of cardinality 2.)

Definition 5.2. A 1-dimensional submanifold $S \subset W$ satisfying $S \pitchfork \partial W$ and $\partial S = S \cap \partial W$ is said to be *adapted to a pattern* (f, φ) if for every $x \in P_f$ there exists a component S_x of S with boundary $\partial S_x = \{x, \varphi(x)\}$. By a *realization* of the pattern (f, φ) we mean a generic extension $F\colon W \to \mathbb{R}^2$ of $f|_{[0,\varepsilon') \times \partial W}$ for some $\varepsilon' \in (0, \varepsilon)$ such that the singular locus $S(F)$ is non-empty and adapted to φ.

Remark 5.3. Requiring that $S(F) \neq \emptyset$ for realizations F of the pattern (f, φ) excludes the case that F is an immersion. The extension problem for stable immersions $\partial W \to \mathbb{R}^2$ to immersions $W \to \mathbb{R}^2$ has been characterized combinatorially in [6].

Fix a pattern (f, φ). A map $\iota\colon P_f \to \{-1, +1\}$ is defined by assigning to every point $x \in P_f$ a positive or a negative sign according to whether a non-zero tangent vector $u_x \in T_x S(f)$ which points into $(0, \varepsilon) \times \partial W$ determines the canonical orientation of $S(F)$ at x (see Definition 2.3) or its opposite orientation. If F is a realization of (f, φ), then $S(F)$ is adapted to φ, and $\iota(x) \neq \iota(\varphi(x))$ for all $x \in P_f$.

Proposition 5.4. *The pattern* (f, φ) *admits a realization if and only if*

(i) $\iota(x) \neq \iota(\varphi(x))$ *for all* $x \in P_f$, *and*
(ii) *there exists a submanifold* $S \subset W$ *which is adapted to* (f, φ).

Proof. It suffices to show that a realization F of (f, φ) exists under the given conditions. First, we produce a realization F' of some pattern (f, φ') by extending the germ of f at ∂W generically over W in an arbitrary way. Then, for every point $x \in P_f$, we use (C) once to create a new pair of cusps sufficiently close to x on $S(F) \setminus \partial W$. Next, for every pair $\{x, y\} \subset P_f$ of the partition φ we eliminate the cusp of F which is closest to x and the cusp of F which is closest to y by applying general cusp elimination (Proposition 2.4) along a path in $W \setminus \partial W$ that connects these two cusps, and whose interior intersects $S(F)$ transversely and only in fold points of F. As a result, in view of assumption (i), the singular locus of the modified generic map $\widetilde{F}$ will contain a component with boundary $\{x, y\}$, and which contains no fold points. Finally, note that assumption (ii) makes sure that we can perform all these cusp eliminations independently of each other along pairwise disjoint paths in $W \setminus \partial W$. □

6. Proof of Theorem 1.1

Part (a) follows immediately from the following

Proposition 6.1. *Given an orientation* σ *of* $P = \partial W$, *the number of cusps of any realization* $F\colon W \to \mathbb{R}^2$ *of a given pattern* (f, φ) *on* W *has the same parity as the quantity*

$$\Gamma^\sigma := \chi(W) + |P_f|/2 + \omega_\sigma(f).$$

Proof. Let $F\colon W \to \mathbb{R}^2$ be an extension of f. First we show in part (1) of the proof that Γ^σ is even when F has an even number of cusps. This will then be used in part (2) of the proof to show that Γ^σ is odd when F has an odd number of cusps.

(1). Suppose that F has an even number of cusps.

First suppose that W is orientable. Then, we may assume by Proposition 3.8 that W is oriented in such a way that σ is the induced orientation of P. As F has an even number of cusps, we may assume by Proposition 2.4 that F has no cusps. Then, the claim that Γ^σ is even follows from Proposition 4.2.

Now suppose that W is non-orientable. Let $C \subset \operatorname{int} W$ be an embedded circle whose normal line bundle $\nu\colon N \to C$ is non-trivial, where $N \subset W$ denotes a tubular neighborhood of C in W. Without loss of generality we may assume that C is transverse to $S(F)$, and avoids the cusps of F. If we cut W along C, then the resulting manifold W_0 has one additional boundary component C_0. By construction there is a quotient map $\pi\colon W_0 \to W$ that restricts to a diffeomorphism $W_0 \setminus C_0 \xrightarrow{\cong} W \setminus C$, and to a covering map $\gamma\colon C_0 \to C$ of degree 2. The pullback bundle of $\nu\colon N \to C$ under $\gamma\colon C_0 \to C$ is the trivial line bundle over C_0, and γ can be covered by a canonical bundle map $\widetilde{\gamma}\colon \mathbb{R}\times C_0 \to N$ such that $\widetilde{\gamma}(r,-x) = -\widetilde{\gamma}(r,x)$, where $-x$ denotes the unique element in the fiber $\gamma^{-1}(\gamma(x))$ that is distinct from $x \in C_0$. Moreover, there exists a collar $[0,\infty) \times C_0 \subset W_0$ of C_0 in W_0 such that $\pi\colon W_0 \to W$ and $\widetilde{\gamma}\colon \mathbb{R} \times C_0 \to N$ restrict to the same map $[0,\infty) \times C_0 \to N$. Let ρ denote an orientation of C_0. Then, writing $g := F \circ \widetilde{\gamma}\colon \mathbb{R} \times C_0 \to \mathbb{R}^2$, it follows from Lemma 3.7 that $\omega_\rho(g)$ has the same parity as the cardinality of $C \cap S(F)$. (In fact, this can be shown by choosing an embedding $\alpha\colon C \to N$ for which the lift of $\alpha \circ \gamma$ to a map $\tilde{\alpha}\colon C_0 \to \mathbb{R}\times C_0$ is a (ρ, g)-adapted embedding in the sense of Definition 3.2.) The quotient map $\pi\colon W_0 \to W$ restricts to a diffeomorphism $\pi_0\colon P_0 = \pi^{-1}(P) \xrightarrow{\cong} P$. Let σ_0 denote the orientation of P_0 induced by σ via π_0. The collar $[0,\infty) \times P \subset W$ induces a collar $[0,\infty) \times P_0 \subset W_0$ by means of $(t,x) \mapsto \pi^{-1}(t, \pi_0(x))$. Let $f_0 = f \circ \pi|_{[0,\infty)\times P_0}$. Then,

$$\omega_{\sigma_0 \sqcup \rho}(f_0 \sqcup g) = \omega_{\sigma_0}(f_0) + \omega_\rho(g) \equiv \omega_\sigma(f) + |C \cap S(F)| \bmod 2.$$

From now on, we assume that C is chosen in such a way that W_0 is orientable. Then, since $F_0 := F \circ \pi$ is an extension of $f_0 \sqcup g$ to W_0 with an even number of cusps, we have already shown above that

$$\chi(W_0) + |(P_0 \sqcup C_0)_{f_0 \sqcup g}|/2 + \omega_{\sigma_0 \sqcup \rho}(f_0 \sqcup g) \equiv 0 \bmod 2.$$

By construction, $\chi(W_0) = \chi(W)$. Moreover, note that we have

$$|(P_0 \sqcup C_0)_{f_0 \sqcup g}| = |P_f| + 2|C \cap S(F)|.$$

All in all, this completes the proof of part (1).

(2). Suppose that F has an odd number of cusps, and let c denote a cusp of F. Let $U \subset \operatorname{int} W$ be a small open disc centered at c in which F looks like the stable Whitney cusp discussed in Example 3.6. Recall that $\omega_\rho(f_{\mathrm{cusp}}) = \pm 1$, where ρ denotes an orientation of $\partial U \cong S^1$. For $W_0 = W \setminus U$ and the restriction $F_0 = F|_{W_0}$ we know by part (1) that

$$\chi(W_0) + |(\partial W_0)_{f \sqcup f_{\mathrm{cusp}}}|/2 + \omega_{\sigma \sqcup \rho}(f \sqcup f_{\mathrm{cusp}}) \equiv 0 \bmod 2.$$

By construction, $\chi(W) = \chi(W_0)+1$, $|(\partial W_0)_{f \sqcup f_{\mathrm{cusp}}}| = |P_f|+2$, and $\omega_{\sigma \sqcup \rho}(f \sqcup f_{\mathrm{cusp}}) = \omega_\sigma(f) \pm 1$. Thus, the claim follows. $\square$

Remark 6.2. Compare Proposition 6.1 to the Fukuda-Ishikawa theorem (see Corollary 1.2 in [7, p. 377]), where the target manifold is a compact surface N, and the extension $F\colon W \to N$ of boundary conditions restricts to a Morse function $\partial W \to \partial N$. In this context, in place of the boundary turning invariant, one has to use the related notion of degree of a continuous map $S^1 \to S^1$.

As for the proof of part (b) of Theorem 1.1, let $F\colon W \to \mathbb{R}^2$ be a realization of the pattern (f, φ). An inspection of the proof of Proposition 5.4 shows that F can be modified on $W \setminus \partial W$ by a finite sequence of (E) and (C) moves to obtain a realization F_1 of (f, φ) whose cusps lie all on loops of F_1. Let $T \subset W$ denote the union of the components of $S(F_1)$ that are diffeomorphic to the unit interval. Using general cusp elimination (Proposition 2.4), we can modify F_1 on $W \setminus (T \cup \partial W)$ by a finite sequence of (E) and (C) moves to obtain a realization F_2 of (f, φ) such that each component of $W \setminus T$ contains at most one cusp of F_2. If $T = \emptyset$, then F_2 has by construction at most one cusp as desired. Otherwise, fix a component J of T. Note that every cusp x of F_2 is the unique cusp on some loop of F_2, say C_x. We may also assume that C_x is trivial (see Definition 2.1). (In fact, if the normal line bundle of $C_x \subset W$ is non-trivial, then we apply (C) once at a fold point of F_2 on C_x to create a new pair of cusps, and then we apply (E) to a joining curve between these two cusps that goes around once in a small tubular neighborhood of C_x. The resulting new loop containing the cusp will then be trivial.) By a finite iteration of the loop tunneling move (Proposition 2.8) we can modify F_2 on $W \setminus (J \cup \partial W)$ by a finite sequence of (E) and (C) moves to obtain a realization F_3 of (f, φ) such that every cusp c of F_3 is the unique cusp on some loop of F_3, and there exists a path $\alpha_c\colon [0,1] \to W \setminus \partial W$ such that $\alpha_c^{-1}(S(F_3)) = \{0,1\}$ and $\alpha_c(0) = c$, $\alpha_c(1) \in J$. Furthermore, we may assume that for every cusp c of F_3 the vector $\alpha_c'(0)$ is pointing downward at c (see Section 2). (In fact, this can always be achieved by one (C) move followed by one (E) move near c.) Two cusps c and d of F_3 for which $\alpha_c'(1)$ and $\alpha_d'(1)$ point to different sides of J can be eliminated by first applying (C) to generate a new cusp pair on J, and then applying (E) twice along joining curves given by suitable perturbations of α_c and α_d. Repeating this process until no such cusps pairs are left, the remaining cusps lie all in the same component of $W \setminus T$, and can thus be eliminated by means of general cusp elimination (Proposition 2.4) up to at most one cusp.

This completes the proof of Theorem 1.1.

7. Realizations and Number of Loops

Throughout the present section, we suppose that W is orientable, and that σ denotes an orientation of $P = \partial W$ which is induced by one of the two orientations of W. The statement of our main result Theorem 1.2(b) involves the following quantities derived from a pattern (f, φ) on W and the fixed orientation σ of ∂W:

- $n_\sigma(f)$ denotes the number of components C of ∂W with the property that f restricts to an orientation preserving immersion $(-\varepsilon', \varepsilon') \times C \to \mathbb{R}^2$ for some $\varepsilon' \in (0, \varepsilon)$. (Here, we assume that $(-\varepsilon', \varepsilon') \times C$ is oriented in such a way that the inclusion $[0, \varepsilon') \times C \subset W$ is orientation preserving. Thus, $(-\varepsilon', \varepsilon') \times C$ is equipped with the orientation opposite to the product orientation.)
- $c_\sigma(f, \varphi)$ denotes the number of components of the closed 1-manifold $Z_\sigma(f, \varphi)$ obtained as follows. Let $\pi_\sigma(f)$ be the fixed-point free involution of $P_f = (\{0\} \times P) \cap S(f)$ that corresponds to the partition of P_f into pairs of the form $P_f \cap \overline{V}$, where V runs through the components of $(-\varepsilon, \varepsilon) \times \partial W \setminus S(f)$ on which f is an orientation preserving immersion. Now, $Z_\sigma(f, \varphi)$ is obtained from the finite set P_f (considered as a 0-dimensional CW-complex) by attaching for every member $\{x, y\}$ of the partitions of P_f induced by φ and $\pi_\sigma(f)$ a 1-cell with endpoints x and y.

  The quantity $c_\sigma(f, \varphi)$ captures the combinatorial interplay between φ and $S(f)$ for given σ.

By construction we have the following

Proposition 7.1. *Let (f, φ) be a pattern on W, and let F be a realization of (f, φ) without cusps. Then, the number of boundary components of W_F^σ (see Definition 4.1) equals $c_\sigma(f, \varphi) + l_F + n_\sigma$, where l_F is the number of loops of F. Moreover, there exists an integer $h_F^\sigma \geq 0$ such that*

$$\chi(W_F^\sigma) = c_\sigma(f, \varphi) + l_F + n_\sigma - 2h_F^\sigma.$$

Furthermore, if $P_f \neq \emptyset$, then there exists a closed one-dimensional submanifold $Z \subset W_F^\sigma \setminus \partial W_F^\sigma$ such that the number of components of Z is $h_F^\sigma - n_\sigma$ (≥ 0), and such that every component of $W_F^\sigma \setminus Z$ has nonempty intersection with $S(F)$.

Proof. It follows from the definitions of $c_\sigma(f, \varphi)$ and n_σ that the number of boundary components of W_F^σ is $c_\sigma(f, \varphi) + l_F + n_\sigma$. To get the formula for

$\chi(W_F^\sigma)$, choose h_F^σ to be the number of surgeries on embedded 0-spheres S^0 that are necessary to obtain W_F^σ from the disjoint union of $c_\sigma(f,\varphi)+l_F+n_\sigma$ copies of D^2, the closed unit disc in $\mathbb{R}^2$.

It remains to construct the desired submanifold $Z \subset W_F^\sigma \setminus \partial W_F^\sigma$ under the assumption that $P_f \neq \emptyset$. For this purpose, note that $c_\sigma(f,\varphi)+l_F$ is the number of boundary components of W_F^σ which have nonempty intersection with $S(F)$. Let X and Y denote the disjoint unions of $c_\sigma(f,\varphi)+l_F$ and n_σ copies of D^2, respectively. Recall from the construction of h_F^σ that there exist pairwise disjoint embeddings

$$\varphi_i \colon S^0 \hookrightarrow (X \setminus \partial X) \sqcup (Y \setminus \partial Y), \qquad i = 1, \ldots, h_F^\sigma,$$

such that W_F^σ can be obtained from the disjoint union $X \sqcup Y$ by means of surgeries on the embeddings φ_i, and ∂X becomes the union of those boundary components of W_F^σ that have nonempty intersection with $S(F)$. Since W is connected and $P_f \neq \emptyset$ by assumption, every component of W_F^σ must contain at least one boundary component which has nonempty intersection with $S(F)$. Thus, by moving some of the handles $S^1 \times [-1,1] \subset W_F^\sigma$ associated with the embeddings φ_i over each other, we may achieve that for every component D of Y, there is an index i_D such that the embedding φ_{i_D} maps one point of S^0 to X and the other point to D. In particular, $h_F^\sigma - n_\sigma \geq 0$, and we may take Z to be the union of the circles $S^1 \times \{0\} \subset S^1 \times [-1,1]$ in the handles associated with the remaining $h_F^\sigma - n_\sigma$ embeddings. □

The statement of our main result Theorem 1.2 involves the quantity

$$\Delta^\sigma := \Gamma^\sigma/2 - c_\sigma(f,\varphi) + n_\sigma(f),$$

which also depends on Γ^σ as introduced in Proposition 6.1. Note that Proposition 4.2 and Proposition 7.1 together imply

Corollary 7.2. *If F is a realization of a pattern (f,φ) of W without cusps, then*

$$\Delta^\sigma = l_F - 2h_F^\sigma + 2n_\sigma.$$

8. Proof of Theorem 1.2

First suppose that W is non-orientable. In view of loop generation (Proposition 2.2(i)) and loop reduction (Proposition 2.7) it suffices to show that F can be modified on $W \setminus \partial W$ by a finite number of (E) and (C) moves to obtain a realization $\widetilde{F}$ of (f,φ) without cusps in such a way that the number of loops of F does not have the same parity as the number of loops of $\widetilde{F}$. For this purpose, let $C \subset W \setminus \partial W$ be an embedded circle which intersects $S(F)$ transversely, say in the points $p_1, \dots, p_r$. We modify F by applying (C) once to create a pair of cusps at p_1. Next, we modify the resulting realization F_1 of (f, φ) by applying general elimination of cusps (Proposition 2.4) to a curve α that connects the two new cusps by going once around in a tubular neighborhood of C while intersecting $S(F_1)$ in $r-1$ fold points. Note that the total number of (E) moves involved in the modification of F_1 equals r. As a result, we obtain an extension F_2 of f without cusps. However, the (E) moves we used to modify F_1 may have changed the singular pattern, so that F_2 is not necessarily a realization of the pattern (f, φ). Nevertheless, we can modify F_2 further to obtain the desired realization $\widetilde{F}$ of the pattern (f, φ) as follows. Every (E) move we have used can be considered as the first step of loop generation (Proposition 2.2(ii)), and we perform the remaining moves. All in all, we obtain a realization $\widetilde{F}$ of the pattern (f, φ) without cusps, but the number of loops has increased by r due to r-fold application of loop generation. Hence, our claim follows from the fact that r will be odd whenever C is chosen such that its normal line bundle is non-trivial. This completes the proof of part (a).

As for the proof of part (b), suppose that W is orientable. Let σ denote an orientation of ∂W which is induced by one of the two orientations of W. Moreover, let $P_f \neq \emptyset$.

(i) $\Rightarrow$ (ii). Let F be a realization of the pattern (f, φ) which has no cusps and l_F loops. Then, Corollary 7.2 implies that $\Delta^\rho \equiv l_F \bmod 2$ and $\Delta^\rho = l_F - 2(h_F^\rho - n_\rho) \leq l_F$ for $\rho \in \{-\sigma, \sigma\}$, where $h_F^\rho \geq n_\rho$ by Proposition 7.1. Consequently, $l_F \in \mathbb{N} \cap (\Delta^\sigma + 2\mathbb{N}) \cap (\Delta^{-\sigma} + 2\mathbb{N})$.

(ii) $\Rightarrow$ (i). Let $l \in \mathbb{N} \cap (\Delta^\sigma + 2\mathbb{N}) \cap (\Delta^{-\sigma} + 2\mathbb{N})$. In view of loop generation (Proposition 2.2(i)) it suffices to show that F can be modified on $W \setminus \partial W$ by a finite sequence of (E) and (C) moves to a realization of (f, φ) which has no cusps and $\min(\mathbb{N} \cap (\Delta^\sigma + 2\mathbb{N}) \cap (\Delta^{-\sigma} + 2\mathbb{N}))$ loops. This can be achieved as follows. Since $P_f \neq \emptyset$, we can use loop simplification (Proposition 2.6) to achieve that all loops of F are contractible. Using tunneling (Proposition 2.8) and loop reduction (Proposition 2.7), one can in addition achieve that the remaining loops of F are boundary components of the same component of $W \setminus S(F)$, say a component V of W_F^ρ for suitable orientation $\rho \in \{\pm\sigma\}$. Since all loops of F are contractible, it follows that every loop bounds a contractible component of $W_F^{-\rho}$. In this situation, an argument analogous to the last part of the proof of Proposition 7.1 implies that

$$l_F + n_\rho \leq h_F^\rho.$$

(In fact, note that $c_\rho(f, \varphi)$ is an upper bound for the number of components of W_F^ρ because all loops of F lie in the boundary of V.) Hence, by Corollary 7.2,

$$\Delta^\rho = l_F - 2h_F^\rho + 2n_\rho \leq 2(l_F - h_F^\rho + n_\rho) \leq 0.$$

Therefore, using $\Delta^\rho \equiv l_F \equiv \Delta^{-\rho} \bmod 2$, we obtain

$$\min(\mathbb{N} \cap (\Delta^\sigma + 2\mathbb{N}) \cap (\Delta^{-\sigma} + 2\mathbb{N})) = \min(\mathbb{N} \cap (\Delta^{-\rho} + 2\mathbb{N})).$$

Finally, we modify F by means of balancing (Proposition 2.9) to reduce the number of loops to $\min(\mathbb{N}\cap(\Delta^{-\rho}+2\mathbb{N}))$ as follows. By Proposition 7.1, there is a closed one-dimensional submanifold $Z \subset W_F^{-\rho} \setminus \partial W_F^{-\rho}$ that consists of $b := h_F^{-\rho} - n_{-\rho} \geq 0$ components, and such that every component of $W_F^{-\rho} \setminus Z$ has nonempty intersection with $S(F)$. For every component H of Z we may then choose an embedding $\alpha_H \colon [0, 1] \to W \setminus \partial W$ with the following properties:

- The endpoints of α_H satisfy $\alpha_H(0), \alpha_H(1) \in W_F^\rho \setminus (\partial W \cup S(F))$.
- α_H intersects $S(F)$ transversely, and $\alpha_H^{-1}(S(F)) = \{t_0^H, t_1^H\}$ for some $0 < t_0^H < t_1^H < 1$.
- α_H intersects H transversely, and $\alpha_H^{-1}(H) = \{s^H\}$ for some $s^H \in (t_0^H, t_1^H)$.
- $\alpha_H(t_0^H)$ and $\alpha_H(t_1^H)$ are fold points of F that do not lie on the loops of F. (Note that the components of $W_F^{-\rho}$ that contain a loop of F in the boundary are disjoint to H because the loops of F are all contractible.)
- If H and H' are different components of Z, then $\alpha_H([0, 1]) \cap \alpha_{H'}([0, 1]) = \emptyset$.

If $\Delta^{-\rho} > 0$, i.e, $l_F > 2(h_F^{-\rho} - n_{-\rho}) = 2b$, then we perform b (≥ 0) balancing moves to reduce the number of loops to $\Delta^{-\rho} = \min(\mathbb{N} \cap (\Delta^{-\rho} + 2\mathbb{N}))$ as required. (More precisely, for every component H of Z we first apply tunneling (Proposition 2.8) to move an unused pair (C_0^H, C_1^H) of loops of F to the endpoints of α_H, that is, $\alpha_H^{-1}(C_i^H) = \{i\}$ for $i = 0, 1$. This is always possible since W is connected, and all loops of F are contractible, and hence trivial. The desired balancing moves can then be performed because the assumptions of Proposition 2.9 are satisfied for our choices of H and α_H.) If, however, $\Delta^{-\rho} \leq 0$, i.e, $l_F \leq 2(h_F^{-\rho} - n_{-\rho}) = 2b$, then let $\varepsilon \in \{0, 1\}$ such that $\varepsilon \equiv l_F \bmod 2$. Then, we can analogously perform $(l_F - \varepsilon)/2$ balancing moves to reduce the number of loops to $\varepsilon = \min(\mathbb{N} \cap (\Delta^{-\rho} + 2\mathbb{N}))$.

This completes the proof of Theorem 1.2.

Remark 8.1. Our approach ignores the complexity of the *apparent contour* $F(S(F))$ of a generic map $F\colon W \to \mathbb{R}^2$. In particular, when $\partial W = \emptyset$ and F is *stable* (i.e., the immersion $F|\colon S(F) \to \mathbb{R}^2$ self-intersects only at fold points of F, and all self-intersections are nodes, that is, transverse double points), results including the number of nodes of F can for instance be found in [16]. It might be interesting to think about a version of Theorem 1.2 for stable maps of surfaces with boundary into the plane (compare [3]) that also includes the number of nodes. Note that the number of nodes cannot directly be controlled by our algorithm since the (E) and (C) moves easily produce new nodes.

9. Applications

Throughout this section we suppose W to be orientable (the non-orientable case could be handled similarly). Let σ denote an orientation of $P = \partial W$ which is induced by one of the two orientations of W.

In this section the spirit of the gluing principle is reflected in our applications of Theorem 1.2 to pseudo-immersions (see Proposition 9.1 and Example 9.2) and cusp counting (see Proposition 9.4).

Proposition 9.1. *Let F be a realization of the pattern (f, φ) without cusps. Then, the following statements are equivalent for any integer $l \geq 1$:*

(i) *F can be modified on $W \setminus \partial W$ by a finite sequence of* (E) *and* (C) *moves to a realization of (f, φ) which has no cusps and l loops.*
(ii) $l \in 1 + \mathbb{N} \cap (\Delta^{\sigma} - 1 + 2\mathbb{N}) \cap (\Delta^{-\sigma} - 1 + 2\mathbb{N})$.

Proof. We present the proof of implication (i) $\Rightarrow$ (ii) in detail – the proof of the converse implication is similar.

Let G be a realization of (f, φ) which has no cusps and l loops. Let C be the boundary of a small open disc U centered at a fold point on a loop of G such that C intersects $S(G)$ transversely in precisely two points. Then, the surface $V := W \setminus U$ has boundary $\partial V = P \sqcup C$, which we equip with an orientation ρ such that $\rho|_{\partial W} = \sigma$, and such that ρ is induced by an orientation of V. (If $\partial W \neq \emptyset$, then ρ is uniquely determined by σ.) Now $H := G|_V$ is a fold map on V with $l-1$ loops. If we equip C with a suitable collar neighborhood in V, then H can be considered as a realization of the pattern $(f \sqcup f_{\text{fold}}, \varphi \sqcup \varphi_2)$, where f_{fold} is taken from Example 3.6, and φ_2 denotes the unique fixed-point free involution on $C \cap S(G)$. Hence, using $\chi(W) = \chi(V) + 1$, $|(P \sqcup C)_{f \sqcup f_{\text{fold}}}| = |P_f| + 2$, and $\omega_{\pm\rho}(f \sqcup f_{\text{fold}}) = \omega_{\pm\sigma}(f)$ (see Example 3.6), we obtain

$$\Gamma_V^{\pm\rho} := \chi(V) + |(P \sqcup C)_{f \sqcup f_{\text{fold}}}|/2 + \omega_{\pm\rho}(f \sqcup f_{\text{fold}}) = \Gamma^{\pm\sigma}.$$

Moreover, $n_{\pm\rho}(f \sqcup f_{\text{fold}}) = n_{\pm\sigma}(f)$ and $c_{\pm\rho}(f \sqcup f_{\text{fold}}, \varphi \sqcup \varphi_2) = c_{\pm\sigma}(f, \varphi) + 1$ imply

$$\Delta_V^{\pm\rho} := \Gamma_V^{\pm\rho}/2 - c_{\pm\rho}(f \sqcup f_{\text{fold}}, \varphi \sqcup \varphi_2) + n_{\pm\rho}(f \sqcup f_{\text{fold}}) = \Delta^{\pm\sigma} - 1.$$

Finally, the implication (i) $\Rightarrow$ (ii) of Theorem 1.2(b) implies that $l - 1 \in \mathbb{N} \cap (\Delta^{\sigma} - 1 + 2\mathbb{N}) \cap (\Delta^{-\sigma} - 1 + 2\mathbb{N})$. □

Example 9.2 (Pseudo-immersions). An important special case of Proposition 9.1 arises when f restricts to an immersion $(-\varepsilon', \varepsilon') \times \partial W \to \mathbb{R}^2$ for some $\varepsilon' \in (0, \varepsilon)$. In this case, realizations F of the pattern $(f, \varphi_\emptyset)$ (where $\varphi_\emptyset$ denotes the unique permutation of the empty set $\emptyset$) are called *pseudo-immersions* (note that $S(F) \neq \emptyset$ by Remark 5.3). The case that ∂W is connected has been considered in [15, Theorems 1.2 and 1.3], and Proposition 9.1 can be reduced to it as follows.

The assumption on f implies that $P_f = \emptyset$ and $\omega_\sigma(f) = W(F|_{\partial W})$, the winding number (see Definition 3.5). Hence, $\Gamma^\sigma = \chi(W) + W(F|_{\partial W})$ by Proposition 6.1. Moreover, $n_\sigma(f) = 1$ (for suitable σ), and $c_\sigma(f, \varphi_\emptyset) = 0$. Thus, we obtain

$$\Delta^\sigma = (\chi(W) + W(F|_{\partial W}))/2 + 1 = ((\chi(W) + 1) + (W(F|_{\partial W}) + 1))/2.$$

On the other hand, observe that $\omega_{-\sigma}(f) = -\omega_\sigma(f) = -W(F|_{\partial W})$, $n_{-\sigma}(f) = 0$, and $c_{-\sigma}(f, \varphi_\emptyset) = 0$, which yields

$$\Delta^{-\sigma} = (\chi(W) - W(F|_{\partial W}))/2 = ((\chi(W) + 1) - (W(F|_{\partial W}) + 1))/2.$$

Altogether, $\mathfrak{m} := \max\{\Delta^\sigma, \Delta^{-\sigma}\} = (\chi(W) + 1 + |W(F|_{\partial W}) + 1|)/2$, and by Proposition 9.1, in accordance with [15], the set of possible numbers of loops of F is

$$1 + \mathbb{N} \cap (\mathfrak{m} - 1 + 2\mathbb{N}) = \begin{cases} \max(\mathfrak{m}, 2) + 2\mathbb{N}, & \text{if } \chi(W) - W(F|_{\partial W}) \equiv 0 \bmod 4, \\ \max(\mathfrak{m}, 1) + 2\mathbb{N}, & \text{if } \chi(W) - W(F|_{\partial W}) \equiv 2 \bmod 4. \end{cases}$$

Remark 9.3. Note that the proof of [15, Theorem 1.3] relies on a theorem due to Eliashberg [4] and Francis [5] (see [15, Theorem 3.2, p. 1330]). On the other hand, our approach is purely based on (E) and (C) moves, and we can easily deduce the Eliashberg-Francis theorem from Proposition 9.1.

Proposition 9.4. *Let F be a realization of the pattern (f, φ). If $P_f \neq \emptyset$, then the following statements are equivalent for any integers $c > 0$ and l:*

(i) *F can be modified on $W \setminus \partial W$ by a finite sequence of* (E) *and* (C) *moves to a realization of (f, φ) which has c cusps and l loops.*

(ii) $l \in \bigcup_{w \in \{-c,-c+2,\ldots,c-2,c\}} \mathbb{N} \cap (\Delta^{\sigma} + w/2 + 2\mathbb{N}) \cap (\Delta^{-\sigma} - w/2 + 2\mathbb{N})$.

Proof. The proof is based on similar ideas as the proof of Proposition 9.1.

(i) $\Rightarrow$ (ii). Let G be a realization of (f, φ) which has $c \geq 1$ cusps and l loops. Let $U \subset W$ be the union of c small open discs each of which is centered at a different cusp of G. By choosing those small discs appropriately, we may assume that the boundary C of U is diffeomorphic to the disjoint union of c circles each of which intersects $S(G)$ transversely in precisely two points. Then, the surface $V := W \setminus U$ has boundary $\partial V = P \sqcup C$, which we equip with the unique orientation ρ such that $\rho|_{\partial W} = \sigma$ (where note that $\partial W \neq \emptyset$), and such that ρ is induced by an orientation of V. Now $H := G|_V$ is a fold map on V with $l - d$ loops, where d denotes the number of loops of G that contain at least one cusp of G. If we equip C with a suitable collar neighborhood in V, then H is near every component of C an extension of the boundary condition f_{cusp} considered in Example 3.6. Thus, H can be considered as a realization of the pattern $(f \sqcup g, \psi)$, where g denotes the disjoint union of c copies of f_{cusp}, and ψ denotes the partition of $\partial V \cap S(G)$ into those subsets of cardinality 2 which arise as the boundary points of some component of $S(H)$. Hence, using $\chi(W) = \chi(V) + c$, $|(P \sqcup C)_{f \sqcup g}| = |P_f| + 2c$, and $\omega_{\pm\rho}(f \sqcup g) = \omega_{\pm\sigma}(f) \pm w$ for some integer w with the same parity as c, and such that $|w| \leq c$, we obtain

$$\Gamma_V^{\pm\rho} := \chi(V) + |(P \sqcup C)_{f \sqcup g}|/2 + \omega_{\pm\rho}(f \sqcup g) = \Gamma^{\pm\sigma} \pm w.$$

Moreover, $n_{\pm\rho}(f \sqcup g) = n_{\pm\sigma}(f)$ and $c_{\pm\rho}(f \sqcup g, \psi) = c_{\pm\sigma}(f, \varphi) + d$ imply

$$\Delta_V^{\pm\rho} := \Gamma_V^{\pm\rho}/2 - c_{\pm\rho}(f \sqcup g, \psi) + n_{\pm\rho}(f \sqcup g) = \Delta^{\pm\sigma} \pm w/2 - d.$$

Finally, to conclude that (ii) holds, we apply the implication (i) $\Rightarrow$ (ii) of Theorem 1.2(b) to the realization H of the pattern $(f \sqcup g, \psi)$ to obtain

$$l - d \in \mathbb{N} \cap (\Delta^{\sigma} + w/2 - d + 2\mathbb{N}) \cap (\Delta^{-\sigma} - w/2 - d + 2\mathbb{N}).$$

(ii) $\Rightarrow$ (i). We may choose an integer $w \in \{-c, -c+2, \ldots, c-2, c\}$ such that

$$l \in \mathbb{N} \cap (\Delta^{\sigma} + w/2 + 2\mathbb{N}) \cap (\Delta^{-\sigma} - w/2 + 2\mathbb{N}).$$

We modify F as follows. After possibly creating a number of new pairs of cusps for F by means of (C), we can choose cusps $x_1, \ldots, x_c$ of F admitting pairwise disjoint small open disc neighborhoods $U_1, \dots, U_c$, respectively, where the boundary circle C_i of each U_i intersects $S(F)$ transversely in precisely two points, and the following property holds. If f_i denotes the restriction of F to some fixed collar neighborhood of C_i in $V := W \setminus \bigsqcup_i U_i$, and ρ denotes the unique orientation of $\partial V = \partial W \sqcup \bigsqcup_i C_i$ that is induced by an orientation of V in such a way that $\rho|_{\partial W} = \sigma$ (where note that $\partial W \neq \emptyset$), then

$$w = \sum_{i=1}^{c} \omega_{\rho|_{C_i}}(f_i).$$

(Indeed, since $w \in \{-c, -c+2, \dots, c-2, c\}$ can clearly be written as the sum of c summands of the form ± 1, we only have to apply the following observation.

If we create a new pair (x, x') of cusps for F by means of (C) and choose sufficiently small open disc neighborhoods U and U' of x and x', respectively, whose respective boundary circles C and C' intersect $S(F)$ transversely in precisely two points, then the following property holds. If g and g' denote the restrictions of F to some fixed collar neighborhoods of C and C' in $Y = W \setminus (U \sqcup U')$, respectively, and τ denotes the unique orientation of $\partial Y = \partial W \sqcup C \sqcup C'$ that is induced by an orientation of Y in such a way that $\tau|_{\partial W} = \sigma$, then $\omega_{\tau|_C}(g), \omega_{\tau|_{C'}}(g') \in \{\pm 1\}$ and $\omega_{\tau|_C}(g) = -\omega_{\tau|_{C'}}(g')$.

In fact, the statement $\omega_{\tau|_C}(g), \omega_{\tau|_{C'}}(g') \in \{\pm 1\}$ follows because we can choose the neighborhoods U and U' so that C and C' have collar neighborhoods in $Y = W \setminus (U \sqcup U')$, respectively, in which g and g' look like copies of the boundary condition f_{cusp} considered in Example 3.6. Moreover, the statement $\omega_{\tau|_C}(g) = -\omega_{\tau|_{C'}}(g')$ can be deduced as follows. Since the pair (x, x') of cusps has been created by means of (C), we can choose the closures of U and U' to lie in a small closed disc $Z \subset W \setminus \partial W$ such that the boundary circle ∂Z intersects $S(F)$ transversely in precisely two points, and the restriction of F to some fixed collar neighborhood of ∂Z in $W \setminus Z$ looks like the boundary condition f_{fold} considered in Example 3.6. Then the claim follows by applying Proposition 4.2 to $F|_{Z \setminus (U \sqcup U')}$.)

The restriction $H := F|_V$ is a generic map on V that can be considered as a realization of a pattern of the form (h, ψ), where $h = f \sqcup \bigsqcup_i f_i$, and where ψ denotes the partition of $\partial V \cap S(G)$ into those subsets of cardinality 2 which arise as the boundary points of some component of $S(H)$. Hence, using $\chi(W) = \chi(V) + c$, $|(\partial V)_h| = |P_f| + 2c$, as well as

$\omega_{\pm\rho}(h) = \omega_{\pm\sigma}(f) \pm w$, we obtain

$$\Gamma_V^{\pm\rho} := \chi(V) + |(\partial V)_h|/2 + \omega_{\pm\rho}(h) = \Gamma^{\pm\sigma} \pm w.$$

Without loss of generality, we may assume that the cusps $x_1, \ldots, x_c$ lie on components of $S(F)$ that have nonempty intersection with ∂W. Consequently, $c_{\pm\rho}(h, \psi) = c_{\pm\sigma}(f, \varphi)$. Using also that $n_{\pm\rho}(h) = n_{\pm\sigma}(f)$, we obtain

$$\Delta_V^{\pm\rho} := \Gamma_V^{\pm\rho}/2 - c_{\pm\rho}(h, \psi) + n_{\pm\rho}(h) = \Delta^{\pm\sigma} \pm w/2.$$

Note that Γ_V^ρ is even because $\Delta_V^\rho = \Delta^\sigma + w/2$ is an integer by assumption. Hence, Proposition 6.1 implies that H has an even number of cusps. Thus, according to Theorem 1.1(b), H can be modified on $V \setminus \partial V$ by a finite sequence of (E) and (C) moves to obtain a realization H_0 of (h, ψ) without cusps. Since

$$l \in \mathbb{N} \cap (\Delta_V^\rho + 2\mathbb{N}) \cap (\Delta_V^{-\rho} + 2\mathbb{N}),$$

we may apply the implication (ii) $\Rightarrow$ (i) of Theorem 1.2(b) to modify the realization H_0 of the pattern (h, ψ) by a finite sequence of (E) and (C) moves to a realization of (h, ψ) which has no cusps and l loops. The above modifications of H and H_0 can be understood as modifications of F on $W \setminus \partial W$ by a finite sequence of (E) and (C) moves that turn F into a realization of (f, φ) which has c cusps and l loops. □

Remark 9.5. The work on this paper has been motivated by Banagl's recent construction of TFT-type invariants [2]. In fact, Banagl has introduced the notion of a positive TFT as a convenient framework for constructing TFT-type invariants based on *semirings* rather than on rings. Following the essential feature of a TFT, the state sum (or partition function) of a positive TFT is required to satisfy a gluing axiom, which forces the information it detects to be local to a certain extent. As for the technical implementation of Banagl's positive TFT based on fold maps, the combinatorial information captured by singular patterns of fold maps is incorporated into the morphisms of a category. The resulting *Brauer category* is a strict monoidal category that is constructed as categorification of the Brauer algebras classically known from representation theory of the orthogonal group $O(n)$. By means of Theorem 1.2 we are able to list those integers l for which a given singular pattern on W admits a realization having no cusps and l loops. From the perspective of Banagl's positive TFT based on fold maps, this leads to the computation of the state sum invariant on all 2-dimensional cobordisms (see [14] for details).

References

[1] F. Aicardi, T. Ohmoto, *First order local invariants of apparent contours*, Topology **45** (2006), 27–45.

[2] M. Banagl, *Positive topological quantum field theories*, Quantum Topology **6** (2015), no. 4, 609–706.

[3] J. W. Bruce, P. J. Giblin, *Projections of surfaces with boundary*, London Math. Society (3) **60** (1990), 392–416.

[4] Y. Eliashberg *On singularities of folding type*, Math. USSR Izv. **4** (1970), 1119–1134.

[5] G. K. Francis, *The folded ribbon theorem: a contribution to the study of immersed circles*, Trans. Amer. Math. Soc. **141** (1969), 271–303.

[6] G. K. Francis, *Assembling compact Riemann surfaces with given boundary curves and branch points on the sphere*, Illinois J. Math. **20** (1976), 198–217.

[7] T. Fukuda, G. Ishikawa, *On the number of cusps of stable perturbations of a plane-to-plane singularity*, Tokyo J. Math. **10** (1987), 375–384.

[8] A. Haefliger, *Quelques remarques sur les applications différentiables d'une surface dans le plan*, Ann. Inst. Fourier Grenoble **10** (1960), 47–60.

[9] D. Hacon, C. Mendes de Jesus, M. C. Romero Fuster, *Fold maps from the sphere to the plane*, Experimental Math. **15** (2006), 491–497.

[10] T. Kálmán, *Stable maps of surfaces into the plane*, Topology Appl. **107** (2000), 307–316.

[11] H. I. Levine, *Elimination of cusps*, Topology **3**, Suppl. 2 (1965), 263–296.

[12] H. Whitney, *On regular closed curves in the plane*, Compositio Mathematica, **4** (1937), 276–284.

[13] H. Whitney, *On singularities of mappings of euclidean spaces, I: Mappings of the plane into the plane*, Ann. of Math. (2) **62** (1955), 374–410.

[14] D. J. Wrazidlo, *Fold maps and positive topological quantum field theories*, Dissertation, Heidelberg (2017), `http://nbn-resolving.de/urn:nbn:de:bsz:16-heidok-232530`.

[15] M. Yamamoto, *Pseudo-immersions of oriented surfaces with one boundary component into the plane*, Proceedings of the Royal Society of Edinburgh **139A** (2009), 1327–1335.

[16] T. Yamamoto, *Number of singularities of stable maps on surfaces*, Pacific Journal of Mathematics **280**, no. 2 (2016), 489–510.

Motivic Milnor–Hirzebruch classes revisited

Shoji Yokura

Department of Mathematics and Computer Science, Faculty of Science, Kagoshima University, 21-35 Korimoto 1-chome, Kagoshima 890-0065, Japan, E-mail: yokura@sci.kagoshima-u.ac.jp

In this paper we give a survey about what is called Milnor–Hirzebruch class, which is the difference of the motivic Hirzebruch class (of a local complete intersection X in a smooth variety) and the Poincaré dual of the Hirzebruch cohomology class of the virtual tangent bundle. For that we give a quick review of the motivic Hirzebruch class, and Verdier–Riemann–Roch formula, which is a motivation behind defining the so-called Milnor class. In the final section we discuss how to capture the Milnor–Hirzebruch class as a special value of a natural transformation.

Keywords: Hirzebruch class, Milnor class, motivic characteristic class, Verdier–Riemann–Roch

2010 MSC: 13D15, 14C17, 14C40, 14F25, 14F45, 16E20, 32S30, 32S35, 32S55

1. Introduction

What is called Milnor class is defined for a local complete intersection X in a smooth variety M and it is a homology class supported on the singular locus X_{sing} of X, defined as follows. For such a local complete intersection variety X a normal bundle N_X in M is defined and via the tangent bundle TM of M the virtual tangent bundle $T_X^{\rm vir} := TM|_X - N_X M$ is a well-defined element of the Grothendieck group $K^0(X)$ of complex vector bundles. Then Fulton–Johnson's class of X (see [27] and [26]) is defined by

$$c_*^{FJ}(X) := c(T_X^{\rm vir}) \cap [X] = \frac{c(TM|_X)}{c(N_X M)} \cap [X].$$

Here $c(-)$ is the total Chern class of the (virtual) bundle $T_X^{\rm vir}, TM|_X, N_X M$. Another distinguished Chern homology class of a possibly singular variety X is the Chern–Schwartz–MacPherson class $c_*(X) := c_*(1\!\!1_X)$, where $c_* : F(-) \to H_*(-)$ is MacPherson's Chern class transformation [33]. Then the Milnor class $\mathcal{M}(X)$ is defined as the difference of these two classes with a sign

$$\mathcal{M}(X) := (-1)^{\dim X} \left(c_*^{FJ}(X) - c_*(X) \right).$$

In a very simple case when X is a hypersurface with only isolated singularities in a smooth variety, the Milnor class turns out to be the sum of the Milnor numbers of these isolated singular points (e.g., see [2], [42], [45], [61]). For the Milnor class, e.g., also see [8, 9, 13, 35, 41, 50].

Remark 1.0.1. The Milnor class $\mathcal{M}(X)$ (also see [68], which is the extended version of the unpublished preprint [65]) is just the difference of the above two classes (up to sign), but in fact the author was naturally led to considering this difference of these two classes in such a special case while he was looking for a Verdier-type Riemann–Roch theorem [67] for MacPherson's Chern class transformation $c_* : F(-) \to H_*(-)$, which will be recalled in a bit more details in §3. Due to these preceding results that the difference of two classes in the case of isolated singularities is the sum of the well-known Milnor numbers of the singularities, it was natural for the author to call the difference $(-1)^{\dim X} \left(c_*^{FJ}(X) - c_*(X) \right)$ the *Milnor class*, denoted $\mathcal{M}_*(X)$.

Motivated by the formations of MacPherson's Chern class (natural) transformation $c_* : F(-) \to H_*(-)$, Baum–Fulton–MacPheson's Todd class (natural) transformation $td_* : G_0(-) \to H_*(-) \otimes \mathbb{Q}$ and Cappell–Shaneson's homology L-class (natural) transformation $L_* : \Omega(-) \to H_*(-) \otimes \mathbb{Q}$ (in which Goresky–MacPherson's homology L-class [29] is used) and also for the "unification[1]" of these three distinguished theories of characteristic homology classes for possibly singular varieties, J.-P. Brasselet, J. Schürmann and the author have constructed the motivic Hirzebruch class (transformation) $T_{y*} : K_0(\mathcal{V}/-) \to H_*^{\mathrm{BM}}(-) \otimes \mathbb{Q}[y]$ (see [10]). This is the unique natural transformation from the covariant functor of relative Grothendieck group $K_0(\mathcal{V}/X)$ of complex algebraic varieties over X to the covariant homology functor with the coefficients $\mathbb{Q}[y]$ such that for a smooth variety X the distinguished value $T_{y*}([X \xrightarrow{\mathrm{id}_X} X])$ is equal to the Poincaré dual of the total Hirzebruch cohomology class $T_y(TX)$ of the

[1]At the very end of MacPherson's survey article [34] (cf. [66] where an explicit problem is posed) he wrote "*It remains to be seen whether there is a unified theory of characteristic classes of singular varieties like the classical one outlined above.*" At that time Goresky–MacPherson's homology L-class was not available yet and it was defined only after the theory of Intersection Homology [29] was invented by M. Goresky and R. MacPherson in 1980.

tangent bundle TX (see §2 below):

$$T_{y*}([X \xrightarrow{\mathrm{id}_X} X]) = T_y(TX) \cap [X] \in H_*(X) \otimes \mathbb{Q}[y].$$

Whether X is singular or nonsingular, the homology class $T_{y*}(X) := T_{y*}([X \xrightarrow{\mathrm{id}_X} X])$ is called the motivic Hirzebruch class of X.

Due to a very naive motivation of trying to capture the Milnor class $\mathcal{M}(X)$ as the distinguished value $\mathcal{M}([X \xrightarrow{\mathrm{id}_X} X])$ of a certain natural transformation $\mathcal{M} : K_0(\mathcal{V}/-) \to H_*(-)$ and also of trying to get a "T_{y*}"-analogue of the Milnor class, in [72] we introduce a modified version $K^{Prop}_{\ell.c.i.}(\mathcal{V}/X)$ of the relative Grothendieck group $K_0(\mathcal{V}/X)$ of complex algebraic varieties over X and we define *a motivic Milnor–Hirzebruch class* as a natural transformation

$$\mathscr{M}T_{y*} : K^{Prop}_{\ell.c.i.}(\mathcal{V}/-) \to H_{**}(-) \otimes \mathbb{Q}[y],$$

where $H_{**}(X)$ is the twisted Borel–Moore homology theory, i.e., the Borel-Moore homology theory $H_*(X)$ with the pushforward $f_{**} : H_*(X) \to H_*(Y)$ for a proper map $f : X \to Y$ defined by $f_{**} := (-1)^{\dim Y - \dim X} f_*$. In particular, the special value $\mathscr{M}T_{y*}(X) := \mathscr{M}T_{y*}([X \xrightarrow{\mathrm{id}_X} X])$ for a local complete intersection X is equal to

$$\mathscr{M}T_{y*}(X) = (-1)^{\dim X}\Big(T_y(T_X^{vir}) \cap [X] - T_{y*}(X)\Big),$$

which is called the motivic Milnor–Hirzebruch class of X, because in the special case when $y = -1$ we have $\mathscr{M}T_{-1*}(X) = \mathcal{M}(X) \otimes \mathbb{Q}$, the rationalized Milnor class of X.

In [21] S. E. Cappell, L. G. Maxim, Schürmann and J. L. Shaneson independently consider the difference of these two classes of a hypersurface X without the sign:

$$\mathcal{M}T_{y*}(X) := T_y(T_X^{vir}) \cap [X] - T_{y*}(X),$$

which they call *Hirzebruch–Milnor class.* Furthermore they capture it as the canonical homology class of the vanishing cycle in the level of (the Grothendieck group of the Abelian category of) mixed Hodge modules:

$$\mathcal{M}T_{y*}(X) = MHT_{y*}(\Phi'^H_f([\mathbb{Q}^H_M])) \in H_*(X) \otimes \mathbb{Q}[y].$$

Here $K_0(MHM(X))$ is the Grothendieck group of mixed Hodge modules on X and $MHT_{y*} : K_0(MHM(X)) \to H_*(X) \otimes \mathbb{Q}[y, y^{-1}]$ is the mixed Hodge module version of the motivic Hirzebruch class (transformation) $T_{y*} : K_0(\mathcal{V}/-) \to H_*(-) \otimes \mathbb{Q}[y]$. $\Phi'^H_f := \Phi^H_f[-1]$ is the shifted vanishing cycle functor. Their approach is quite natural. The Milnor class $\mathcal{M}(X)$ of a global hypersurface $X := f^{-1}(0)$ in a complex manifold M defined by a holomorphic function $f : M \to \mathbb{C}$ is captured as

$$\mathcal{M}(X) = (-1)^{\dim X} c_*(\phi_f(1\!\!1_M)),$$

where $\phi_f : F(M) \to F(X)$ is the constructible function version of the vanishing cycle functor[2]. Thus $\Phi'^H_f([\mathbb{Q}^H_M])$ is the lift of the vanishing cycle $\phi_f(1\!\!1_M)$ to the level of mixed Hodge modules.

In the final section, via the above Cappell–Maxim–Schürmann–Shaneson's description of the Milnor–Hirzebruch class by the vanishing cycle functor, we capture the Hirzebruch–Milnor class of a hypersurface as a *motivic* one, namely *as a natural transformation* $\mathcal{M}T_{y*} : K_0(\mathcal{V}/-) \to H_*(-) \otimes \mathbb{Q}[y]$ *such that the distinguished value* $\mathcal{M}T_{y*}([X \xrightarrow{\mathrm{id}_X} X])$ *is equal to* $\mathcal{M}T_{y*}(X)$, which we call a *"motivicfication"* of the Milnor–Hirzebruch class, abusing words.

2. Motivic Hirzebruch classes

In this section we give a very quick review of *the motivic Hirzebruch class* [10].

2.1. *Hirzebruch χ_y-genera $\chi_y(X)$ of a smooth variety X and Hirzebruch classes $T_y(E)$ of a complex vector bundle E*

Definition 2.1.1. For a compact complex algebraic manifold X the *Hirzebruch χ_y-genus $\chi_y(X)$* is defined by

$$\chi_y(X) := \sum_{p\geq 0} \chi(X, \Lambda^p T^*X) y^p = \sum_{p\geq 0} \Big(\sum_{i\geq 0} (-1)^i \dim_{\mathbb{C}} H^i(X, \Lambda^p T^*X) \Big) y^p \,.$$

If $\chi^p(X) := \chi(X, \Lambda^p T^*X)$ then $\chi_y(X) = \sum_{p\geq 0} \chi^p(X) y^p$.

More generally we have

[2]Recall that if $\psi_f : F(M) \to F(X)$ is the nearby cycle "functor", then we have $\phi_f = \psi_f - i_X^*$ where $i_X : X \to M$ is the inclusion. Hence $\phi_f(\mathbf{1}_M) = \psi_f(\mathbf{1}_M) - i_X^*(\mathbf{1}_M) = \psi_f(\mathbf{1}_M) - \mathbf{1}_X$ and $\phi_f(\mathbf{1}_M)(x) = \chi(F_{f,x}) - 1$ where $F_{f,x}$ is the Milnor fiber of f at the point $x \in X$. The Milnor number $\mu(x)$ of a hypersruface X with isolated singular point x is $\mu(x) = (-1)^{\dim X}(\chi(F_{f,x}) - 1)$. $c_*(\psi_f(\mathbf{1}_M)) = c(T_X^{vir}) \cap [X]$, thus we have $\mathcal{M}(X) = (-1)^{\dim X} c_*(\phi_f(\mathbf{1}_M))$.

Definition 2.1.2. For E a holomorphic vector bundle over X, the *Hirzebruch χ_y-genus of E* is defined by

$$\begin{aligned}\chi_y(X,E) &:= \sum_{p\geq 0} \chi(X, E\otimes \Lambda^p T^*X) y^p \\ &= \sum_{p\geq 0}\Big(\sum_{i\geq 0}(-1)^i \dim_{\mathbb{C}} H^i(X, E\otimes \Lambda^p T^*X)\Big) y^p.\end{aligned}$$

Theorem 2.1.3 (The generalized Hirzebruch–Riemann–Roch theorem).

$$\chi_y(X,E) = \int_X T_y(TX)\cdot ch_{(1+y)}(E)\cap [X] \quad \in \mathbb{Q}[y]. \tag{2.1}$$

$$ch_{(1+y)}(E) := \sum_{j=1}^{\operatorname{rank} E} e^{\beta_j(1+y)}, \quad T_y(TX) := \prod_{i=1}^{\dim X}\Big(\frac{\alpha_i(1+y)}{1-e^{-\alpha_i(1+y)}} - \alpha_i y\Big),$$

where β_j are the Chern roots of E, α_i are the Chern roots of the tangent bundle TX.

The special case when $y=0$ is the following well-known theorem:

Theorem 2.1.4 (Hirzebruch–Riemann–Roch theorem).

$$\chi(X,E) = \int_X td(TX)\cdot ch(E)\cap [X], \tag{2.2}$$

$ch(E) = \sum_{j=1}^{\operatorname{rank} E} e^{\beta_j}$ *is the Chern character and* $td(TX) = \prod_{i=1}^{\dim X}\Big(\frac{\alpha_i}{1-e^{-\alpha_i}}\Big)$ *is the Todd class.*

Remark 2.1.5. For a complex vector bundle E the above class

$$T_y(E) := \prod_{i=1}^{\operatorname{rank} E}\Big(\frac{\alpha_i(1+y)}{1-e^{-\alpha_i(1+y)}} - \alpha_i y\Big)$$

with α_i the Chern roots of the bundle E unifies the following three important characteristic cohomology classes of E:

(1) $(y=-1)$: $c(E) = \prod_{i=1}^{\operatorname{rank} E}(1+\alpha_i)$ the total Chern class,
(2) $(y=\ \ 0)$: $td(E) = \prod_{i=1}^{\operatorname{rank} E}\frac{\alpha_i}{1-e^{-\alpha_i}}$ the total Todd class,
(3) $(y=\ \ 1)$: $L(E) = \prod_{i=1}^{\operatorname{rank} E}\frac{\alpha_i}{\tanh\alpha_i}$ the total Thom–Hirzebruch L-class.

$T_y(E)$ is called *the Hirzebruch class* of the complex vector bundle E and $T_y(TX)$ is called *the cohomology Hirzebruch class* of the algebraic manifold X.

2.2. Hirzebruch χ_y-genus $\chi_y(X)$ and motivic Hirzebruch class $T_{y*}(X)$ for a singular variety X

The above Hirzebruch χ_y-genus of a smooth compact variety is extended to the case of singular varieties, using Deligne's mixed Hodge structures, as follows:

Definition 2.2.1 (Hodge–Deligne polynomial). For a mixed Hodge structure $(H_*(X), W^\bullet, F_\bullet)$ where $W^\bullet$ is the weight filtration and $F_\bullet$ is the Hodge filtration, the Hodge–Deligne polynomial[3] $\chi_{u,v}(X)$ is defined to be

$$\chi_{u,v}(X) := \sum_{i,p,q\geq 0} (-1)^i(-1)^{p+q} \dim_{\mathbb{C}}(Gr_F^p Gr_{p+q}^W H_c^i(X,\mathbb{C}))u^p v^q.$$

$\chi_{u,v}$ satisfies the following properties:

(1) $X \cong X'$ (isomorphism) $\Longrightarrow \chi_{u,v}(X) = \chi_{u,v}(X')$,
(2) $\chi_{u,v}(X) = \chi_{u,v}(X \setminus Y) + \chi_{u,v}(Y)$ for a closed subvariety $Y \subset X$,
(3) $\chi_{u,v}(X \times Y) = \chi_{u,v}(X) \cdot \chi_{u,v}(Y)$,
(4) $\chi_{u,v}(pt) = 1$.

Let $\mathrm{Iso}(\mathcal{V})$ be the free abelian group generated by the isomorphism classes $[X]$ of varieties X. Then the property (1) implies that the homomorphism $\chi_{u,v} : \mathrm{Iso}(\mathcal{V}) \to \mathbb{Z}[u,v]$ defined by $\chi_{u,v}([X]) := \chi_{u,v}(X)$ is well-defined. The subgroup of $\mathrm{Iso}(\mathcal{V})$ generated by elements of the form $[X] - [Y] - [X \setminus Y]$ with Y a closed subvariety of X shall be denoted by $\{[X] - [Y] - [X \setminus Y]\}$ and the quotient

$$K_0(\mathcal{V}) := \mathrm{Iso}(\mathcal{V})/\{[X] - [Y] - [X \setminus Y]\}$$

is called the *Grothendieck group of varieties*. The equivalence class of $[X]$ in $K_0(\mathcal{V})$ is still denoted by $[X]$ for the sake of simplicity. The property (2) implies that the homomorphism $\chi_{u,v} : \mathrm{Iso}(\mathcal{V}) \to \mathbb{Z}[u,v]$ induces the following finer homomorphism

$$\chi_{u,v} : K_0(\mathcal{V}) \to \mathbb{Z}[u,v].$$

In [32] E. Looijenga defines the relative Grothendieck group $K_0(\mathcal{V}/X)$ analogously as the free abelian group generated by the isomorphism classes $[V \xrightarrow{h} X]$ of a morphism $f : V \to X$ modulo the relation

$$[V \xrightarrow{h} X] = [W \xrightarrow{h|_W} X] + [V \setminus W \xrightarrow{h|_{V \setminus W}} X]$$

[3] It is sometimes called the E-polynomial and denoted by $E_{u,v}(X)$. However, as remarked above, for a smooth variety, $E_{y,-1}(X) = E_{-1,y}(X)$ is equal to the Hirzebruch χ_y-genus, thus we use the symbol $\chi_{u,v}$ instead of $E_{u,v}$.

for a closed subvariety $W \subset V$. Note that $K_0(\mathcal{V}/pt) = K_0(\mathcal{V})$ and for a morphism $f : X \to Y$ the pushforward $f_* : K_0(\mathcal{V}/X) \to K_0(\mathcal{V}/Y)$ is defined by $f_*([V \xrightarrow{h} X]) := [V \xrightarrow{f \circ h} Y]$ and clearly $(g \circ f)_* = g_* \circ f_*$, namely $K_0(\mathcal{V}/-)$ is a covariant functor. The pullback $f^* : K_0(\mathcal{V}/Y) \to K_0(\mathcal{V}/X)$ is defined by $f^*([V \xrightarrow{h} Y]) := [V \times_Y X \xrightarrow{\tilde{h}} X]$ where we use the following fiber product:

$$\begin{array}{ccc} V \times_Y X & \xrightarrow{\tilde{h}} & X \\ {\scriptstyle \tilde{h}}\downarrow & & \downarrow{\scriptstyle f} \\ V & \xrightarrow[h]{} & Y. \end{array}$$

Here is a question of wether or not one can claim *Grothendieck–Riemann–Roch type theorem for* $\chi_{u,v} : K_0(\mathcal{V}) \to \mathbb{Z}[u,v]$, namely asking if there exists a natural transformation

$$\tau : K_0(\mathcal{V}/-) \to H_*(-) \otimes \mathbb{Z}[u,v],$$

where $H_*(-) := H_{2*}^{\mathrm{BM}}(-)$ is the Borel–Moore homology group of even degree, such that if $X = pt$ is a point, then $\tau : K_0(\mathcal{V}/-) \to H_*(-) \otimes \mathbb{Z}[u,v]$ is nothing but the above homomorphism $\chi_{u,v} : K_0(\mathcal{V}) \to \mathbb{Z}[u,v]$. We impose one more requirement ("smooth condition" [4]) that there exists a multiplicative characteristic cohomology class $c\ell$ such that if X is smooth, then

$$\tau([X \xrightarrow{\mathrm{id}_X} X]) = c\ell(TX) \cap [X].$$

This "smooth condition" implies that $(u+1)(v+1) = 0$, i.e., $u = -1$ or $v = -1$, therefore we only have to consider $\chi_{u,-1}(X)$ or $\chi_{-1,v}(X)$. $\chi_{u,v}(X)$ is in fact symmetric with respect to (u,v), i.e., $\chi_{u,v}(X) = \chi_{v,u}(X)$, thus $\chi_{u,-1}(X) = \chi_{-1,u}(X)$. We observe that $\chi_{u,-1}(X)$ is nicer than $\chi_{-1,u}(X)$ since $\chi_{u,-1}(X)$ involves only the Hodge filtration as follows: (changing u to y)

$$\chi_y(X) := \chi_{y,-1}(X) = \sum_{i,p \geq 0} (-1)^i \dim_{\mathbb{C}} Gr_F^p \left(H_c^i(X, \mathbb{C})\right)(-y)^p.$$

When X is non-singular and compact, it is equal to the original Hirzebruch χ_y-genus $\chi_y(X)$, therefore the above $\chi_y(X)$ is still called the Hirzebruch χ_y-genus, whether X is smooth or singular. Now let us consider the following

[4]This "smooth condition" is motivated by similar properties which the three well-known theories of characteristic classes for singular varieties, which will be recalled later, satisfy.

commutative diagram for $[\mathbb{P}^n \xrightarrow{\mathrm{id}_{\mathbb{P}^n}} \mathbb{P}^n]$:

$$\begin{array}{ccc} K_0(\mathcal{V}/X) & \xrightarrow{\tau} & H_*(X) \otimes \mathbb{Z}[y] \\ {\scriptstyle (a_X)_*}\downarrow & & \downarrow{\scriptstyle (a_X)_*} \\ K_0(\mathcal{V}) & \xrightarrow[\chi_y]{} & \mathbb{Z}[y], \end{array}$$

then we have

$$\chi_y (a_{\mathbb{P}^n})_*([\mathbb{P}^n \xrightarrow{\mathrm{id}_{\mathbb{P}^n}} \mathbb{P}^n]) = (a_{\mathbb{P}^n})_* \tau([\mathbb{P}^n \xrightarrow{\mathrm{id}_{\mathbb{P}^n}} \mathbb{P}^n]).$$

Then the left hand side of the above equality is equal to $\chi_y(\mathbb{P}^n)$ and the right hand side is equal to $\int_{\mathbb{P}^n} c\ell(T\mathbb{P}^n) \cap [\mathbb{P}^n]$. Thus $\chi_y(\mathbb{P}^n) = \int_{\mathbb{P}^n} c\ell(T\mathbb{P}^n) \cap [\mathbb{P}^n]$. Since $\chi_y(\mathbb{P}^n) = 1 - y + y^2 + \cdots + (-1)^n y^n$, we get

$$\int_{\mathbb{P}^n} c\ell(T\mathbb{P}^n) \cap [\mathbb{P}^n] = 1 - y + y^2 + \cdots + (-1)^n y^n.$$

In [30] F. Hirzebruch proved that such a characteristic class $c\ell$ has to be the Hirzebruch class T_y.

In our previous paper [10] (see also [52], [57] and [71]), using Saito's theory of mixed Hodge modules [48] we gave a positive answer to the above question of the existence of Grothendieck–Riemann–Roch type theorem for $\chi_y : K_0(\mathcal{V}) \to \mathbb{Z}[y]$:

Theorem 2.2.2. *There exists a unique natural transformation (called "motivic Hirzebruch class")*

$$T_{y_*} : K_0(\mathcal{V}/X) \to H_*(X) \otimes \mathbb{Q}[y],$$

such that

(1) *for a nonsingular variety X, $T_{y_*}([X \xrightarrow{\mathrm{id}_X} X]) = T_y(TX) \cap [X]$. Here $T_y(TX)$ is the Hirzebruch class of the tangent bundle TX,*
(2) *for a point $X = pt$, $T_{y_*} : K_0(\mathcal{V}/pt) = K_0(\mathcal{V}) \to H_*(pt) \otimes \mathbb{Q}[y] = \mathbb{Q}[y]$ is equal to the homomorphism $\chi_y : K_0(\mathcal{V}) \to \mathbb{Z}[y]$ followed by the inclusion $\mathbb{Z}[y] \hookrightarrow \mathbb{Q}[y]$.*

Definition 2.2.3. For any variety X, singular or nonsingular, $T_{y*}(X) := T_{y*}([X \xrightarrow{\mathrm{id}_X} X])$ is also called the motivic Hirzebruch class of X.

When we consider the natural transformation $T_{y_*} : K_0(\mathcal{V}/X) \to H_*(X) \otimes \mathbb{Q}[y]$ for the three special cases of $y = -1, 0, 1$, since $T_{-1}(TX) = c(TX)$ the total Chern class, $T_0(TX) = td(TX)$ the total Todd class and $T_1(TX) = L(TX)$ the total Thom–Hirzebruch class, we get the following:

(1) $(y = -1)$: There exists a unique natural transformation

$$T_{-1*} : K_0(\mathcal{V}/X) \to H_*(X) \otimes \mathbb{Q},$$

such that for a nonsingular variety X, $T_{-1*}([X \xrightarrow{\mathrm{id}_X} X]) = c(TX) \cap [X]$.

(2) $(y = \ \ 0)$: There exists a unique natural transformation

$$T_{0*} : K_0(\mathcal{V}/X) \to H_*(X) \otimes \mathbb{Q},$$

such that for a nonsingular variety X, $T_{0*}([X \xrightarrow{\mathrm{id}_X} X]) = td(TX) \cap [X]$.

(3) $(y = \ \ 1)$: There exists a unique natural transformation

$$T_{1*} : K_0(\mathcal{V}/X) \to H_*(X) \otimes \mathbb{Q},$$

such that for a nonsingular variety X, $T_{1*}([X \xrightarrow{\mathrm{id}_X} X]) = L(TX) \cap [X]$.

The three well-known natural transformations similar to the above are respectively as follows:

(1) MacPherson's Chern class [33]: *There exists a unique natural transformation*

$$c_* : F(X) \to H_*(X),$$

such that for a nonsingular variety X, $c_*(1\!\!1_X) = c(TX) \cap [X]$. Here F is the covariant functor assigning to X the abelian group $F(X)$ of constructible functions on X. For a variety X, singular or nonsingular, the value $c_*(X) := c_*(1\!\!1_X)$ is called the Chern–Schwartz–MacPherson class of X [5].

(2) Baum–Fulton–MacPherson's Todd class [7]: *There exists a unique natural transformation*

$$td_* : G_0(X) \to H_*(X) \otimes \mathbb{Q},$$

such that for a nonsingular variety X, $td_*(\mathcal{O}_X) = td(TX) \cap [X]$. Here G_0 is the covariant functor assigning to X the Grothendieck group $G_0(X)$ of coherent sheaves on X.

(3) Cappell–Shaneson's L-class [14] (also see [64]), which extends Goresky–MacPherson's L-class [29] as a natural transformation: *For a compact variety* X *there exists a unique natural transformation*

$$L_* : \Omega(X) \to H_*(X) \otimes \mathbb{Q},$$

[5]Here we remark that Brasselet and M.-H. Schwartz [11] (see also [3]) showed that MacPherson's Chern class $c_*(\mathbf{1}_X)$ corresponds to the Schwartz class $c^S(X) \in H^*_X(M) = H^*(M, M\setminus X)$ (see [58–60]) by Alexander duality for X embedded in the smooth complex manifold M. That is why the total homology class $c_*(X) = c_*(\mathbf{1}_X)$ is called the Chern–Schwartz–MacPherson class of X.

such that if X is nonsingular, then $L_*(\mathbb{Q}_X[2\dim X]) = L(TX) \cap [X]$. Here Ω is the covariant functor assigning to X the abelian group $\Omega(X)$ of cobordism classes of self-dual complexes sheaves on X (cf. Schürmann–Woolf's very recent paper [54]).

Our $T_{y*} : K_0(\mathcal{V}/X) \to H_*(X) \otimes \mathbb{Q}[y]$ "unifies" the above three characteristic classes c_*, td_*, L_* in the sense that we have the following commutative diagrams:

$$\begin{array}{ccc} K_0(\mathcal{V}/X) & & \\ {\scriptstyle \epsilon}\downarrow & \searrow^{T_{-1*}} & \\ F(X) & \xrightarrow[c_* \otimes \mathbb{Q}]{} & H_*(X) \otimes \mathbb{Q}, \end{array} \qquad \begin{array}{ccc} K_0(\mathcal{V}/X) & & \\ {\scriptstyle \Gamma}\downarrow & \searrow^{T_{0*}} & \\ G_0(X) & \xrightarrow[td_*]{} & H_*(X) \otimes \mathbb{Q}, \end{array}$$

$$\begin{array}{ccc} K_0(\mathcal{V}/X) & & \\ {\scriptstyle \omega}\downarrow & \searrow^{T_{1*}} & \\ \Omega(X) & \xrightarrow[L_*]{} & H_*(X) \otimes \mathbb{Q}. \end{array}$$

Here $\epsilon : K_0(\mathcal{V}/-) \to F(-)$, $\Gamma : K_0(\mathcal{V}/-) \to G_0(-)$ and $sd : K_0(\mathcal{V}/-) \to \Omega(-)$ are unique natural transformations such that for X smooth, $\epsilon([X \xrightarrow{\mathrm{id}_X} X]) = 1\!\!1_X$, $\Gamma([X \xrightarrow{\mathrm{id}_X} X]) = [\mathcal{O}_X]$ and $sd([X \xrightarrow{\mathrm{id}_X} X]) = [\mathbb{Q}_X[2\dim X]]$ [6].

Remark 2.2.4. This "unification" can be considered as a positive answer to a remark (see Footnote 1 on page 266) stated at the end of MacPherson's survey article [34] (cf. [66]). The above three theories

$$c_* : F(-) \to H_*(-), \; td_* : G_0(-) \to H_*(-) \otimes \mathbb{Q}, \; L_* : \Omega(X) \to H_*(X) \otimes \mathbb{Q}$$

have different covariant functors $F(-), G_0(-), \Omega(-)$ as the source functor. This fact is an obstacle to coming up with a reasonable covariant functor for a reasonable theory unifying these three theories c_*, td_*, L_*. What was in the author's mind at the beginning was that there would be a certain covariant functor $\mathfrak{M}(-)$ as a sort of 'motif' of these three functors:

$$\begin{array}{ccccc} & & \mathfrak{M}(-) & & \\ & \swarrow & \downarrow & \searrow & \\ F(-) & & G_0(-) & & \Omega(-) \end{array}$$

[6] $\epsilon : K_0(\mathcal{V}/-) \to F(-)$ is simply defined by $\epsilon([V \xrightarrow{f} X]) := f_* 1\!\!1_V$, where $(f_* 1\!\!1_V)(x) := \chi_c(f^{-1}(x) \cap V)$. However, as to the other two Γ (for which see §2.3) and sd are not simply defined.

and a certain natural transformation $\mathfrak{N}_{c\ell} : \mathfrak{M}(-) \to H_*(-) \otimes \Lambda_{c\ell}$ (which is related to the characteristic cohomology class $c\ell$), from which the above three characteristic homology classes could be incarnated:

$$\mathfrak{N}_{c\ell} : \mathfrak{M}(-) \to H_*(-) \otimes \Lambda_{c\ell}$$
$$\Longrightarrow \begin{cases} c_* : F(-) \to H_*(-), \ c\ell = \text{Chern class} \\ td_* : G_0(-) \to H_*(-) \otimes \mathbb{Q}, \ c\ell = \text{Todd class} \\ L_* : \Omega(-) \to H_*(-) \otimes \mathbb{Q}, \ c\ell = \text{L-class.} \end{cases}$$

In [74] the author defines a toy model of a "unified" theory of these three characteristic classes.

2.3. *A proof of T_{y*} by Saito's theory of mixed Hodge modules*

Our motivic Hirzebruch class transformation $T_{y_*} : K_0(\mathcal{V}/X) \to H_*(X) \otimes \mathbb{Q}[y]$ is obtained as the composite $T_{y_*} = \widetilde{td_{*(y)}} \circ \Lambda_y^{mot}$ of the following natural transformations:

$$\Lambda_y^{mot} : K_0(\mathcal{V}/X) \to G_0(X) \otimes \mathbb{Z}[y],$$
$$\widetilde{td_{*(y)}} : G_0(X) \otimes \mathbb{Z}[y] \to H_*(X) \otimes \mathbb{Q}[y, (1+y)^{-1}].$$

Here

$$\widetilde{td_{*(y)}} := \sum_{i \geq 0} \frac{1}{(1+y)^i} td_{*i}$$

and

$$td_{*i} : G_0(X) \xrightarrow{td_*} H_*(X) \otimes \mathbb{Q} \xrightarrow{\text{projection}} H_i(X) \otimes \mathbb{Q}.$$

$\Lambda_y^{mot} : K_0(\mathcal{V}/X) \to G_0(X) \otimes \mathbb{Z}[y]$ is a key ingredient and in our paper [10] it is denoted by mC_* and called the *motivic Chern class.* In this paper, we use this symbol because of the following property:

Theorem 2.3.1. *("motivic" λ_y-class transformation) There exists a unique natural transformation*

$$\Lambda_y^{mot} : K_0(\mathcal{V}/X) \to G_0(X) \otimes \mathbb{Z}[y]$$

satisfying the "smooth condition" that for smooth X

$$\Lambda_y^{mot}([X \xrightarrow{\text{id}} X]) = \lambda_y(\Omega_X) := \sum_{p \geq 0}^{\dim X} [\Omega_X^p] y^p = \lambda_y(T^*X) \otimes \mathcal{O}_X.$$

(Here $\otimes\mathcal{O}_X : K^0(X) \cong G_0(X)$ is the isomorphism for smooth X, i.e., taking the sheaf of local sections.)

Then the composite $\widetilde{td_{*(y)}} \circ \Lambda_y^{mot}$ satisfies that for X smooth, $\widetilde{td_{*(y)}} \circ \Lambda_y^{mot}([X \xrightarrow{\mathrm{id}} X]) = T_y(TX) \cap [X]$ (e.g., see [71]).

Remark 2.3.2. Note that even though the target of $\widetilde{td_{*(y)}} : G_0(X) \otimes \mathbb{Z}[y] \to H_*(X) \otimes \mathbb{Q}[y,(1+y)^{-1}]$ is $H_*(X) \otimes \mathbb{Q}[y,(1+y)^{-1}]$, the image of $T_{y*} = \widetilde{td_{*(y)}} \circ \Lambda_y^{mot}$ is in $H_*(X) \otimes \mathbb{Q}[y]$.

Now, in order to define Λ_y^{mot}, first we recall the following four things:

MHM1 : To X one can associate an abelian category of *mixed Hodge modules* $MHM(X)$, together with a functorial pullback f^* and pushforward f_* on the level of bounded derived categories $D^b(MHM(X))$ for any (not necessarily proper) map. These natural transformations are functors of triangulated categories.

MHM2 : Let $i : Y \to X$ be the inclusion of a closed subspace, with open complement $j : U := X \backslash Y \to X$. Then one has for $M \in D^bMHM(X)$ a distinguished triangle

$$j_!j^*M \to M \to i_!i^*M \xrightarrow{[1]}$$

Hence, by the definition of the Grothendieck group $K_0(D^bMHM(X))$ of the derived category $D^bMHM(X)$, in $K_0(D^bMHM(X))$ we have the following equality

$$[M] = [j_!j^*M] + [i_!i^*M].$$

MHM3 : For all $p \in \mathbb{Z}$ one has a "filtered De Rham complex" functor of triangulated categories

$$gr_p^F DR : D^b(MHM(X)) \to D^b_{coh}(X)$$

commuting with proper pushforward. Here $D^b_{coh}(X)$ is the bounded derived category of sheaves of $\mathcal{O}_X$-modules with coherent cohomology sheaves. Moreover, $gr_p^F DR(M) = 0$ for almost all p and $M \in D^bMHM(X)$ fixed.

MHM4 : There is a distinguished element $\mathbb{Q}^H_{pt} \in MHM(\{pt\}/k)$ such that

$$gr^F_{-p} DR(\mathbb{Q}^H_X) \simeq \Omega^p_X[-p] \in D^b_{coh}(X)$$

for X smooth and pure dimensional. Here $\mathbb{Q}^H_X := \pi_X^*\mathbb{Q}^H_{pt}$ for $\pi_X : X \to pt$ a constant map, with $\mathbb{Q}^H_{pt}$ viewed as a complex concentrated in degree zero.

Remark 2.3.3. The above transformations are functors of triangulated categories, thus they induce functors even on the level of *Grothendieck groups of triangulated categories*, which we denote by the same name. Note that for these *Grothendieck groups* we have isomorphisms

$$K_0(D^bMHM(X)) \simeq K_0(MHM(X)) \quad \text{and} \quad K_0(D^b_{coh}(X)) \simeq G_0(X)$$

by associating to a complex its alternating sum of cohomology objects.

Definition 2.3.4. We define
(1) $mH : K_0(\mathcal{V}/X) \to K_0(MHM(X))$ by $mH([V \xrightarrow{f} X]) := [f_!\mathbb{Q}^H_V]$.
(2) $gr^F_{-*}DR : K_0(MHM(X)) \to G_0(X) \otimes \mathbb{Z}[y, y^{-1}]$ by

$$gr^F_{-*}DR([M]) := \sum_p [gr^F_{-p}DR(M)] \cdot (-y)^p.$$

(3) $\Lambda^{mot}_y := gr^F_{-*}DR \circ mH : K_0(\mathcal{V}/X) \xrightarrow{mH} K_0(MHM(X)) \xrightarrow{gr^F_{-*}DR} G_0(X) \otimes \mathbb{Z}[y]$.

Remark 2.3.5. By (MHM4), for X smooth and pure dimensional we have that

$$gr^F_{-*}DR \circ mH([X \xrightarrow{\mathrm{id}_X} X]) = \sum_{p \geq 0}^{\dim X} [\Omega^p_X] \cdot y^p \in G_0(X) \otimes \mathbb{Z}[y]\,.$$

Remark 2.3.6. If $y = 0$, then $T_{0*} = \widetilde{td_{*(0)}} \circ \Lambda^{mot}_0 = td_* \circ \Lambda^{mot}_0$ and $\Lambda^{mot}_0([V \xrightarrow{f} X]) = gr^F_0 DR([f_!\mathbb{Q}^H_V])$. Hence $\Gamma : K_0(\mathcal{V}/X) \to G_0(X)$ is nothing but $\Lambda^{mot}_0 : K_0(\mathcal{V}/X) \to G_0(X)$.

Remark 2.3.7. For more recent works on T_{y*}, e.g. see [22–24, 36–40]. As to recent applications of motivic Chern class Λ^{mot}_y in Schubert calculus, see [25, 47] (cf. [4–6]).

3. Motivic Milnor–Hirzebruch classes

3.1. *Milnor class*

The aforementioned well-known three theories of characteristic classes of singular varieties are natural transformations from covariant functors to the homology covariant functors. As to a *contravariant aspect* of Baum–Fulton–MacPherson's Todd classes, they conjectured that for a local complete intersection morphism $f : X \to Y$, $td_*(X) = td(T_f) \cap f^*(td_*(Y))$, where T_f is the virtual tangent bundle of f (see [26, B.7.6]). In [62] Verdier proved it affirmatively, by showing the following more general statement, which is called Verdier–Riemann–Roch:

Theorem 3.1.1. *For a local complete intersection morphism $f : X \to Y$, the following diagram commutes:*

$$\begin{array}{ccc} G_0(Y) & \xrightarrow{td_*} & H_*(Y)\otimes \mathbb{Q} \\ {\scriptstyle f^*}\downarrow & & \downarrow{\scriptstyle td(T_f)\cap f^*} \\ G_0(X) & \xrightarrow[td_*]{} & H_*(X)\otimes \mathbb{Q}. \end{array}$$

Here $f^* : G_0(Y) \to G_0(X)$ is defined by $f^*\mathcal{F} = \sum_i (-1)^i \operatorname{Tor}_i^{\mathcal{O}_Y}(\mathcal{F}, \mathcal{O}_X)$ and $f^* : H_*(Y)\otimes\mathbb{Q} \to H_*(X)\otimes\mathbb{Q}$ is the Gysin homomorphism (see [26]).

In [67] we showed a similar Verdier–Riemann–Roch type theorem for MacPherson's Chern class transformation $c_* : F(-) \to H_*(-)$ *for a smooth morphism*:

Theorem 3.1.2 (VRR for MacPherson's Chern class transformation c_*). *For a smooth morphism $f : X \to Y$, the following diagram commutes:*

$$\begin{array}{ccc} F(Y) & \xrightarrow{c_*} & H_*(Y) \\ {\scriptstyle f^*}\downarrow & & \downarrow{\scriptstyle c(T_f)\cap f^*} \\ F(X) & \xrightarrow[c_*]{} & H_*(X). \end{array} \tag{3.1}$$

In a general case when $f : X \to Y$ is a local complete intersection morphism, the above Verdier–Riemann–Roch formula (3.1) does not hold and there is some defect for the commutativity. In the simplest case when X is a local complete intersection variety in a smooth variety M the map to a point $a_X : X \to pt$ is a local complete intersection morphism. Consider the above diagram for this map $a_X : X \to pt$:

$$\begin{array}{ccc} F(pt) & \xrightarrow{c_*} & H_*(pt) \\ {\scriptstyle (a_X)^*}\downarrow & & \downarrow{\scriptstyle c(T_{a_X})\cap (a_X)^*} \\ F(X) & \xrightarrow[c_*]{} & H_*(X). \end{array} \tag{3.2}$$

For the characteristic function $1\!\!1_{pt}$ we have

$$(c_* \circ (a_X)^*)(1\!\!1_{pt}) = c_*((a_X)^*(1\!\!1_{pt})) = c_*(1\!\!1_X) = c_*(X)$$

which is Chern–Schwartz–MacPherson's class and

$$(c(T_{a_X}) \cap (a_X)^* \circ c_*)(\mathbb{1}_{pt}) = c(T_{a_X}) \cap (a_X)^*([pt]) = c(T_X^{vir}) \cap [X]$$

which is Fulton-Johnson's class $c_*^{FJ}(X)$. Certainly, if $a_X : X \to pt$ is smooth, i.e., X is smooth, thus $T_X^{vir} = T_X$ and $c_*(X) = c(T_X) \cap [X]$, namely we have the above diagram is commutative. Thus the defect of the commutativity of the diagram (3.2), i.e., the difference $c_* \circ (a_X)^* - c(T_{a_X}) \cap (a_X)^* \circ c_*$ evaluated on the generator $\mathbb{1}_{pt}$ of $F(pt)$ is nothing but (up to sign) the Milnor class $\mathcal{M}(X)$.

Remark 3.1.3. Here we remark that a general Verdier–Riemann–Roch type theorem for MacPherson's Chern class, i.e., a formula for the defect of the commutativity of the diagram (3.1) for a local complete intersection morphism $f : X \to Y$ has been obtained by Schürmann in [49].

Remark 3.1.4. As to Cappell–Shaneson's L-class $L_* : \Omega(-) \to H_*(-) \otimes \mathbb{Q}$, we do not know if Verdier–Riemann–Roch type formula (for a local complete intersection morphism) holds or not and also we do not know if it holds for a smooth morphism.

The motivic Hirzebruch class $T_{y*} : K_0(\mathcal{V}/-) \to H_*(-) \otimes \mathbb{Q}[y]$ does satisfy a Verdier–Riemann–Roch for a smooth morphism $f : X \to Y$, i.e., the following diagram is commutative (see [10]):

$$\begin{array}{ccc} K_0(\mathcal{V}/Y) & \xrightarrow{T_{y*}} & H_*(Y) \otimes \mathbb{Q}[y] \\ {\scriptstyle f^*}\downarrow & & \downarrow{\scriptstyle T_y(T_f) \cap f^*} \\ K_0(\mathcal{V}/X) & \xrightarrow[T_{y*}]{} & H_*(X) \otimes \mathbb{Q}[y]. \end{array} \tag{3.3}$$

Problem 3.1.5. *For a local complete intersection morphism $f : X \to Y$, identify the defect of the commutativity of the above diagram (3.3) for a local complete intersection morphism $f : X \to Y$ (possibly in a similar way as done in Schürmann's paper [49]).*

As in the case of MacPherson's Chern class c_*, if we consider the above diagram (3.3) for the above map $a_X : X \to pt$, then the defect of the commutativity of the diagram (3.3), i.e., the difference $T_{y*} \circ (a_X)^* - T_y(T_{a_X}) \cap (a_X)^* \circ T_{y*}$ evaluated on the distinguished element $[pt \xrightarrow{\mathrm{id}_{pt}} pt]$ of $K_0(\mathcal{V}/pt)$ is nothing but (up to sign)

$$T_{y*}(X) - T_y(T_X^{vir}) \cap [X],$$

which is what we discuss below. Namely, this is a kind of T_{y*}-version of the Milnor class $\mathcal{M}(X)$.

In [72] we captured the Milnor class as a natural transformation from a "pre-motivic" covariant functor to the homology theory, which is quickly recalled below[7].

3.2. *The Grothendieck group* $K^{\mathcal{P}rop}_{\ell.c.i}(\mathcal{V}/X \xrightarrow{h} S)$

Let S be a complex algebraic variety and fixed. Let $\mathcal{V}_S$ be the category of S-varieties, i.e., an object is a morphism $h : X \to S$ and a morphism from $h : X \to S$ to $k : Y \to S$ is a morphism $f : X \to Y$ such that the following diagram commutes:

$$\begin{array}{ccc} X & \xrightarrow{f} & Y \\ & {\scriptstyle h}\searrow \quad \swarrow{\scriptstyle k} & \\ & S. & \end{array}$$

Definition 3.2.1. Let $M^{\mathcal{P}rop}_{\ell.c.i}(\mathcal{V}/X \xrightarrow{h} S)$ be the monoid consisting of isomorphism classes $[V \xrightarrow{p} X]$ of proper morphisms $p : V \to X$ such that the composite $h \circ p : V \to S$ is an $\ell.c.i.$ morphism, with the addition $(+)$ and zero (0) defined by

- $[V \xrightarrow{h} X] + [V' \xrightarrow{h'} X] := [V \sqcup V' \xrightarrow{h+h'} X]$,
- $0 := [\phi \to X]$.

Then we define

$$K^{\mathcal{P}rop}_{\ell.c.i}(\mathcal{V}/X \xrightarrow{h} S)$$

to be the Grothendieck group of the monoid $M^{\mathcal{P}rop}_{\ell.c.i}(\mathcal{V}/X \xrightarrow{h} S)$.

Lemma 3.2.2.

(1) *The Grothendieck group* $K^{\mathcal{P}rop}_{\ell.c.i}(\mathcal{V}/X \xrightarrow{h} S)$ *is a covariant functor with pushforwards for proper morphisms, i.e., for a proper morphism* $f : X \to Y \in \mathcal{V}_S$

$$\begin{array}{ccc} X & \xrightarrow{f} & Y \\ & {\scriptstyle h}\searrow \quad \swarrow{\scriptstyle k} & \\ & S, & \end{array}$$

[7] For a more general work, see [73] and [56].

the pushforward $f_* : K^{\mathcal{P}rop}_{\ell.c.i}(\mathcal{V}/X \xrightarrow{h} S) \to K^{\mathcal{P}rop}_{\ell.c.i}(\mathcal{V}/Y \xrightarrow{k} S)$ *defined by* $f_*([V \xrightarrow{p} X]) := [V \xrightarrow{f\circ p} Y]$ *is covariantly functorial.*

(2) *The Grothendieck group* $K^{\mathcal{P}rop}_{\ell.c.i}(\mathcal{V}/X \xrightarrow{h} S)$ *is a contravariant functor with pullbacks for smooth morphisms, i.e., for a smooth morphism* $f : X \to Y \in \mathcal{V}_S$*, the pullback* $f^* : K^{\mathcal{P}rop}_{\ell.c.i}(\mathcal{V}/Y \xrightarrow{k} S) \to K^{\mathcal{P}rop}_{\ell.c.i}(\mathcal{V}/X \xrightarrow{h} S)$ *defined by* $f^*([W \xrightarrow{p} Y]) := [W' \xrightarrow{p'} X]$ *is contravariantly functorial. Here we consider the following commutative diagrams whose top square is a fiber square:*

$$\begin{array}{ccccc} W' & & \xrightarrow{\quad f' \quad} & & W \\ {\scriptstyle p'}\downarrow & & & & \downarrow{\scriptstyle p} \\ X & & \xrightarrow{\quad f \quad} & & Y \\ & {\scriptstyle h}\searrow & & \swarrow{\scriptstyle k} & \\ & & S. & & \end{array}$$

3.3. *Generalized motivic Milnor–Hirzebruch classes*

Proposition 3.3.1. *Let* $c\ell : K^0(-) \to H^*(-) \otimes R$ *be a characteristic class of complex vector bundles with a suitable coefficients* R*. Then on the category* $\mathcal{V}_S$ *there exists a unique natural transformation*

$$\gamma_{c\ell *} : K^{\mathcal{P}rop}_{\ell.c.i}(\mathcal{V}/X \xrightarrow{h} S) \to H_*(X) \otimes R$$

such that for a local complete intersection morphism $h : X \to S$

$$\gamma_{c\ell *}([X \xrightarrow{\mathrm{id}_X} X]) = c\ell(T_h) \cap [X].$$

Namely, for a morphism $f : X \to Y$*, i.e., for the following commutative diagram*

$$\begin{array}{ccccc} X & & \xrightarrow{\quad f \quad} & & Y \\ & {\scriptstyle h}\searrow & & \swarrow{\scriptstyle k} & \\ & & S, & & \end{array}$$

the following diagram commutes:

$$\begin{array}{ccc} K^{\mathcal{P}rop}_{\ell.c.i}(\mathcal{V}/X \xrightarrow{h} S) & \xrightarrow{\gamma_{c\ell *}} & H_*(X) \otimes R \\ {\scriptstyle f_*}\downarrow & & \downarrow{\scriptstyle f_*} \\ K^{\mathcal{P}rop}_{\ell.c.i}(\mathcal{V}/Y \xrightarrow{k} S) & \xrightarrow[\gamma_{c\ell *}]{} & H_*(Y) \otimes R. \end{array}$$

If S is a point and $c\ell = c$ the Chern class, then for a local complete intersection variety X in a smooth manifold, we have that $\gamma_{c*}([X \xrightarrow{\mathrm{id}_X} X]) = c(T_X^{\mathrm{vir}}) \cap [X]$, which is Fulton–Johnson's class $c_*^{FJ}(X)$. Thus the natural transformation $\gamma_{c\ell *} : K_{\ell.c.i}^{\mathcal{P}rop}(\mathcal{V}/X \xrightarrow{h} S) \to H_*(X) \otimes R$ is a generalization of Fulton–Johnson's class as *a natural transformation*. It is called *a motivic Fulton-Johnson-type $c\ell$ class*, denoted by $c\ell_*^{FJ}$, since it is modeled after Fulton–Johnson's class c_*^{FJ}.

If we consider the Hirzebruch class T_y for the characteristic class $c\ell$ and we use the motivic Hirzebruch class $T_{y*} : K_0(\mathcal{V}/X) \to H_*(X) \otimes \mathbb{Q}[y]$, then the natural transformation

$$\gamma_{T_{y*}} : K_{\ell.c.i}^{\mathcal{P}rop}(\mathcal{V}/X \xrightarrow{h} S) \to H_*(X) \otimes \mathbb{Q}[y]$$

is called *the motivic Fulton–Johnson-type Hirzebruch class* and denoted by T_{y*}^{FJ}.

The Borel–Moore homology theory with the twisted pushforward $f_\star = (-1)^{\dim Y - \dim X} f_* : H_*(X) \to H_*(Y)$ is still a covariant functor and shall be denoted by $H_\star(X)$.

Theorem 3.3.2. *We define $\mathscr{M}T_{y*} : K_{\ell.c.i}^{\mathcal{P}rop}(\mathcal{V}/X \xrightarrow{h} S) \to H_\star(X) \otimes \mathbb{Q}[y]$ by*

$$\mathscr{M}T_{y*}([V \xrightarrow{p} X]) := (-1)^{\dim V} \left(T_{y*}^{FJ} - T_{y*}\right)([V \xrightarrow{p} X]).$$

$\mathscr{M}T_{y*} : K_{\ell.c.i}^{\mathcal{P}rop}(\mathcal{V}/- \xrightarrow{h} S) \to H_\star(-) \otimes \mathbb{Q}[y]$ is a unique natural transformation such that for a local complete intersection morphism $h : X \to S$ it satisfies that

$$\begin{aligned}\mathscr{M}T_{y*}([X \xrightarrow{\mathrm{id}_X} X]) &= (-1)^{\dim X} \left(T_{y*}^{FJ} - T_{y*}\right)([X \xrightarrow{\mathrm{id}_X} X]) \\ &= (-1)^{\dim X} \Big(T_y(T_X^{\mathrm{vir}}) \cap [X] - T_{y*}(X)\Big).\end{aligned}$$

Let S be a point. Then the above motivic natural transformation

$$\mathscr{M}T_{y*} : K_{\ell.c.i}^{\mathcal{P}rop}(\mathcal{V}/X) \to H_\star(X) \otimes \mathbb{Q}[y]$$

is called *a motivic Milnor–Hirzebruch class*, even though $K_{\ell.c.i}^{\mathcal{P}rop}(\mathcal{V}/X)$ is not the motivic group $K_0(\mathcal{V}/X)$, but because it is a subgroup of $K_0(\mathcal{V}/X)$ and it is defined by using the motivic Hirzebruch class $T_{y*} : K_0(\mathcal{V}/X) \to H_*(X) \otimes \mathbb{Q}[y]$ and also because, if we specialize $\mathscr{M}T_{y*}$ to the case when $y = -1$ and X is a local complete intersection variety in a smooth manifold,

we have

$$\begin{aligned}\mathscr{M}T_{-1_*}([X \xrightarrow{\mathrm{id}} X]) &= (-1)^{\dim X}\left\{T_{-1}(T_X^{\mathrm{vir}}) \cap [X] - T_{-1*}([X \xrightarrow{\mathrm{id}} X])\right\}\\ &= (-1)^{\dim X}\left(c_*^{FJ}(X)\otimes\mathbb{Q} - c_*(X)\otimes\mathbb{Q}\right)\\ &= \left((-1)^{\dim X}\left(c_*^{FJ}(X) - c_*(X)\right)\right)\otimes\mathbb{Q},\end{aligned}$$

which is the Milnor class $\mathcal{M}(X) = (-1)^{\dim X}\left(c_*^{FJ}(X) - c_*(X)\right)$ of X *modulo torsion.* Thus $\mathscr{M}T_{-1_*} : K_{\ell.c.i}^{\mathcal{P}rop}(\mathcal{V}/X) \to H_\star(X)\otimes\mathbb{Q}$ is called *the motivic Milnor class* and the more general one $\mathscr{M}T_{y_*} : K_{\ell.c.i}^{\mathcal{P}rop}(\mathcal{V}/X \xrightarrow{h} S) \to H_\star(X)\otimes\mathbb{Q}[y]$ *a generalized motivic Milnor–Hirzebruch class.*

Proposition 3.3.3. *Let $f : X \to Y$ be a smooth morphism in the category $\mathcal{V}_S$:*

$$\begin{array}{ccc} X & \xrightarrow{f} & Y \\ & {}_h\searrow \quad \swarrow_k & \\ & S. & \end{array}$$

Then the following diagram commutes:

$$\begin{array}{ccc} K_{\ell.c.i}^{\mathcal{P}rop}(\mathcal{V}/Y \xrightarrow{k} S) & \xrightarrow{c\ell_*^{FJ}} & H_*(Y)\otimes R \\ {\scriptstyle f^*}\downarrow & & \downarrow{\scriptstyle c\ell(T_f)\cap f^*} \\ K_{\ell.c.i}^{\mathcal{P}rop}(\mathcal{V}/X \xrightarrow{h} S) & \xrightarrow[c\ell_*^{FJ}]{} & H_*(X)\otimes R. \end{array}$$

In [69, Theorem 2.2] we obtained a Verdier-type Riemann–Roch formula of the Milnor class in a special case. The following Verdier-type Riemann–Roch formula of the motivc Milnor–Hirzebruch class is a generalization of this result:

Theorem 3.3.4. *For a smooth morphism $f : X \to Y$, the twisted Gysin pullback homomophism $f^\star : H_*(Y) \to H_*(X)$ is defined by $f^\star = (-1)^{\dim X - \dim Y} f^*$. For a smooth morphism $f : X \to Y$ in the category $\mathcal{V}_S$ as in Proposition 3.3.3, the following diagram commutes:*

$$\begin{array}{ccc} K_{\ell.c.i}^{\mathcal{P}rop}(\mathcal{V}/Y \xrightarrow{k} S) & \xrightarrow{\mathscr{M}T_{y_*}} & H_\star(Y)\otimes\mathbb{Q}[y] \\ {\scriptstyle f^*}\downarrow & & \downarrow{\scriptstyle T_y(T_f)\cap f^\star} \\ K_{\ell.c.i}^{\mathcal{P}rop}(\mathcal{V}/X \xrightarrow{h} S) & \xrightarrow[\mathscr{M}T_{y_*}]{} & H_\star(X)\otimes\mathbb{Q}[y]. \end{array}$$

4. Cappell–Maxim–Schürmann–Shaneson's Hirzebruch-Milnor class via the vanishing cycle functor

4.1. *Hirzebruch–Milnor class via the vanishing cycle functor*

In this section we give a very quick recall of Cappell–Maxim–Schürmann–Shaneson's result on the Milnor–Hirzebruch class. In [21], for a hypersurface X, globally defined as the zero-set $X = f^{-1}(0)$ of an algebraic function $f : M \to \mathbb{C}$ on a complex algebraic manifold M, the following homology class without the sign $(-1)^{\dim X}$

$$T_y(T_X^{\mathrm{vir}}) \cap [X] - T_{y*}(X)$$

is denoted by $\mathcal{M}T_{y*}(X)$ and they called it the Milnor–Hirzebruch class.

In [51] Schürmann proves the following theorem, which is a generalization of Verdier's result [62] on the specialization of MacPherson's Chern class transformation:

Theorem 4.1.1 (Schürmann). *Let the situation be as above and let $i : X \hookrightarrow M$ be the inclusion. Then $MHT_{y*} : K_0(MHM(-)) \to H_*(-) \otimes \mathbb{Q}[y, y^{-1}]$ commutes with the specialization, namely the following diagram commutes:*

$$\begin{array}{ccc} K_0(MHM(M)) & \xrightarrow{MHT_{y*}} & H_*(M) \otimes \mathbb{Q}[y, y^{-1}] \\ {\scriptstyle \Psi'^H_f}\downarrow & & \downarrow{\scriptstyle i^!} \\ K_0(MHM(X)) & \xrightarrow[MHT_{y*}]{} & H_*(X) \otimes \mathbb{Q}[y, y^{-1}]. \end{array}$$

Here $\Psi'^H_f = \Psi^H_f[-1] : K_0(MHM(M)) \to K_0(MHM(X))$ is the shifted nearby cycle functor.

Using this specialization theorem, in [21] Cappell–Maxim–Schürmann–Shaneson prove the following theorem:

Theorem 4.1.2. *Let the situation be as above.*

$$\mathcal{M}T_{y*}(X) = MHT_{y*}(\Phi'^H_f([\mathbb{Q}^H_M])).$$

Here $\Phi'^H_f = \Phi^H_f[-1]$ is the shifted one of the vanishing cycle functor $\Phi^H_f : K_0(MHM(M)) \to K_0(MHM(X))$ defined by $\Phi^H_f := \Psi^H_f - i^*$.

Here we remark that the following diagrams commute:

$$
\begin{array}{ccccc}
K_0(\mathcal{V}/M) & \longrightarrow & K_0(\mathcal{V}/X) & \xrightarrow{\quad T_{y*}\quad} & H_*(X)\otimes\mathbb{Q}[y] \\
\downarrow{\scriptstyle \chi_{\mathrm{Hodge}}} & & \downarrow{\scriptstyle \chi_{\mathrm{Hodge}}} & & \downarrow \\
K_0(MHM(M)) & \xrightarrow[\Psi_f]{\Phi_f} & K_0(MHM(X)) & \xrightarrow{\quad MHMT_{y*}\quad} & H_*(X)\otimes\mathbb{Q}[y,y^{-1}] \\
\downarrow{\scriptstyle \mathrm{rat}} & & \downarrow{\scriptstyle \mathrm{rat}} & & \downarrow{\scriptstyle y=-1} \\
K_0(D^b_c(M)) & \xrightarrow[\Psi_f]{\Phi_f} & K_0(D^b_c(X)) & \xrightarrow{\quad c_*\otimes\mathbb{Q}\quad} & H_*(X)\otimes\mathbb{Q} \\
\downarrow{\scriptstyle \chi_{\mathrm{stalk}}} & & \downarrow{\scriptstyle \chi_{\mathrm{stalk}}} & \nearrow{\scriptstyle c_*\otimes\mathbb{Q}} & \\
F(M) & \xrightarrow[\Psi_f]{\Phi_f} & F(X). & &
\end{array}
$$

Here we have

- $\chi_{\mathrm{Hodge}} : K_0(\mathcal{V}/Z) \to K_0(MHM(Z))$ is defined by $\chi_{\mathrm{Hodge}}([V \xrightarrow{f} Z]) := f_*\mathbb{Q}^H_Z$ and $\mathbb{Q}^H_Z := (\mathcal{O}_Z, F, \mathbb{Q}_Z, W)$ with the Hodge filtration F such that $gr^i_F = 0$ $(i \neq 0)$ and the pure weight filtration W.
- rat denotes the forgetful functor which assigns to a complex of mixed Hodge modules the underlying rational constructible complex.
- $\chi_{\mathrm{stalk}} : K_0(D^b_c(Z)) \to F(Z)$ is defined by, for a constructible sheaf complex $\mathcal{F}_\bullet$,
$$(\chi_{\mathrm{stalk}}(\mathcal{F}_\bullet))(x) := \chi_x(\mathcal{F}_\bullet) := \sum_j (-1)^j \dim H^j(\mathcal{F}_\bullet)_x.$$
- $c_* \otimes \mathbb{Q} : K_0(D^b_c(X)) \to H_*(X) \otimes \mathbb{Q}$ is defined by the composite of $\chi_{\mathrm{stalk}} : K_0(D^b_c(X)) \to F(X)$ and the rationalized MacPherson's Chern class homomorphism $c_* \otimes \mathbb{Q} : F(X) \to H_*(X) \otimes \mathbb{Q}$.

The case of the Milnor class is the following:

$$\mathcal{M}(X) = c_*(\Phi_f(\mathbb{1}_M)).$$

In [53] Schürmann extends this result to the case of global complete intersections. Let X be a global complete intersection in a smooth variety M defined by $f_1, f_2, \cdots, f_n : M \to \mathbb{C}$, i.e., $X = (f_1, f_2, \cdots, f_n)^{-1}((0, 0, \cdots, 0))$ for $(f_1, f_2, \cdots, f_n) : M \to \mathbb{C}^n$. Here we emphasize that the ordering of these maps f_i's does not matter. When it comes to the nearby and vanishing cycles, the orderings does matter as done by Schürmann:

Definition 4.1.3 (Nearby and vanishing cycles for an ordered tuple). Let $(f) := (f_1, \cdots, f_n) : X \to \mathbb{C}^n$ be an ordered n-tuple of complex algebraic functions on X, with $X_0 := \{f = 0\} = \{f_1 = 0, \cdots f_n = 0\}$ the zero-fiber of (f). Then nearby cycles of $(f) := (f_1, \cdots, f_n)$ are defined by iteration as

$$\Psi_{(f)} := \Psi_{f_1} \circ \cdots \circ \Psi_{f_n} : F(X) \to F(X_0). \tag{4.1}$$

By linearity, $\Psi_{(f)}$ is uniquely defined by the convention that for a closed complex algebraic subspace $Z \subset Y$ the value

$$\Psi_{(f)}(1\!\!1_Z)(x) := \chi(H^*(F_{(f)|_Z,x}; \mathbb{Q})) \tag{4.2}$$

is just the Euler-Poincaré characteristic of a local Milnor fiber $F_{(f)|_Z,x}$ of $(f)|_Z$ at x. Here this local Milnor fiber of (f) at x is given by

$$F_{(f)|_Z,x} := Z \cap B_\epsilon(x) \cap \{f_1 = y_1, \cdots f_n = y_n\}, \tag{4.3}$$

with $0 < |y_n| << \cdots << |y_1| << \epsilon << 1$ and $B_\epsilon(x)$ an open (or closed) ball of radius ϵ around x (in some local coordinates).

The corresponding vanishing cycles of (f) are defined by

$$\Phi_{(f)} := \Psi_{(f)} - i^* : F(X) \to F(X_0), \tag{4.4}$$

with $i : X_0 \hookrightarrow X$ the inclusion. By linearity, $\Phi_{(f)}$ is uniquely defined by the convention that for a closed complex algebraic subspace $Z \subset Y$ the value

$$\Phi_{(f)}(1\!\!1_Z)(x) := \chi(H^*(F_{(f)|_Z,x}; \mathbb{Q})) - 1 = \chi(\tilde{H}^*(F_{(f)|_Z,x}; \mathbb{Q})) \tag{4.5}$$

is the reduced Euler–Poincaré characteristic of a local Milnor fiber $F_{(f)|_Z,x}$ of $(f)|_Z$ at x.

Theorem 4.1.4 (Schürmann). *If $X = \{f = 0\} = \{f_1 = 0, \cdots, f_n = 0\}$ is a global complete intersection in a smooth manifold M, defined by the zero-fiber of an ordered n-tuple of complex algebraic functions $(f) := (f_1, \cdots, f_n) : M \to \mathbb{C}^n$, then we have*

$$c_*^{FJ}(X) - c_*(X) = c_*(\Phi_{(f)}(1\!\!1_M)).$$

Remark 4.1.5. The above argument works for extending it to the case of T_{y*}, i.e., we get $\mathcal{M}T_{y*}(X) = MHT_{y*}(\Phi'_{(f)}(\mathbb{Q}^H_M))$ (see [21, Remark 1.6]).

Remark 4.1.6. Although these vanishing cycles $\Phi_{(f)}(1\!\!1_M)$ do depend on the order of n-tuple $\{f_1, f_2, \cdots, f_n\}$, but after evaluating them by MacPherson's Chern class transformation c_*, the dependence of ordering disappears and they are all equal to the Milnor class $\mathcal{M}(X)$ (up to sign). Therefore we can say the following. One can define the *averaged* vanishing cycle (which does not depend on the order any more)

$$\widetilde{\Phi}_{(f)}(\mathbb{1}_M) := \frac{1}{n!}\sum_{\sigma \in \mathcal{S}_n} \Phi_{\{f_{\sigma(1)},\cdots,\sigma(n)\}}(\mathbb{1}_M)$$

and we have $c_*(\widetilde{\Phi}_{(f)}(\mathbb{1}_M)) = \mathcal{M}(X)$ (up to sign).

5. A "motivicfication" of Milnor–Hirzebruch class via the vanishing cycle functor

First we consider a "motivicfication" of Chern–Mather class, which was one of the motivations of considering "motivicfication" of homology classes.

5.1. *A "motivicfication" of Chern-Mather class*

The Chern–Schwartz–MacPherson class satisfies the following scissor formula:

$$c_*(X) = c_*(Y) + c_*(X \setminus Y)$$

for a closed subvariety Y of X, since $\mathbb{1}_X = \mathbb{1}_Y + \mathbb{1}_{X \setminus Y}$, thus c_* is *motivic.*

The Chern-Mather class $c_*^{\mathrm{Ma}}(X)$ is a key ingredient in defining the natural transformation $c_* : F(-) \to H_*(-)$, and defined by

$$c_*^{\mathrm{Ma}}(X) := c_*(\mathrm{Eu}_X),$$

where Eu_X is the local Euler obstruction. The Chern–Mather class is not motivic, i.e., does not satisfy the scissor formula: $c_*^{\mathrm{Ma}}(X) \neq c_*^{\mathrm{Ma}}(Y) + c_*^{\mathrm{Ma}}(X \setminus Y)$, simply because

$$\mathrm{Eu}_X \neq \mathrm{Eu}_Y + \mathrm{Eu}_{X \setminus Y}.$$

A very simple example for this is, for example, a singular curve C with only one singular point p which is a double point. Then

$$\mathrm{Eu}_C(x) = \begin{cases} 1, \ x \neq p \\ 2, \ x = p. \end{cases}$$

On the other hand

$$\left(\mathrm{Eu}_p + \mathrm{Eu}_{C \setminus \{p\}}\right)(x) = 1, \ \ \forall x,$$

because p and $C \setminus \{p\}$ are both non-singular, thus $\mathrm{Eu}_p = \mathbb{1}_p$ and $\mathrm{Eu}_{C\setminus\{p\}} = \mathbb{1}_{C\setminus\{p\}}$.

The correct equality for $\mathrm{Eu}_X \neq \mathrm{Eu}_Y + \mathrm{Eu}_{X\setminus Y}$ is the following:

$$\mathrm{Eu}_X = (i_Y)_*\left((i_Y)^*\,\mathrm{Eu}_X\right) + (i_{X\setminus Y})_*\left((i_{X\setminus Y})^*\,\mathrm{Eu}_X\right).$$

The *pullback followed by pushforward* of the local Euler obstruction Eu_X of the target variety X can be generalized to any constructible function on X. This was done in [12].

Lemma 5.1.1. *For a constructible function $\delta_X \in F(X)$, the homomorphism*

$$[\delta_X] : K_0(\mathcal{V}/X) \to F(X)$$

defined by $[\delta_X]([V \xrightarrow{h} X]) := h_*h^*\delta_X$ and extended it linearly is well-defined.

Lemma 5.1.2. *For* any *morphism $g : X \to Y$ and* any *constructible function $\delta_Y \in F(Y)$, the following diagrams commute:*

$$\begin{array}{ccc} K_0(\mathcal{V}/X) & \xrightarrow{[g^*\delta_Y]} & F(X) \\ {\scriptstyle g_*}\downarrow & & \downarrow{\scriptstyle g_*} \\ K_0(\mathcal{V}/Y) & \xrightarrow[{[\delta_Y]}]{} & F(Y), \end{array} \qquad \begin{array}{ccc} K_0(\mathcal{V}/Y) & \xrightarrow{[\delta_Y]} & F(Y) \\ {\scriptstyle g^*}\downarrow & & \downarrow{\scriptstyle g^*} \\ K_0(\mathcal{V}/X) & \xrightarrow[{[g^*\delta_Y]}]{} & F(X). \end{array}$$

Corollary 5.1.3. *Let $c_*^{\delta_X} = c_* \circ [\delta_X] : K_0(\mathcal{V}/X) \to H_*(X)$ be the composite of the above homomorphism $[\delta_X] : K_0(\mathcal{V}/X) \to F(X)$ and MacPherson's Chern class transformation $c_* : F(X) \to H_*(X)$.*

(1) *For a proper morphism $g : X \to Y$ and* any *constructible function $\delta_Y \in F(Y)$, the following diagram commutes:*

$$\begin{array}{ccc} K_0(\mathcal{V}/X) & \xrightarrow{c_*^{g^*\delta_Y}} & H_*(X) \\ {\scriptstyle g_*}\downarrow & & \downarrow{\scriptstyle g_*} \\ K_0(\mathcal{V}/Y) & \xrightarrow[c_*^{\delta_Y}]{} & H_*(Y). \end{array}$$

(2) *For a smooth morphism $g : X \to Y$ and* any *constructible function $\delta_Y \in F(Y)$ the following diagram commutes:*

$$\begin{array}{ccc} K_0(\mathcal{V}/Y) & \xrightarrow{c_*^{\delta_Y}} & H_*(Y) \\ {\scriptstyle g^*}\downarrow & & \downarrow{\scriptstyle c(T_g)\cap g^*} \\ K_0(\mathcal{V}/X) & \xrightarrow[c_*^{g^*\delta_Y}]{} & H_*(X). \end{array}$$

For the homomorphism $c_*^{\delta_X} : K_0(\mathcal{V}/X) \to H_*(X)$, the distinguished value $c_*^{\delta_X}([X \xrightarrow{\mathrm{id}_X} X])$ is $c_*(\delta_X)$. For example, $c_*^{\mathrm{Eu}_X}([X \xrightarrow{\mathrm{id}_X} X]) = c_*(\mathrm{Eu}_X) = c_*^{\mathrm{Ma}}(X)$ is the Chern–Mather class. Although the Chern–Mather class is not motivic as observed above, but due to this homomorphism $c_*^{\mathrm{Eu}_X} : K_0(\mathcal{V}/X) \to H_*(X)$ the Chern–Mather class is captured as *a motivic homology class.* This is what we call *a "motivicfication" of the Chern–Mather class.*

5.2. *A "motivicfication" of Hirzebruch–Milnor class via the vanishing cycle functor*

First we observe the following, which is the mixed Hodge module version of the above Lemma 5.1.1:

Lemma 5.2.1. *For $\mathcal{M} \in K_0(MHM(X))$, the homomorphism*

$$[\mathcal{M}] : K_0(\mathcal{V}/X) \to K_0(MHM(X))$$

defined by

$$[\mathcal{M}]([V \xrightarrow{h} X]) := h_! h^* \mathcal{M}$$

is well-defined.

Proof. It suffices to show that for a morphism $h : V \to X$ and a closed sub variety $W \subset V$, we have

$$h_! h^* \mathcal{M} = (h|_W)_!(h|_W)^* \mathcal{M} + (h|_{V \setminus W})_!(h|_{V \setminus W})^* \mathcal{M}.$$

Let $i : W \hookrightarrow V$ and $j : V \setminus W \hookrightarrow V$ be the inclusions. Then we have that $h|_W = h \circ i$ and $h|_{V \setminus W} = h \circ j$. It follows from the fundamental properties of mixed Hodge modules (see MHM2 in §2.3) that we have

$$h^* \mathcal{M} = j_! j^*(h^* \mathcal{M}) + i_! i^*(h^* \mathcal{M}),$$

which implies that

$$h_!(h^* \mathcal{M}) = h_!\Big(j_! j^*(h^* \mathcal{M})\Big) + h_!\Big(i_! i^*(h^* \mathcal{M})\Big),$$

i.e.,

$$h_! h^* \mathcal{M} = (h \circ i)_!(h \circ i)^* \mathcal{M} + (h \circ j)_!(h \circ j)^* \mathcal{M}.$$

Hence $h_! h^* \mathcal{M} = (h|_W)_!(h|_W)^* \mathcal{M} + (h|_{V \setminus W})_!(h|_{V \setminus W})^* \mathcal{M}$. Therefore we have

$$[\mathcal{M}]\Big([V \xrightarrow{h} X] - [W \xrightarrow{h|_W} X] - [V \setminus W \xrightarrow{h|_{V \setminus W}} X]\Big) = 0,$$

hence the homomorphism $[\mathcal{M}] : K_0(\mathcal{V}/X) \to K_0(MHM(X))$ is well-defined. $\square$

Definition 5.2.2. For a hypersurface X, globally defined as the zero-set $X = f^{-1}(0)$ of an algebraic function $f : M \to \mathbb{C}$ on a complex algebraic manifold M, we define

$$\mathcal{M}T_{y*} : K_0(\mathcal{V}/X) \to H_*(X) \otimes \mathbb{Q}[y]$$

by

$$\mathcal{M}T_{y*}([V \xrightarrow{h} X]) := MHT_{y*}\Big(h_! h^*(\Phi'_f(\mathbb{Q}^H_M))\Big).$$

Note that $\mathcal{M}T_{y*}([X \xrightarrow{\mathrm{id}_X} X]) = MHT_{y*}(\Phi'_f(\mathbb{Q}^H_M)) = \mathcal{M}T_{y*}(X)$.

Theorem 5.2.3. *For a proper morphism $\rho : N \to M$ of complex manifolds and for an algebraic function $f : M \to \mathbb{C}$, we define $Y := (f \circ \rho)^{-1}(0) = \rho^{-1}(X)$ and consider the restriction map $\rho_Y : Y \to X$. Then we have the following commutative diagram:*

$$\begin{array}{ccc} K_0(\mathcal{V}/Y) & \xrightarrow{\mathcal{M}T_{y*}} & H_*(Y) \otimes \mathbb{Q}[y] \\ {\scriptstyle (\rho|_Y)_*}\downarrow & & \downarrow{\scriptstyle (\rho|_Y)_*} \\ K_0(\mathcal{V}/X) & \xrightarrow[\mathcal{M}T_{y*}]{} & H_*(X) \otimes \mathbb{Q}[y]. \end{array}$$

Note that this can be extended to the case of global complete intersections just like Theorem 4.1.4.

Acknowledgements: The author would like to thank the referee for his/her thorough reading and useful comments. This work was supported by JSPS KAKENHI Grant Number JP16H03936.

References

[1] P. Aluffi, *Singular schemes of hypersurfaces*, Duke Math. J., **80**, No. 10 (1995), 325–351.

[2] P. Aluffi, *Chern classes for singular hypersurfaces*, Trans. Amer. Math. Soc., **351**, No. 10 (1999), 3989–4026.

[3] P. Aluffi, and J.-P. Brasselet, *Une nouvelle preuve de la concordance des classes définies par M.-H. Schwartz et par R. MacPherson*, Bull. Soc. Math. France **13** (2008), 159–166.

[4] P. Aluffi and L. C. Mihalcea, *Chern classes of Schubert cells and varieties*, J. Algebraic Geom. **18** (2009), 63–100.

[5] P. Aluffi and L. C. Mihalcea, *Chern–Schwartz–MacPherson classes for Schubert cells in flag manifolds*, Compositio Mathematica, **152** (2016), 2603–2625.

[6] P. Aluffi, L. C. Mihalcea, J. Schürmann and C. Su, *Shadows of characteristic cycles, Verma modules, and positivity of Chern-Schwartz-MacPherson classes of Schubert cells*, arXiv:1709.08697.

[7] P. Baum, W. Fulton and R. MacPherson, *Riemann–Roch for singular varieties*, Publ. Math. I.H.E.S. **45** (1975), 101–145.

[8] J.-P. Brasselet, D. Lehmann, J. Seade and T. Suwa, *Milnor classes of local complete intersections*, Trans. Amer. Math. Soc., **354** (2002), 1351–1371.

[9] J.-P. Brasselet, J. Seade and T. Suwa, *Vector fields on singular varieties*, Lecture Notes in Mathematics, vol. **1987**, Springer-Verlag, Berlin, 2009.

[10] J.-P. Brasselet, J. Schürmann and S. Yokura, *Hirzebruch classes and motivic Chern classes for singular spaces*, Journal of Topology and Analysis, Vol. **2**, No. 1 (2010), 1–55.

[11] J.-P. Brasselet and M.-H. Schwartz, *Sur les classes de Chern d'une ensemble analytique complexe*, Astérisque **82–83** (1981), 93–148.

[12] V. Bussi and S. Yokura, *Naive motivic Donaldson-Thomas type Hirzebruch classes and some problems*, Journal of Singularities, **10** (2014), 26–53.

[13] R. Callejas-Bedregal, M. F. Z. Morgado and J. Seade, *Lê cycles and Milnor classes*, Invent. Math. **197** (2014),453–482; erratum, 483–489.

[14] S. E. Cappell and J. L. Shaneson, *Stratifiable maps and topological invariants*, J. Amer. Math. Soc. **4** (1991), 521–551.

[15] S. E. Cappell and J. L. Shaneson, *Characteristic classes, singular embeddings, and intersection homology*, Proc. Natl. Acad. Sci. USA, **84** (1991), 3954–3956.

[16] S. E. Cappell and J. L. Shaneson, *Singular spaces, characteristic classes, and intersection homology*, Ann. of Math. **134** (1991), 325–374.

[17] S. E. Cappell, A. Libgober, L. G. Maxim and J. L. Shaneson, *Hodge genera of algebraic varieties, II*, Math. Ann. **345**, No. 4 (2009), 925–972.

[18] S. E. Cappell, A. Libgober, L. G. Maxim and J. L. Shaneson, *Hodge genera and characteristic classes of complex algebraic varieties*, Electron. Res. Announc. Math. Sci. **15** (2008), 1–7.

[19] S. E. Cappell, L. G. Maxim and J. L. Shaneson, *Euler characteristics of algebraic varieties*, Comm. Pure Appl. Math. **61**, no. 3 (2008), 409–421.

[20] S. E. Cappell, L. G. Maxim and J. L. Shaneson, *Hodge genera of algebraic varieties, I*, Comm. Pure Appl. Math. **61**, no. 3 (2008), 422–449.

[21] S. Cappell, L. Maxim, J. Schürmann and J.Shaneson, *Characteristic classes of complex hypersurfaces*, Adv. Math. **225** (2010), no. 5, 2616–2647.

[22] S. Cappell, L. Maxim, J. Schürmann and J.Shaneson, *Equivariant characteristic classes of complex algebraic varieties*, Comm. Pure Appl. Math. **65** (2012), 1722–1769.

[23] S. Cappell, L. Maxim, T. Ohmoto, J. Schürmann and S. Yokura, *Characteristic classes of Hilbert schemes of points via symmetric products*, Geometry & Topology **17** (2013), 1165–1198.

[24] S. Cappell, L. Maxim, J. Schürmann, J.Shaneson and S. Yokura, *Charac-*

teristic classes of symmetric products of complex quasi-projective varieties, J. Reine Angew. Math. **728** (2017), 35–63.
[25] L. Fehér, R. Rimányi and A. Weber, *Motivic Chern classes and K-theoretic stable envelopes*, arXiv:1802.01503v2.
[26] W. Fulton, *Intersection Theory*, Springer Verlag, 1981.
[27] W. Fulton and K. Johnson, *Canonical classes on singular varieties*, Manus. Math. **32** (1980), 381–389.
[28] W. Fulton and R. MacPherson, *Categorical frameworks for the study of singular spaces*, Memoirs of Amer. Math. Soc. **243**, 1981.
[29] M. Goresky and R. MacPherson, *Intersection homology theory*, Topology **149** (1980), 155–162.
[30] F. Hirzebruch, *Topological Methods in Algebraic Geometry, 3rd ed. (1st German ed. 1956)*, Springer-Verlag, 1966.
[31] F. Hirzebruch, T. Berger and R. Jung, *Manifolds and Modular Forms*, Vieweg, 1992.
[32] E. Looijenga, *Motivic measures*, Séminaire Bourbaki, Astérisque **276** (2002), 267–297.
[33] R. MacPherson, *Chern classes for singular algebraic varieties*, Ann. of Math. **100** (1974), 423–432.
[34] R. MacPherson, *Characteristic Classes for Singular Varieties*, Proceedings of the 9-th Brazilian Mathematical Colloquium (Poços de Caldas 1973) Vol. II, Instituto de Matemática Pura e Aplicada, São Paulo, (1977), 321–327.
[35] L. Maxim, *On Milnor classes of complex hypersurfaces*, in "Topology of Stratified Spaces", MSRI Publications **58**, Cambridge Univ. Press (2010), 161–175.
[36] L.Maxim, M. Saito and J. Schürmann, *Hirzebruch-Milnor classes of complete intersections*, Adv. Math. **241** (2013), 220–245.
[37] L.Maxim, M. Saito and J. Schürmann, *Hirzebruch-Milnor classes and Steenbrink spectra of certain projective hypersurfaces*, Arbeitstagung Bonn 2013, Progress in Mathematics 319 (2016) , 265–287.
[38] L. Maxim and J. Schürmann, *Characteristic classes of mixed Hodge modules and applications*, in *"Schubert Varieties, Equivariant Cohomology and Characteristic Classes (IMPANGA 15)"*, EMS Series of Congress Reports 2017, 159–202,
[39] L. Maxim and J. Schürmann, *Characteristic classes of singular toric varieties*, Comm. Pure Appl. Math. **68** (2015), 2177–2236.
[40] L. Maxim and J. Schürmann, *Equivariant characteristic classes of external and symmetric products of varieties*, Geometry & Topology **22** (2018), 471–515.
[41] T. Ohmoto and S. Yokura, *Product formulas for the Milnor class*, Bull. Polish Acad. Sci. Math. **48** (2000), 387–401.
[42] A. Parusiński, *A generalization of the Milnor number*, Math. Ann. **281** (1988), 247–254.
[43] A. Parusiński, *Multiplicity of the dual variety*, Bull. London Math. Soc. **23** (1991), 429–436.
[44] A. Parusiński and P. Pragacz, *Chern–Schwartz–MacPherson classes and the*

Euler characteristic of degeneracy loci and special divisors, J. Amer. Math. Soc. **8** (1995), 793–817.

[45] A. Parusiński and P. Pragacz, *A formula for the Euler characteristic of singular hypersurfaces*, J. Algebraic Geometry **4** (1995), 337–351.

[46] A. Parusiński and P. Pragacz, *Characteristic classes of hypersurfaces and characteristic cycles*, J. Algebraic Geometry **10** (2001), 63–79.

[47] R. Rimányi, *Motivic characteristic classes in cohomological Hall algebras*, arXiv:1808.05654v1.

[48] M. Saito, *Mixed Hodge Modules*, Publ. RIMS., Kyoto Univ. **26** (1990), 221–333.

[49] J. Schürmann, *A generalized Verdier-type Riemann–Roch theorem for Chern-Schwartz-MacPherson classes*, math.AG/ 0202175.

[50] J. Schürmann, *Topology of singular spaces and constructible sheaves*, Mathematics Institute of the Polish Academy of Sciences. Mathematical Monographs (New Series), vol. **63**, Birkhäuser Verlag, Basel, 2003.

[51] J. Schürmann, *Specialization of motivic Hodge–Chern classes*, arXiv:0909.3478 [math.AG].

[52] J. Schürmann, *Characteristic classes of mixed Hodge modules*, MSRI Publications **58**, Cambridge Univ. Press (2010), 419–470.

[53] J. Schürmann, *Nearby cycles and characteristic classes of singular spaces*, in "Singularities in Geometry and Topology, Strasbourg 2009", IRMA Lectures in Mathematics and Theoretical Physics **20** (2012), 181–205.

[54] J. Schürmann and J. Woolf, *Witt groups of abelian categories and perverse sheaves*, Ann. K-Theory **4** (2019), 621–670.

[55] J. Schürmann and S. Yokura, *A survey of characteristic classes of singular spaces*, in "Singularity Theory" (Denis Cheniot et al, ed.), Dedicated to Jean-Paul Brasselet on his 60th birthday, Proceedings of the 2005 Marseille Singularity School and Conference, World Scientific (2007), 865–952,

[56] J. Schürmann and S. Yokura, *Grothendieck groups and a categorification of additive invariants*, International J. Math. **23** (2012), 1–37.

[57] J. Schürmann and S. Yokura, *Motivic bivariant characteristic classes*, Adv. Math. **250** (2014), 611–649.

[58] M.-H. Schwartz, *Classes caractéristiques définies par une stratification d'une variétié analytique complex*, C. R. Acad. Sci. Paris **t. 260** (1965), 3262–3264, 3535–3537

[59] M.-H. Schwartz, *Classes et caractères de Chern des espaces linéaires*, C. R. Acad. Sci. Paris Sér. I. Math. **295** (1982), 399–402.

[60] M.-H. Schwartz, *Champs radiaux sur une stratification analytique*, Travaux en cours, **39** Hermann, Paris, 1991.

[61] T. Suwa, *Classes de Chern des intersections complétes locales*, C. R. Acad. Sci. Paris Sér. **324** (1996), 76–70.

[62] J.-L. Verdier, *Le théorème de Riemann-Roch pour les variétés algébriques éventuellement singulières*, Astérisque **1974-1975**, No.464 (1975), 159–175.

[63] J.-L. Verdier, *Spécialisation de classes de Chern*, Astérisque **82-83** (1981), 149–159.

[64] S. Yokura, *On Cappell–Shaneson's homology L-class of singular algebraic*

varieties, Trans. Amer. Math. Soc. **347** (1995), 1005–1012.
[65] S. Yokura, *On a Milnor class*, preprint (unpublished), 1997.
[66] S. Yokura, *A singular Riemann–Roch theorem for Hirzebruch characteristics*, Banach Center Publ. **44** (1998), 257–268.
[67] S. Yokura, *On a Verdier-type Riemann–Roch for Chern-Schwartz-MacPherson class*, Topology and its Applications **94** (1999), 315–327.
[68] S. Yokura, *On characteristic classes of complete intersections*, Algebraic Geometry: Hirzebruch 70 (Warsaw, 1998), Contemporary Math., **241** (1999), 349–369.
[69] S. Yokura, *An application of bivariant theory to Milnor classes*, Topology and its Applications **115** (2001), 43–61.
[70] S. Yokura, *Oriented bivariant theory, I*, International J. Math. **20**, No. 10 (2009), 1305–1334.
[71] S. Yokura, *Motivic characteristic classes*, in "Topology of Stratified Spaces", MSRI Publications **58**, Cambridge Univ. Press (2010), 375–418.
[72] S. Yokura, *Motivic Milnor classes*, Journal of Singularities **1** (2010), 39–459.
[73] S. Yokura, *Natural transformations associated to additive homology classes*, Kyoto University RIMS. Kokyuroku **1764** (2011), 30–47.
[74] S. Yokura, *"Counting" spaces and related topics*, Topology and its Applications **225** (2017), 206–219.

Participants

of the 5th Franco-Japanese-Vietnamese Symposium on Singularities (FJV2017 Kagoshima)

Takuro Abe – Kyushu, Japan
Tadashi Ashikaga – Tohoku, Japan
Pauline Bailet – Bremen, Germany
Jean-Paul Brasselet – Marseille, France
Dang Tuan Hiep – Taiwan
Dang Van Doat – Hanoi, Vietnam
Christophe Eyral – Warsaw, Poland
Goulwen Fichou – Rennes, France
Maria R. Gonzalez-Dorrego – Madrid, Spain
Vincent Grandjean – Ceara, Brazil
Weili Guo – Hokkaido, Japan
Ha Huy Vui – Hanoi, Vietnam
Kazumasa Inaba – Tohoku, Japan
Hiroaki Ishida – Kagoshima, Japan
Shihoko Ishii – Tokyo, Japan
Masaharu Ishikawa – Tohoku, Japan
Ilia Itenberg – Paris, France
Naohiko Kasuya – Kyoto, Japan
Hiraku Kawanoue – Kyoto, Japan
Masayuki Kawashima – Okayama, Japan
Toshitake Kohno – Tokyo, Japan
Wojciech Kucharz – Krakow, Poland
Krzysztof Kurdyka – Chambery, France
Le Cong Trinh – Qui Nhon, Vietnam
Ye Liu – Hokkaido, Japan
Ursula Ludwig – Duisburg-Essen, Germany
Yukio Matsumoto – Tokyo, Japan
Todor Milanov – IPMU, Tokyo, Japan
Kimio Miyajima – Kagoshima, Japan

Jean-Philippe Monnier	– Angers, France
Hiroyuki Nakaoka	– Kagoshima, Japan
Nguyen Tat Thang	– Hanoi, Vietnam
Nguyen Thi Thao	– Hanoi, Vietnam
Nguyen Viet Dung	– Hanoi, Vietnam
Toru Ohmoto	– Hokkaido, Japan
Mutsuo Oka	– Tokyo, Japan
Tomohiro Okuma	– Yamagata, Japan
Adam Parusinski	– Nice, France
Pham Minh Quy	– Dalat, Vietnam
Pham Tien Son	– Dalat, Vietnam
Phan Phien	– Nha Trang, Vietnam
Pho Duc Tai	– Hanoi, Vietnam
Anne Pichon	– Marseille, France
Delphine Pol	– Hokkaido, Japan
Fabien Priziac	– Marseille, France
Oswald Riemenschneider	– Hamburg, Germany
Guillaume Rond	– Marseille, France
Osamu Saeki	– Kyushu, Japan
Kyoji Saito	– IPMU, Tokyo, Japan
Takahiro Saito	– Tsukuba, Japan
Mark Spivakovsky	– Toulouse, France
Tatsuo Suwa	– Hokkaido, Japan
Homare Tadano	– Tokyo, Japan
Takuhiro Takahashi	– Tohoku, Japan
Kiyoshi Takeuchi	– Tsukuba, Japan
Susumu Tanabe	– Istanbul, Turkey
Tadashi Tomaru	– Gunma, Japan
Tran Nhat Tan	– Hokkaido, Japan
Trinh Duc Tai	– Dalat, Vietnam
Asahi Tsuchida	– Hokkaido, Japan
Kazushi Ueda	– Tokyo, Japan
Michel Vaquie	– Toulouse, France
Zhenjian Wang	– Cote d'Azur, France
Dominik Wrazidlo	– Kyushu, Japan
Toshihiro Yamaguchi	– Kochi, Japan
Takehiko Yasuda	– Osaka, Japan
Yimu Yin	– Los Angeles, U.S.A.
Shoji Yokura	– Kagoshima, Japan
Masahiko Yoshinaga	– Hokkaido, Japan

Lightning Source UK Ltd.
Milton Keynes UK
UKHW020031220620
365311UK00001B/44

9 789811 206023